Management Essentials for Civil Engineers

Management Essentials for Civil Engineers

A Practical Guide to Business, Communication, Ethics, and Risk

CODY A. PENNETTI
C. KAT GRIMSLEY
BRIAN M. GRINDALL

WILEY

Published by John Wiley & Sons, Inc., Hoboken, New Jersey.
Published simultaneously in Canada.

For general information on our other products and services or for technical support, please contact our Customer Care Department within the United States at (800) 762-2974, outside the United States at (317) 572-3993 or fax (317) 572-4002.

Wiley also publishes its books in a variety of electronic formats. Some content that appears in print may not be available in electronic formats. For more information about Wiley products, visit our web site at www.wiley.com.

Library of Congress Cataloging-in-Publication Data is applied for:

Hardback ISBN: 9781119851608

Cover Image: Cody A. Pennetti
Cover Design: Wiley

SKY10082010_081424

To those who are determined to build a better future for our communities, and to those who support them in making such achievements possible.

Together, may you continue to inspire, innovate, and lead with integrity.

Contents

Acknowledgments

From Cody A. Pennetti
I am immensely grateful for the opportunity to have collaborated with numerous individuals who exemplify authentic leadership. This includes my early mentors in civil engineering, Chris dePascale, Tim Culleiton, and Daniela Medek, whose guidance remains the foundation for my work ethic. My proficiency in project planning and risk management has been influenced by my work with the distinguished faculty at the University of Virginia, especially Dr. James H. Lambert. Additionally, countless insights have been gained from my colleagues, students, and clients throughout the years. I sincerely appreciate my family's contribution to this effort, especially my wife, whose unwavering support and grounding principles inspire my dedication to teaching and supporting emerging leaders.

From Kat C. Grimsley
My heartfelt appreciation goes out to all my friends and colleagues in industry for their input and support on this and many other projects over the years. I am also tremendously grateful to my academic peers across many distinguished institutions, but especially the Housing Economics and Real Estate Sector Research Group at the University of Alicante in Spain, who hosted me during the development of this project. And, of course, I am thankful for the unwavering support of my family, who have always come with me on projects that have taken us around the world.

From Brian M. Grindall
I am deeply grateful to the many people who have made invaluable contributions to pursuing this endeavor. First and foremost, I would like to thank my family for their enduring support in promoting the pursuit of curiosity and understanding – especially Everett Grindall, Henry Grindall, Theo Grindall, Colin Grindall, Harry Grindall, Karen Grindall, and Sean Grindall. Thank you to the folks at Georgetown University and George Mason University for their continued support of innovative approaches to teaching and learning the principles of real estate, law, and finance. Finally, thank you to my colleagues in the practice of law who continue to develop new approaches to age-old challenges.

CHAPTER 1

Introduction to Management Essentials

Where to begin.

CHAPTER OUTLINE

Management Essentials for Civil Engineers: A Practical Guide to Business, Communication, Ethics, and Risk, First Edition. Cody A. Pennetti, C. Kat Grimsley, and Brian M. Grindall.

This chapter introduces management terminology, including how project success is measured. It also outlines the communication, ethics, and legal aspects of engineering, with more detail provided throughout the book.

1.1 Introduction to Project Management

Civil infrastructure and real estate development projects are complex, long-term projects that require the coordination and contribution of a wide range of *stakeholders*. The Project Management Institute (PMI) defines a project as a "temporary endeavor undertaken to create a unique product, service, or result." Civil engineering projects fall squarely within this definition, often following a predictive workflow of conception, design, permitting, and construction. Other operations will occur before the project starts (e.g., procurement) and after construction (e.g., maintenance and operations). These projects require years of planning, design, and construction, leading to changes in the built and natural environment that will last for decades. Each civil engineering project is unique due to the geographic properties (each project location) and the current political, environmental, and community perception during the project execution.

Intuitively, civil engineering projects require team members with technical expertise in engineering, science, economics, business, law, and architecture. Civil engineers are equipped with a solid foundation in science, mathematics, and design, enabling them to create new products and services. Within the realm of infrastructure and real estate development, civil engineers tackle a wide array of project types and navigate diverse stakeholder perspectives. Succeeding under these conditions demands more than just technical expertise.

This book is a project management resource tailored for civil engineering projects, which is derived from professional experiences and the methodologies defined by the Project Management Institute (PMI), American Society of Civil Engineers (ASCE), International Council on Systems Engineering (INCOSE), and others. In addition to core management concepts, this book includes design narratives and detailed cases to highlight the ambiguity and complexity of these project types. This book will guide new and experienced engineers as the concepts are adapted to fit each engineer's personal and organizational parameters.

This book includes project management topics for the feasibility, procurement, planning, design, and permitting of civil development projects. The content in this book presents these topics through the lens of the engineering team, setting it apart from construction or development management viewpoints. It's common for engineers to transition into project management roles after first working on purely technical production tasks. During this transition, engineers must learn to balance inward-facing duties, like leading design production and understanding the inner workings of their firm, with outward-facing responsibilities, such as coordinating with various stakeholders for procurement and work validation. This book includes information about the value of stakeholder perspectives to inform management and decision-making for engineers.

1.2 Organization and Terminology of This Book

This book will frequently reference information across different chapters. For example, it is difficult to consider the implications of project schedules without also considering the scope, resources, and cost – each of which is identified in different chapters. While the content in the book does not require sequential progression, the general order is as follows: the first chapters focus on pre-project planning and procurement; the middle chapters cover topics of management processes that include scope, quality, schedule, cost, and risk management; and the later chapters focus on core concepts of communication and leadership.

1.2.1 Organization by Chapter

Each chapter includes principles of various management topics, ethical considerations, and a scenario relevant to project management topics.

Chapter 1 Introduction to Project Management

- This chapter provides an overview of the book and project management principles, success measures, communication, legal aspects, and ethics.

Chapter 2 Origins and the Initiation of Projects

- This chapter describes the origin of a project, including early planning and project objectives that extend to requests for proposals of engineering services.

Chapter 3 Project Pursuit Processes

- This chapter describes pursuit planning to develop and submit proposals for services.

Chapter 4 Contractual Frameworks and Liability

- This chapter describes agreements and contracts pertaining to project work and principles of liability.

Chapter 5 Scope Definition and Quality Management

- This chapter describes how project work is defined and includes topics of quality management based on the scope of work.

Chapter 6 Project Planning and Scheduling

- This chapter includes a sample project plan and describes project schedules, including task dependencies and schedule modifications.

Chapter 7 Cost Management and Monitoring

- This chapter describes the fundamentals of project expenses and the methods to monitor and control project costs from the perspective of the engineering team.

Chapter 8 Risk Management

- This chapter describes the principles of risk management, specifically how they pertain to project operations and objectives.

Chapter 9 Principles of Effective Communication

- This chapter describes various forms of communication and best practices for communicating technical engineering work to diverse audiences.

Chapter 10 Leadership Dynamics and Stakeholder Relationships

- This chapter describes various forms of leadership and power along with topics of stakeholder management.

Appendix A *The Millbrook Logistics Park*

- This narrative complements the textbook by offering 10 distinct scenarios, each corresponding to a different chapter. These scenarios are presented in a story format, effectively illustrating the application of various project management principles.

1.2.2 Terminology

This book references industry standards of project management when defining management topics. These terms are based on those from the project management industry (e.g., PMI, INCOSE). For clarity, a few initial terms are defined in this section. In practice, several disparate terms may describe team member roles and stakeholders. For this book, these roles focus on the scope associated with the design production and permitting documents for civil engineering projects.

As an initial distinction, there is a difference between a project role, an organizational title, and a certification. An individual's project *role* is most important when defining project responsibilities. Confusingly, some organizations may assign employee *titles* such as *senior project manager*. The title is used with internal reporting and hierarchical structures; however, titles are almost arbitrary when addressing the role and responsibility of an individual on a project. There is only one project manager for each project, regardless of organizational title.

Roles and responsibilities should be clearly defined at the start of the project. This is critical when a project operates across multiple disciplines and companies, as the authority, accountability, responsibility, and obligations are convoluted. For example, suppose a design conflict is identified where the building utilities do not match the site utilities' location (or size). In that case, there needs to be a defined process to resolve the conflict. Who is responsible for identifying conflicts and documenting change requests? Should the architecture team or the site-civil team change their design? Do all team members understand the cost and schedule impacts of the change? Who does the client contact to discuss this issue? For this reason, it is good practice to refer to the industry-standard designations and define the terms to ensure consistency across all team members and to publish organizational charts with each project.

This section includes terminology for project management tailored to civil engineering projects.

Project Manager

The project manager is responsible for leading the team and meeting project objectives. The project manager is accountable for the project's success and the team's actions. Each project has one project manager. In this book, the *project manager* refers to a civil engineer serving in this role for the engineering consulting work associated with a civil infrastructure or real estate development project.

A project manager's operating policies will vary by organizational structure. On large projects, a project manager is primarily focused on communication, documentation, and team leadership. Small projects may require a project manager to hold both managerial and production responsibilities. Engineers must recognize that clients typically have their own operational procedures for projects and often appoint their own project manager to oversee the development scope of work. However, within the context of the engineering aspects of the project, the engineer assumes the role of project manager. All references to the project manager in this book pertain to the project manager handling the engineering scope of work (not the client's project manager overseeing the entire development operation).

Functional Manager

An organization will have several people in positions of authority that will support a project and control resources. These staff supervisors serve as functional managers to the project manager. *Functional managers* will monitor task budgets allocated to their work but do not oversee the project operations the way the project manager does.

For example, suppose an engineering firm has multiple departments, including transportation engineering, landscape architecture, land survey, and stormwater management. Each department has personnel that supervise its staff and often hold a managerial title. For a highway design project, the project manager will likely be someone from the transportation engineering department while the supervisors in other departments (landscape, survey, and stormwater) will operate as functional managers and coordinate with the project manager to assign personnel to work on the project. While there is only one project manager for each project, functional managers are often involved in coordinating resource demands. Functional managers' level of involvement and authority will vary based on the project team structure, as described in Chapter 6 (Section 6.2.2).

Project Teams

The project team size, hierarchy, and positions will vary by the organizational structure. From the engineer's perspective, the team will include a project manager and production staff. Other team members, such as attorneys, economists, brokers, and developers, are part of the larger, overall project team but may have disparate organizational structures. In this book, the *project team* refers to the staff that operates under the direction of the project manager.

Client / Owner

In this book, the term *client* is used to refer to the party responsible for initiating and driving the project, which may include private developers, landowners, public agencies, and similar entities.

It's important to understand that the *client* can differ depending on the contract. Typically, the project contract outlines who the engineering firm's client is, which may not always be the direct project owner. For example, if an architect subcontracts a civil engineering firm, then the architect is considered the engineer's client. Meanwhile, the architect directly serves the project owner and developer (the project client). Similarly, when construction contractors subcontract civil engineers, those contractors are the engineers' clients, even as they themselves serve the client of the project.

Although *owner* is another common term referring to the individual or entity with investment control, decision-making authority, and overall project oversight, this text will consistently use *client* instead. This choice emphasizes the service-oriented relationship that focuses on meeting the needs and expectations of the entity for whom the project is being carried out. Therefore, irrespective of the contractual hierarchy or project structure, the term *client* in this book refers to the project client, or entity for whom the engineering services are ultimately provided.

Engineering Firm / Organization

Engineering operations may be part of a larger consulting firm, an architecture firm, or a public agency division (transportation, schools, municipal, and others). Both the terms *engineering firm* and *organization* are used in this book to describe the operational organization that influences the project policies and processes and includes the engineering project team.

Stakeholders

The project stakeholders may include any of the individuals, community members, and organizations influencing the project. These stakeholders may be internal or external to the project. A *stakeholder* will have varying levels of impact, power, influence, and interest in the project. These parameters can vary throughout the project as stakeholders gain or lose interest and power at different phases. More information about stakeholders is in Chapter 10 (Section 10.7).

Civil Infrastructure and Real Estate Development

This book is tailored to civil infrastructure and real estate development projects. The distinction between these two project types is that *civil infrastructure* projects are often associated with public projects, such as roads, trails, parks, utility systems, stormwater management, and other design services that serve the public. The term *real estate development*, or land development, is often associated with private development projects such as new retail centers, residential communities, commercial offices, industrial centers, and other uses. The principles of this book also apply to projects involving new development, redevelopment, planning, or restoration projects of the built and natural environment.

While the book distinguishes between public and private projects, many initiatives do not fit neatly into these categories. Examples include joint ventures, public-private partnerships, and design-build-operate projects. Despite this variety in terminology and project structure, these projects often share common characteristics in terms of their operational aspects. Therefore, the insights and principles presented here are relevant across a wide spectrum of project types, transcending the basic public-private dichotomy.

Qualifiers in Terminology

In many cases, there are qualifiers for statements or terms, such as *often, generally,* or *good practice*. These qualifiers acknowledge the inherent complexity and unique character of civil infrastructure and real estate development projects and the project teams. The terms used in this book generally align with industry standards and are tailored to civil engineering practices.

1.3 The Purpose of Projects

The core purpose of projects is creating value, which is achieved through an organization's vision. The success of a project requires technical competence and management efficacy. Project value is attained with a disciplined approach to management, which is required to maintain effective communication, prevent cost overruns, reduce rework, manage scope creep, support the project team, and achieve success. The purpose of value creation in projects is consistent with civil infrastructure and real estate development projects, even though they may be completed to meet various objectives. For example, a developer will initiate a project to create a new product (e.g., homes, commercial space, or other) to capture profit from the sale or lease of the resulting property. Or civil infrastructure projects are initiated to meet regulatory requirements, improve safety, satisfy capital infrastructure needs, achieve sustainability goals, improve community resilience, or repair infrastructure systems. Best practices in project management will enable engineers and clients to derive a higher value from a project. In all cases, engineers must recognize that there are objectives beyond the technical solutions, and each possible solution will have multiple perspectives that will define success.

Engineers must recognize that there are objectives beyond the technical solutions, and each possible solution will have multiple perspectives that will define success.

While this book is tailored to the civil engineering perspective, the principles apply to other team members operating in architecture, engineering, and construction (AEC). Similarly, other engineering fields will recognize anecdotes used to describe standard operating conditions and experiences described in the case studies. Fortunately, civil engineering is a tangible and relatable practice. This means the applications and narratives described in this book can easily be transferred to other professional services.

1.3.1 Management Perspectives

This section will introduce two project management perspectives: (1) the role and best practices for project management of the engineer's scope of work and (2) how the engineering decisions can impact an infrastructure or real estate development project, including project outcomes, cost, and schedule.

The engineering project manager focuses on successfully providing business value. The project's scope for the engineer can be considered to include the production of the design documents and any related consulting services required. From an internal perspective, value is driven by managing the project's cost and budget to ensure the work is profitable while satisfying project requirements without jeopardizing the relationships with clients or stakeholders. Engineers will use technical expertise to prioritize compliance and safety in the design solutions. While the client will have similar objectives, the engineer's perspective differs from how a client may view the project – clients are most focused on the final development (the product) achieved through construction, operations, maintenance, and delivered value. From the civil engineer's perspective, project management and success are typically limited to a specific assignment, such as the engineer's responsibility for creating a site layout. However, project management and success are much broader for the client. To fully appreciate and support client goals, engineers must recognize the client's perspective and how the engineer's work affects the comprehensive objectives of a project.

To fully appreciate and support client goals, engineers must recognize the client's perspective and how the engineer's work affects the comprehensive objectives of a project.

Typically, the client's perspective is based on undertaking the entire development process: identifying and acquiring a site, obtaining the necessary entitlements, securing financing, completing the site and building design, finishing construction, and beginning operations. For private-sector real estate developers, the process also includes additional cost- and profit-related activities such as conducting a market analysis, constructing improvements on the land, and selling or leasing the completed asset to capture financial returns.

Development is a lengthy process that can take several years, regardless of project type. During this time, the client incurs tremendous cost without yet realizing any future value and must manage various project risks, evolving stakeholder concerns, and relationships with an extensive array of professional service providers, including coordinating the timing and interaction of the various contracts, schedules, and deliverables. The client's project management efforts extend beyond any single element, such as the civil engineering design. Before and after the civil engineer has completed the design tasks, the client is involved in contractual and noncontractual relationships that shape the project and are often not visible to the civil engineer. Figure 1.1 shows the client's development partners and the civil engineer's relationship with the client in this context.

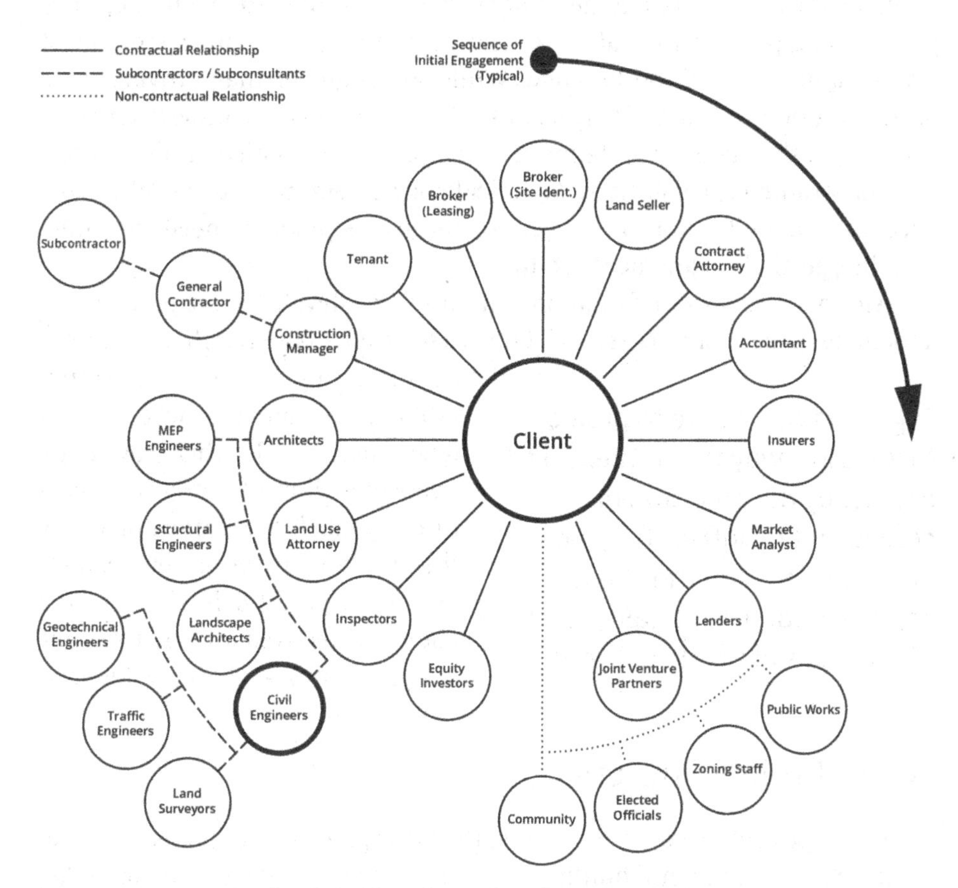

FIGURE 1.1 Example of stakeholder relationships from the client's perspective.

Stakeholder relationships often influence the scope, cost, and schedule throughout the project life cycle. For example, various circumstances can necessitate site layout changes as the client adapts to new constraints and requirements. Clients will monitor the need for scope modifications as they consider the costs associated with additional engineering work to accommodate changed conditions. The client will also consider schedule penalties associated with delays for rework. For example, missing a construction delivery date can impact tenant occupancy and can represent a significant financial burden that jeopardizes project success.

Clients also manage the anticipated operations when establishing project success criteria. As design solutions are established, the client expects that the site layout accommodates operational components such as phasing, tenant or end-user requirements, maintenance, etc. Examples of this might be the location of a large utility box in the front yard of a home, steep grades along pedestrian routes, or locating retaining walls in the middle of a yard rather than at the lot edge. While the engineer may provide these design solutions and satisfy the technical requirements, the project's success is based on the client's validation that the project has satisfied their needs. In these examples, a solution's technical correctness may fail to deliver the necessary value if it does not meet the end-user's practical needs for functionality, aesthetics, and accessibility.

An engineer with a full appreciation of the development process can anticipate these "common sense" choices to build a positive client relationship and improve the firm's reputation as a valuable development team member. In this way, engineers contribute to project success in a way that is measured not just by the precision of designs but by integrating the right solution with the client's objectives for functionality, cost-efficiency, and timely delivery.

Engineers contribute to project success in a way that is measured not just by the precision of designs but by integrating the right solution with the client's objectives for functionality, cost-efficiency, and timely delivery.

1.3.2 Project Success

Achieving project success entails more than compliance with project requirements. Success is about fulfilling project objectives *and* satisfying stakeholder needs. Additionally, successful projects must ensure the well-being of the project team. Measures of success extend beyond the confines of an individual project and team. Recognizing that each project operates within a broader system and contributes to overarching goals is essential. Each project should advance the client and the engineering firm toward their unique strategic objectives and new opportunities. Furthermore, projects should be approached to enhance the firm's reputation and prominence. This mindset creates the stage for project management operations, recognizing the interconnected relationships between the engineering team and stakeholders.

Engineering firms, clients, and other stakeholders will have disparate measures of success. For example, an engineering firm may deem a project successful only if the project is profitable for the engineering firm. From the client's perspective, there is little attention (or visibility) to the engineering firm's financial performance. Instead, a developer's success will likely be measured based on whether the project met the schedule so tenant rent could be collected as early as possible. Alternatively, a transportation agency may track schedule compliance but focus more on project costs to ensure compliance with available budgets. Still other types of clients may measure success by quality, safety, risk mitigation, resilience, life cycle costs, or other criteria. While all factors are important to project objectives, stakeholders will determine the acceptable trade-offs to prioritize the project success criteria.

These priority criteria for each stakeholder and the associated trade-offs operate under the *triple constraints* of scope, schedule, and cost (also referred to as the *project management triangle*) and rely on resources to meet these criteria. These factors are interconnected and interdependent on a project's risk and quality, which vary throughout the project life cycle. A few examples are:

A. Scope vs. Resources:

- An increase in the project's scope without adjusting the timeline might necessitate additional resources, leading to potential cost overruns.

B. Schedule vs. Cost:

- A client's demand to accelerate the schedule would require increased costs to accommodate larger teams (more resources) and additional communication channels.

C. Cost vs. Scope:

- Reducing the budget without altering the scope might extend the project's duration, as resource selection is constrained by price.

D. Quality vs. Cost:

- Cutting costs might lead to compromises in quality, which in the long run could affect the project's sustainability or require corrective actions.

E. Risk vs. Schedule:

- Retroactively addressing risks might require extending the project's schedule or increasing the budget to ensure that the scope and quality are not compromised.

These are just a few examples demonstrating that project success requires balancing, prioritizing, and acknowledging trade-offs. A standard graphic visual representation of the interconnected constraints is shown in Figure 1.2.

Each project's unique goals and priorities must be communicated across the team, accompanied by unambiguous and measurable criteria. While this is intuitive, many project teams erroneously prioritize technical problem-solving before adequately defining the problem. Project objectives should align with the client's vision and the engineering firm's mission of delivering business value. Project-specific objectives, like generating revenue from the design services of a road construction project, can sometimes become overly narrow and unintentionally

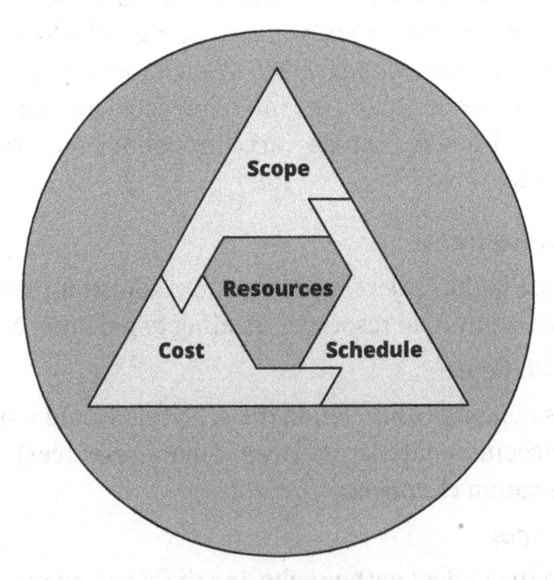

FIGURE 1.2 Project management triangle with a reliance on resources

obscure the organization's broader goals. These broader goals could include building brand recognition, fostering positive relationships with the community, or achieving long-term societal benefits. Understanding how to define and communicate objectives is critical to the measurable success of a project. Other chapters of this book provide greater detail on the principles of scope, quality, schedule, resources, cost, and risk.

Many project teams erroneously prioritize technical problem-solving before adequately defining the problem.

1.3.3 Technical Solutions vs. Client Value

Technical viability alone won't guarantee client satisfaction; success requires aligning the right design solutions with client objectives. Engineers have many technical options for addressing infrastructure and real estate development project challenges. To ensure success, engineers should differentiate between a solution that meets technical requirements and a better alternative that adds value for the client. The distinction lies in the fact that clients might not be satisfied if the solution doesn't align with their explicit or implicit objectives. For example, a structural engineering design that places a column in the middle of a kitchen may satisfy the load-bearing capacity requirements but would likely displease the client by creating an undesirable space for the end-customer and, therefore, reduce the rental or sale income the client can charge for the space. A large concrete trapezoidal ditch may satisfy drainage conveyance requirements but is likely discouraged for a residential backyard. A public park that focuses only on cost savings could lack the expected amenities of recreation facilities and landscaping that provide community value. Engineers must understand their client's (or end users') objectives to succeed.

Technical viability alone won't guarantee client satisfaction; success requires aligning the right design solutions with client objectives.

1.4 Principles of Project Life Cycles

A project has a definitive initiation and closure process – a start and end. Projects progress to completion through planning, executing, and continuously monitoring and controlling the project work. These operations rely on managing project risk, quality, scope, cost, schedule, stakeholders, resources, and communication. Before a project begins, pre-project work often includes the operations necessary to evaluate, fund, scope, prioritize, and select a project. From a client's perspective, this might consist of reviewing a capital infrastructure plan to assess and prioritize the projects that address community needs. From the engineer's perspective, the project begins with a procurement stage as the client advertises (or otherwise communicates) a need for engineering services. This procurement process initiates the project work, often through competitive bidding and then authorization from the client to the engineer to proceed with work (with a notice to proceed, or NTP). Then, there is a constant effort in planning, executing, monitoring, and controlling the project. Understanding that management proficiencies are continuously applied throughout the project is vital. For example, risk and quality management require upfront planning but rely on continuous implementation and updates throughout the project life cycle to maintain relevancy and applicability as project and environmental situations change. Finally, projects require a closure process involving transferring products and documentation to the client, who begins operating and commissioning the project.

For civil infrastructure and real estate development projects, the life cycle encompasses the engineering and design of plans and reports used by the client to construct the project. In many cases, the engineer's role extends from design phases into construction, providing construction administrative services and addressing unforeseen conditions or owner-directed changes. The client may own, operate, maintain, or sell the asset after completing the construction (at the time of project closure). While terminology might differ across industries, Figure 1.3 gives a schematic representation of this life cycle with widely accepted terminology.

There is a crucial distinction between project work and routine operations. A project has a definitive start and end point, involves a change of state, and adds value for the client. An example would be the design and subsequent construction of a new commercial development. The project is initiated, planned, designed, constructed, and then turned over to tenants and operators. Conversely, routine operations like managing that development or an engineering firm's day-to-day activities with accounting or tech support, will influence projects but aren't projects themselves.

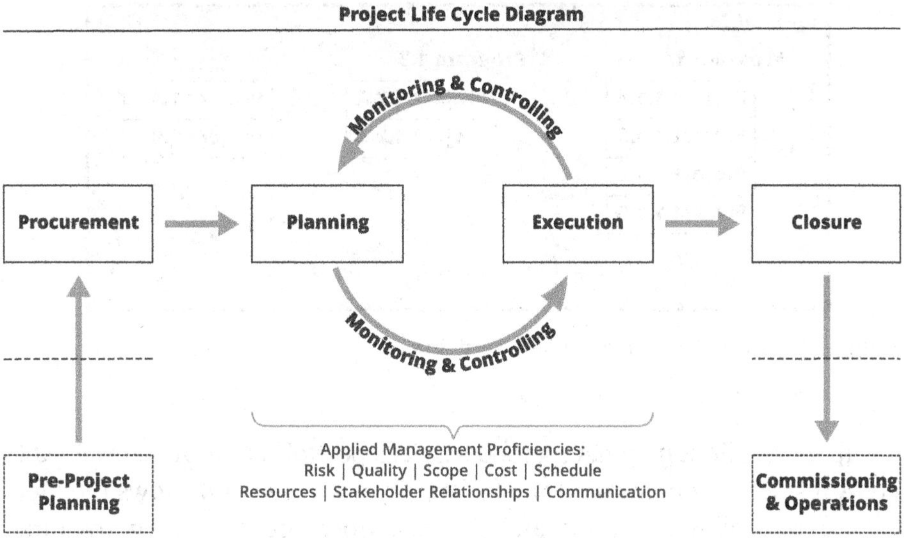

Project Life Cycle Diagram

Monitoring & Controlling
Procurement
Planning
Execution
Closure
Monitoring & Controlling
Applied Management Proficiencies:
Risk | Quality | Scope | Cost | Schedule
Resources | Stakeholder Relationships | Communication
Pre-Project Planning
Commissioning & Operations

FIGURE 1.3 Project life cycle

Further insights into project life cycles, including details on predictive and adaptive life cycles, can be found in Chapter 6 (Section 6.3).

Programs and Portfolios

Projects are evaluated in the larger context of an organization's strategic objectives to ensure consistency in design and monitor interdependencies that focus on supporting the client's multiple endeavors. Multiple projects within an organization are often interconnected and typically organized under a structured hierarchy of programs and portfolios. For example, a retail client may have a project for one new site, which is part of a program focusing on several sites built across the region. Each new site is an independent project under the same program because they share similar resources and stakeholders, and coordinated management activities will benefit the projects. Portfolios are a level higher than programs, which include a collection of programs and projects. With the same

Multiple projects within an organization are often interconnected and typically organized under a structured hierarchy of programs and portfolios.

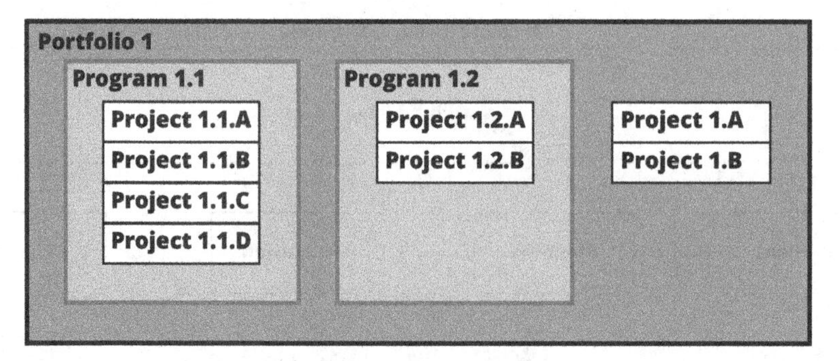

FIGURE 1.4 Hierarchy of programs and portfolios

example, the client may have a food services portfolio that provides for different restaurant programs at the different sites. Figure 1.4 shows the hierarchy of portfolios, programs, and projects. More information on programs and portfolios is included in Chapter 10 (Section 10.2).

1.5 Principles of Communication

In infrastructure and real estate development, successful project outcomes are influenced as much by communication as by technical design. As project managers, civil engineers spend most of their time communicating as they capture new work, meet with clients, present to community members, work with technical consultants, and lead the engineering project team. This section establishes some communication principles and encourages engineers to be deliberate in their communication with team members, clients, and other stakeholders. More information about communication is included in Chapter 9, but this section includes a primer on communication principles.

Communication is a powerful nontechnical skill that must be tailored to various stakeholders and situations. Word choice, for example, is about providing appropriate background or contextual information, controlling emotional responses, and choosing the appropriate style and format to deliver a message. Active listening is a critical part of communication. Listening, when performed to improve understanding, can lead to more effective and creative problem-solving by helping the parties fully appreciate each other's perspectives, needs, and constraints. Communication is ultimately only successful if a clear message is sent, received, and understood (ideally with minimal noise and interference). Depending on the circumstance, acknowledgment of the message and the nature of any resulting action is essential to successful communication.

Ethical considerations related to communication are often linked to what was communicated and when it was communicated. Usually, what is communicated should be fact-based information provided to stakeholders that have a need to know. Just as important is the timing of communication, which should be prompt in delivery and response to inquiries. Much of engineering project work is interdependent, where ambiguous language and delayed responses can have cascading impacts on team members and stakeholders who rely on timely communication to make informed decisions.

Much of engineering project work is interdependent, where ambiguous language and delayed responses can have cascading impacts on team members and stakeholders who rely on timely communication to make informed decisions.

From a legal perspective, the decisions about word choice and what information is included in a message become extremely important in the context of liability. Work emails and messages are often perceived as private communication between the addressed parties; however, the content of any communication, such as an email, is "discoverable" and used as evidence in a legal dispute involving the engineering firm. The need for professionally written emails is critical in the months or years after an email is sent if litigation surrounding a project requires old messages to be produced as part of a lawsuit. For example, an email message could identify when an error was first recognized and how quickly it was addressed compared to the requisite timing established by law or contract language. In all cases, it is important to understand that most forms of communication associated with professional work are discoverable. The legal procedures of an engineering firm will often mandate the nature and duration of data retention for information accessibility.

Engineers should also understand how technical notes on plans may intersect with, or contradict, the project contract. Many engineering notes are concise because the design documents must provide substantial information with limited space. Brevity is encouraged, but notes should be clear to the reader and are best written as complete sentences. While contractually established conditions will govern the project requirements, unclear language with notes will create confusion, particularly during construction operations. For example, the contract language may stipulate that certain published specifications must be adhered to for the project; if the engineer does not understand these project conditions, then the engineering plan may contradict this requirement by providing different specifications.

The following section introduces communication concepts that consider the various audiences encountered with engineering projects and the many forms of communication required throughout the project.

1.5.1 Audiences

Engineers must communicate with various audiences, including their project teams, supervisors and firm principals, clients, government agencies, community members, and others. Each of these groups is a stakeholder because the group has some form of vested interest – or a "stake" – in the outcome of a project. Engineers must understand how to engage in a two-way dialogue with stakeholders rather than simply directing information to them. Successfully interacting with each of the different stakeholders requires different communication strategies. To this end, stakeholders can be categorized in various ways to help guide the appropriate communication approach.

Engineers must understand how to engage in a two-way dialogue with stakeholders rather than simply directing information to them.

Internal versus External

One way to understand different audiences is in relation to the engineer's organization or as internal audiences and external audiences. *Internal audiences* are those people or groups internal to the engineering firm, such as other team members, managers, firm principals, and other firm employees or contractors. Communication with internal stakeholders may be more technical and focused on engineering solutions for a development project. *External audiences* are stakeholders outside the engineer's firm, such as clients, government bodies, and community members. Project team members from other consulting firms are considered part of an external audience, although this may depend on the nature of the contract between the various project firms. It is essential not to share proprietary information with external audiences, and many engineering firms will have policies that dictate what and how information should be shared.

When engaging with external audiences, it is also crucial for engineers to appreciate the context of the client's project objectives beyond their technical considerations. Engineers are just one of the many project team members. There may be larger concerns or considerations that have a significant influence on the overall project. Engineers enhance their value to the client by being knowledgeable allies and actively collaborating with the development team to propose comprehensive solutions that align with the overarching project objectives. Engineers should maintain a high level of professionalism and formality in both written and verbal communication, always respecting the other party.

Engineers enhance their value to the client by being knowledgeable allies and actively collaborating with the development team to propose comprehensive solutions that align with the overarching project objectives.

Technical versus Nontechnical

Another way to understand different audiences is through technical expertise. Communication styles and mediums will vary for technical and nontechnical audiences. It is the engineer's responsibility to understand the audience and tailor the communication style as needed. Technical engineering proficiency can vary within the same group of stakeholders. For example, some clients will have an engineering background instead of (or in addition to) a business administration background. Similarly, community members can include engineers or people with esoteric knowledge about the community settings, politics, environmental conditions, and other elements.

A *technical audience* includes stakeholders with sufficient technical background to understand the engineering issues, constraints, and concepts. This can include sophisticated technical review staff employed by a public agency, certain external project team members, such as architects or construction managers, and some clients. This audience will be familiar with reading engineering plans or reviewing calculations when discussing the project scope. When communicating with a technical audience, engineers should maintain a professional tone. Still, they can generally use technical explanations (and industry jargon) and focus on detailed requirements without fear of alienating the audience.

A *nontechnical audience* includes those stakeholders that do not have a technical background. This may include community members, elected public officials, and some clients. When communicating with nontechnical audiences, it is essential to recognize that these stakeholders are unlikely to comprehend all terms and methods used by engineering professionals. So, communicating in an overly technical way may alienate this stakeholder group and jeopardize the intended communication outcome. Therefore, when engaging with nontechnical audiences, the engineer must consider tone and word choice. Striking the right balance in communication style is crucial; overly technical language can come across as intimidating or pretentious, possibly alienating an audience not versed in industry-specific terminology. On the other hand, oversimplification risks patronizing the audience, which can be equally damaging to the engineer's rapport with them. Engineers must tailor their communication to the audience's level of understanding, ensuring clarity without compromising respect and inclusivity. A lack of shared understanding creates a communication barrier that can frustrate both parties. Maintaining a respectful tone, patience, and emotional control is critical to engaging with nontechnical audiences.

Developing the skill to communicate effectively with stakeholders is learned over time. Aspiring project managers can seek opportunities to shadow more experienced team members or ask to attend public-facing meetings to observe what works – and what does not – when communicating with various audiences.

1.5.2 Forms of Communication

Information can be communicated using a range of different forms: verbal; written; and nonverbal communication, such as body language, gestures, or facial expressions. Engineers need to be cognizant of the diverse communication styles inherent in different cultures and norms, and tailor their approach accordingly to ensure effective and appropriate interactions. Active listening is also a form of non-verbal communication that expresses the participant's willingness to understand the other party and shows respect for their perspective.

Written

Of the different types of communication, technical writing is perhaps the most common form of communication for engineers early in their careers, where a direct style is used to communicate facts. This includes using technical notes on drawings, outlining specifications, or narratives demonstrating compliance with project requirements. However, written communication is used far more broadly than technical documents. Reports, status updates, emails, presentations, requests for information (RFI), and proposal submissions are just a few of the many examples of communication in written form.

The choice of words is important in all types of written communication, mainly because tone and intention are challenging to convey in written form. Word choice can signal a message about the sender's level of professionalism, respect for the recipient, and understanding of the issue. Misunderstandings are often exacerbated because there is no opportunity (or no immediate opportunity) to clarify intent when submitting a written product. Email exchanges may not result in complete understanding if the recipient mistakenly believes they understand the message or escalates the situation without seeking clarification. Table 1.1 provides examples of considerations for written communication.

TABLE 1.1 Guidelines for written communication

Proper formatting
• Use company templates for letterheads, logos, and other content. • Include a cover page, table of contents, and page numbers in reports and similar products. • Be consistent.
Context
• Address communication "gaps" by identifying what information is being assumed, either by yourself or by the recipient, and clarifying. • Include new information that the recipient may not be aware of. • Provide a background or summary to ensure common understanding. • Outline your understanding, or the facts you relied on. • Explain the "why" behind decisions, changes of plan, or unexpected results. • When receiving emails, ask for clarification if there is any doubt regarding the message.

(Continued)

TABLE 1.1 *(Continued)*

Choice of language
• Always avoid slang and jargon. • Use technical language when appropriate, but explain as necessary. • Be aware of inclusivity and professionalism. • When working as part of a team, use terms like "we" rather than "I" to acknowledge the group effort and the engineering firm's support. • Use respectful language.
Visual presentations
• Choose images that help reinforce the message. • Present data clearly and consistently. • Use bullet points or numbered lists, and do not "overload" slides with too many words.

Social Media

Engineering firms often use social media strategically to promote their products and services or signal their support for various causes. Often, sharing updates on project work on social media platforms can advertise the accomplishments of a project team and support a client's objectives. These platforms can also help connect stakeholders and share professional updates.

However, employees also use social media platforms in their personal capacity. Many engineers may not be aware that their firm's policies may contain regulations that have consequences for their out-of-work behavior, which may be seen in social media posts. The professional–personal balance can be further complicated if an engineer is connected to fellow employees on social media platforms, who may feel the need to report certain behaviors. An example of a possible conflict might occur when an employee posts derogatory comments about their employer or client on their personal accounts that could be perceived to harm the firm or client's reputation. Engineers should be cognizant of their social media activity and recognize that personal posts can be recorded and shared, which may not be revealed until much later.

Verbal Communication

Verbal communication occurs during any conversation with team members, upper management, clients, and others. Word choice is often equally important in verbal and written communication, but clarifications are possible during live exchanges. Tone is also easier to express verbally but still requires self-awareness on the speaker's part. While the ability to adjust messaging is an example of the benefits of a live exchange, there are also downsides to verbal communication. Maintaining a calm, impartial communication style during contentious or difficult conversations can be difficult, particularly if the engineer feels like a stakeholder is being unreasonable. Strategies for managing difficult verbal exchanges include:

- Give full attention to the speaker, show genuine interest, and avoid interrupting. Clarify and reflect on what was heard to ensure understanding and to demonstrate empathy.
- Stay composed and avoid emotional responses, even when faced with provocation. Practice techniques such as timed pauses before responding to maintain a professional demeanor.
- Direct the conversation toward constructive solutions rather than dwelling on the problem. Encourage collaborative problem-solving by inviting all parties to contribute their ideas and perspectives.
- When necessary, postpone conversations to a later date or ask for a brief recess; physically leaving the room may help diffuse a challenging discussion.

Nonverbal Communication

Nonverbal cues, such as body language, facial expressions, and other silent gestures, carry significant weight in conveying messages within teams and during stakeholder interactions. These unspoken signals can shape the dynamics of team collaboration, client engagement, and community relations. For instance, a team member might verbally agree to a project's proposed next steps while their body language (e.g., crossed arms, slumped posture, heavy sigh, scowling, etc.) suggests hesitance, potentially indicating deeper concerns. Such concerns could range from team morale issues, an overlooked yet viable alternative solution, interpersonal tensions, or a silent objection from a feeling of not being heard. Recognizing and addressing these nonverbal indicators is crucial for project managers. Ensuring team members feel valued and understood is part of this recognition process. Conversely, signs of inattention, like avoiding eye contact, can unintentionally communicate disinterest. Project managers must be attuned to these subtle signals to effectively lead and maintain a cohesive, open, and responsive environment, both for their internal teams and in other stakeholder interactions.

1.6 Principles of Legal Aspects

It is important for project managers and engineers to understand the law and its implications for the work they do. This book includes information regarding significant legal considerations to support project management in Chapter 4; however, issues involving the law and legal considerations are dealt with throughout this text, and a considerable portion of this book will relate to the law. In order to better understand the role of law in the engineering field and how it relates to project management, it is important to understand that the law essentially consists of a body of rules governing conduct in a society whose rules are prescribed and enforced by an authority. Such rules would include common resources such as constitutions, statutes passed by legislatures, common law precedents from judicial decisions, and rules and regulations promulgated by government agencies. Bodies of rules can also include private legally binding relationships such as contracts.

In order to better understand and apply the role of law to certain engineering and project management issues, it helps to understand that the purpose of the law is to permit societies to organize human affairs. In this context, the term *societies* should be construed broadly to include formal settings – such as nations, states, counties, cities, tribes, etc. – and less formal settings – such as neighborhoods, professional organizations, trade associations, and civic or religious organizations. Further, the effort of "organizing affairs" is also a broad concept including formal regimes – such as constitutions, statutes, regulations, and legally binding covenants – and less formal regimes – societal conventions, commercial customs or norms, social courtesy, or commonly practiced standards. Recognizing the need of societies to organize their affairs, engineers can better understand the significant role of law in the various issues encountered (or avoided).

1.6.1 Legal Concepts and Terminology

Legal concepts inherently involve legal terminology employed in project management, civil infrastructure design, commercial real estate development, and engineering settings. The non-lawyer must understand the blunt tools employed with each legal concept. For example, a relationship between an engineer and a client can be captured in many ways; however, a contract between the client and the engineer establishes a legally binding relationship and elaborates on the professional duties owed by the engineer to the contractual client.

Section 1.2.2 defines the term *client* in this book to reference the project owner and developer. However, it's crucial to understand that this is a broad interpretation. In legal terms, the actual client of an engineering firm could vary, especially when the firm operates as a subcontractor. The contract is the defining document that legally determines who the client is, in a strict sense. Thus, while this book or other project documents might refer to the project owner and developer as the client in a general context, the contractual agreement is the authoritative source that officially specifies the client. This distinction underscores the importance of the contract in legally establishing the identity of the client, which may not always align with the general usage of the term.

The legal concept of contracts involves basic terminology such as covenants, conditions, and remedies, while professional duties implicate legal terminology such as duty, breach, damages, and malpractice. Specific legal terminology will be identified and defined in Chapter 4 and will be applied to specific contexts throughout the book.

1.6.2 Liability and Risk

Quite simply, complying with legal requirements is critical for any successful professional. However, compliance is no simple endeavor. This book's approach to exploring law and legal concepts is rooted in the need to manage liability. Specifically, much of the consideration in this book of the role of law and legal concepts as they relate to engineers and project managers in the civil infrastructure and real estate development perspective is focused on the risk of liability.

The term *liability* essentially consists of a legal obligation – the breach of which would result in legally enforceable consequences. For example, if an engineer agrees to prepare a survey in accordance with specific standards but ultimately fails to adhere to those standards, then the engineer may be held liable for the costs incurred by the client arising out of the engineer's failure. As a matter of self-interest, engineering professionals should better understand the nature of obligations and the likelihood that obligations will arise, be fulfilled, or be breached.

Identifying potential liability is important; however, managing liability is critical. To better assist engineering professionals in managing liability, this book examines liability through the lens of risk management. Risk can be defined in many ways. A finance professional may refer to risk as any uncertainty with respect to an investment that has the potential to negatively impact a financial goal. On the other hand, a construction professional may consider risk to be any exposure to financial loss. A legal professional may identify risk as an outcome arising from failure to comply with statutory or regulatory obligations; an engineering professional may consider risk as the combination of the likelihood of an occurrence and the severity of consequences that may arise from such occurrence. A medical professional may consider risk as a chance or likelihood that something will harm or otherwise affect one's health.

This book adopts a broader perspective on risk that touches on each of these definitions: *risk* consists of the uncertainty of an event occurring (or not occurring) and the associated outcomes, specifically as it changes a project's priorities, constraints, and objectives.

Chapter 8 provides a more detailed explanation of risk and risk assessment in the engineering context, while Chapter 4 describes contracts and liability.

1.7 Engineering Ethics

The study of ethics is a broad field, encompassing both theoretical and applied dimensions related to behavioral standards and the often nuanced, context-dependent nature of moral concepts such as "right" and "wrong." The purpose of this section is not to provide an exhaustive review of ethics but rather to highlight professional, ethical frameworks and demonstrate the relevance – and challenges – of ethical decision-making for civil engineers that may affect their projects, clients, and careers.

As in many industries, professional ethics in the civil engineering and real estate industry centers around the standards and expectations of behavior that affect individual decision-making in various professional circumstances. In straightforward situations, there is often a consensus among professionals on the necessary actions or outcomes in an ethical context. Yet, when it comes to multifaceted projects involving numerous stakeholders and conflicting priorities, ethical dilemmas can arise that don't lend themselves to clear-cut 'right' or 'wrong' answers. The dilemma of choosing a course of action in such circumstances can be exacerbated when there are multiple different solutions, which may be marginally 'better' or 'worse' than others, depending on the stakeholder perspective. Further, different professionals may have legitimate yet different individual interpretations of both a situation and the underlying ethical principles involved. Finally, situations requiring ethical decision-making can be further complicated when a team or individual does not recognize that they are facing a choice with an ethical dimension.

When it comes to multifaceted projects involving numerous stakeholders and conflicting priorities, ethical dilemmas can arise that don't lend themselves to clear-cut 'right' or 'wrong' answers.

Within the technical sectors of the commercial real estate industry, such as engineering and architecture, professional ethical standards are often centered around the broad concept of protecting the public welfare, as defined by the National Society of Professional Engineers (NSPE). Such standards are often designated by industry organizations, which guide best practices and professional, ethical behavior specific to the profession. Such guidance may be disseminated at the chapter, state, or national level. It may be informal (e.g., discussion of ethical topics at conferences or events) or formal (e.g., published and endorsed ethical codes). Indeed, adherence to ethical standards may be obligatory for professionals when such guidance is incorporated as a condition of industry organizational membership. In such instances, the organization may maintain a framework for investigating violations and assigning and imposing sanctions. This can sometimes occur in conjunction with official government bodies, for example, concerning violations connected to professional licensing.

Among the various engineering ethical codes are:

- The American Society of Civil Engineers (ASCE) Code of Ethics
- The National Society of Professional Engineers (NSPE) Code of Ethics for Engineers
- The American Council of Engineering Companies (ACEC) Professional and Ethical Conduct Guidelines

While there is generally an ideological consistency between the various codes, they are not identical. For example, the ASCE Code of Ethics begins with an emphasis on the civil engineer's role that includes sustainable development and environmental stewardship as an ethical imperative, while the NSPE Code of Ethics focuses on safety and health without explicitly mandating sustainability as a guiding principle. Both stances share a common ground in prioritizing the public good, but they approach the engineer's role from slightly different angles, which can influence decision-making in projects with significant environmental impact. This highlights the challenge of determining precisely what constitutes an appropriate ethical response to a particular scenario. The purpose of this section is not to review these standards in detail but to reaffirm that engineers are expected to be aware of the professional, ethical standards that exist within the purview of their practice.

While there are ethical standards specific to the engineering sector of the commercial real estate industry, it is important for civil engineers to recognize that their clients and other development team members also have their own ethical expectations. These may differ materially from the engineer's ethical framework when such standards focus on providing the best outcome for a client, as opposed to an engineering ethical standard that may focus on the best outcome for protecting the public welfare. For instance, a developer might prioritize cost-saving measures to enhance project profitability and meet shareholders' expectations, prompting the choice of less-expensive materials or design shortcuts. Meanwhile, a civil engineer, bound by their professional ethical standards, may argue for more expensive, robust materials or designs that ensure long-term safety and durability. In a project where a residential building is being constructed near a known seismic activity zone, the developer might be inclined to meet the minimum legal seismic design standards in order to control costs while remaining compliant with set safety standards. On the other hand, the civil engineer might advocate for exceeding these minimum requirements, putting enhanced public safety and the structure's long-term resilience at the forefront, even if it comes at a higher initial cost and goes beyond the jurisdictionally mandated requirements. Balancing these different ethical priorities requires clear communication, understanding, and collaboration among all project stakeholders.

It is also important to note that ethical standards are different from legal standards. In other words, certain decisions or actions may be legally permissible yet considered unethical. In this sense, ethics are often considered a more rigorous standard for conduct. Suppose a municipality has a law stating that city officials in public service cannot award contracts to companies in which they have a direct financial interest; however, one of the public officials has a close relative who owns one of the bidding companies. In this case, the law does not explicitly prevent the engineer from being involved in decision-making processes, but ethically, their impartiality could be questioned. Legally, there is no issue, but ethically, the public official would be wise to recuse themselves to avoid even an appearance of bias.

There are many different frameworks of varying complexity that can be used to evaluate an ethical dilemma. In general, the process includes five steps:

1. Recognizing ethical issues
2. Gathering information
3. Applying standards
4. Evaluating recommendations and alternatives
5. Documenting and reporting

The following section includes a basic set of steps engineers can use as a guide to follow when facing a decision that may have ethical implications.

1.7.1 Recognizing Ethical Issues

This first step is incredibly important and easily overlooked in ethical analysis and decision-making. Early in their careers, engineers may assume that ethical issues will present themselves as a clear-cut, up-front problem wherein an egregious situation has obvious ethical implications. In such cases, the need to identify an ethically responsible solution is obvious. However, not all ethical scenarios will be apparent or present themselves as a single, recognizable event. Rather, ethical issues can be introduced subtly or evolve slowly over the project timeline. A series of decisions made by multiple different team members, including the client, can combine over time to create an ethical dilemma. While each individual decision may seem justifiable at the

Not all ethical scenarios will be apparent or present themselves as a single, recognizable event.

time, the eventual outcome may not be. Consider a project where all initial design choices are made to exceed environmental regulations and focus on sustainability. As the project progresses, budget and time constraints lead to minor adjustments. A subcontractor aiming to stay within budget might opt for a slightly cheaper material that still meets the code but has a higher environmental impact. Later, another team member decides to change a drainage design solution to expedite the construction process. While each decision might seem minor and justifiable in isolation, the cumulative effect could result in significant environmental harm. Note that incremental changes obviously need to be monitored to ensure that the project does not ultimately violate the relevant environmental regulations. However, even if it is still compliant by project completion, the once environmentally robust design might have changed substantially and pose ecological threats that none of the team members had initially intended. This introduces an ethical dilemma if there was a commitment to (or by) stakeholders to create and market the project as environmentally sustainable above and beyond regulatory requirements.

It can be difficult to recognize ethical issues for a variety of reasons. For example, an engineer may be too quick to apply judgment based on past experience; however, each project has its own specific context and stakeholders, which may introduce nuances that the engineer needs to consider during the decision-making process to evaluate any new ethical ramifications. Conversely, suppose an ethical issue arises on a topic unfamiliar to engineers, or seemingly outside the scope of a project. In that case, the issue may not be recognized as having ethical implications. In these cases, engineers must practice awareness concerning how engineering assignments are connected to the larger real estate development project. Another example of a situation in which it is challenging to recognize ethical issues is when an engineer works in a highly homogenous environment, where a narrow range of perspectives or dominant conformity of thought culture may hinder the awareness needed to identify an ethical issue.

1.7.2 Gathering Information

Once a situation with potential ethical implications has been identified, the engineer must verify the relevant facts before recommending a course of action. Indeed, the ethical nature of the issue itself must be confirmed.

Civil infrastructure and real estate development projects are complex and multifaceted; this can lead to issues that are not, in fact, ethical dilemmas but rather the result of misunderstanding, poor communication, assumptions on the part of specific team members, use of competing definitions, or a myriad of other possibilities. In addition to gathering facts and clarifying understanding, engineers should also check their internal assumptions about the issue. Finally, it is vital to consider the issue from the perspective of other stakeholders, including the client, who may view it differently or provide new information.

1.7.3 Applying Standards

If a thorough fact-checking exercise suggests a legitimate ethical dilemma, the issue should be clearly outlined. This can be challenging if the issue is nuanced or the context is complex. Nonetheless, stating the issue (and any interrelated issues) is an important exercise, allowing engineers to communicate their position to others as clearly as possible. It also allows the engineer to consider the issue through the lens of any relevant standards, as shown in Figure 1.5, including:

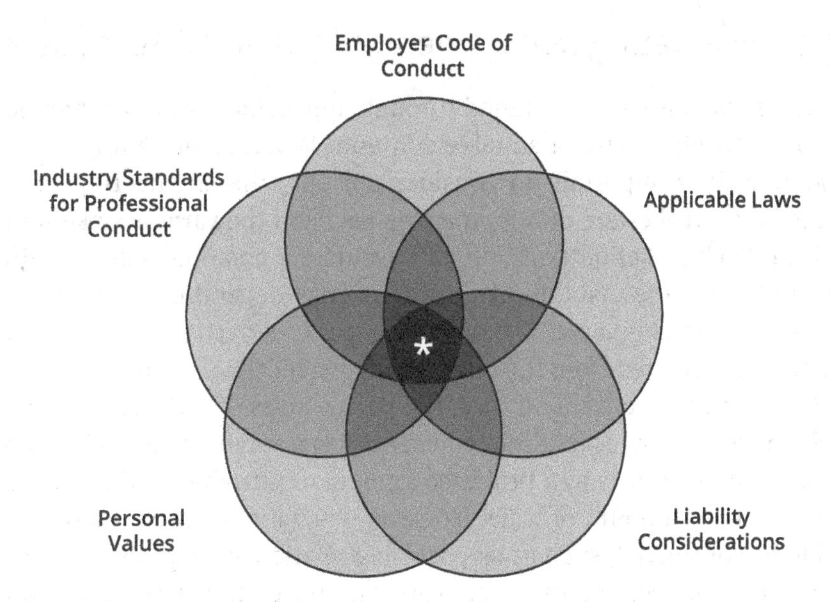

FIGURE 1.5 Standards applicable to ethical evaluations.

- The engineering firm's Code of Conduct or any other firm policy that applies;
- Applicable local or federal laws;
- Contractual obligations or liability considerations that intersect with the issue;
- The engineer's own personal values or moral code;
- Industry standards for professional conduct, which typically include considerations for protecting public well-being.

Note that each applicable standard may treat a particular ethical issue differently than other standards. This is especially likely for nuanced or complex issues that cannot be clearly deemed "right" or "wrong." Beyond these more formal standards, engineers should also apply a basic perception test by imagining how a decision would look if it were brought up at a public hearing or published in a newspaper article about the project. A questionable ethical decision has the potential to harm the firm's reputation or the engineer's own professional reputation. Realistically, not all decisions will be easily defensible, which serves to reinforce the need to articulate an issue clearly, document information thoroughly, and evaluate it through all applicable standards.

1.7.4 Evaluating Recommendations and Alternatives

Once an issue has been examined through applicable standards, engineers can focus on identifying a suitable solution. When formulating a recommendation, it is important to consider not only the engineering project, but the larger real estate development project and the various stakeholders involved. Each stakeholder group will evaluate a potential solution differently and, thus, the viability of a solution may depend on how well it is received by non-engineers. The consequences of a particular solution also need to be considered from the engineering perspective, the larger development project perspective, and from the perspectives of different stakeholders. Such consequences might include easily measurable outcomes, such as increasing project costs by a projected amount or introducing project delays by an estimated amount of time. However, less tangible consequences also need to be considered, such as jeopardizing relationships, project financing implications, project approval implications, future marketing implications, or other potential project outcomes.

After creating a recommendation, engineers should develop a list of alternatives, but should never suggest an option that is infeasible or that they do not feel comfortable supporting. This exercise will help ensure the issue has been considered thoroughly and that no viable alternative has been missed. Indeed, for complex issues, more than one reasonable solution may be possible. The benefits and risks of different potential solutions should also be articulated. This can help different stakeholders evaluate options and may be particularly helpful in communicating the recommendation clearly to the engineer's management team or to the client.

1.7.5 Documenting and Reporting

Engineers will most likely refer high-profile issues to their supervisors or may be asked to participate in discussions with the client to explain the issue. However, many issues do not rise – at least initially – to this level. In instances where an engineer is making ethical decisions at the team-level, it is important to be transparent about the issue and the decision by documenting it in an email, progress report, or by some other formal method. Documenting should include the outline of the issue, considerations, consequences, recommendations, and any other important information. Documenting the issue helps protect the engineer, the team, and the firm by making sure there is a record of the considerations that are factored into the decision. This also gives others the chance to raise concerns or provide other information. Documentation can often help protect engineers should they subsequently face pressure to do something other than what was initially agreed to. Note that this final aspect of ethical decision-making can be tricky if it involves whistleblowing situations. In these cases, engineers should check if their firm or industry organizations have resources or guidance for handling issues that rise to this level.

Despite the complexity of the ethical arena, it is imperative that civil engineers find a way to successfully navigate the challenges they face. Exercising professional integrity, compliance with best practices, and adherence to applicable codes of ethics and regulation

> Engineers should maintain a perspective that extends beyond immediate concerns to consider the long-term implications and sustainability of their decisions.

is paramount for personal career advancement as well as maintaining client relationships. Engineers should maintain a perspective that extends beyond immediate concerns to consider the long-term implications and sustainability of their decisions. Navigating challenging scenarios and engaging in tough conversations can yield enduring benefits, as an engineer's ethical determination will contribute to cultivating their reputation for reliability and integrity.

1.8 Introduction to Management Scenarios

Management and leadership skills are tacit knowledge, often learned through experiences. The narrative included in Appendix A of this book, titled *The Millbrook Logistics Park*, provides anecdotal experiences portraying the nuanced challenges of implementing project management and leadership skills. The narrative includes a scenario for each corresponding chapter of the technical content in this book. Each scenario progresses through the major milestones of a project, from pre-project planning until project completion. The narrative structure mirrors real-world situations, revealing insights that may not be apparent through fact-based exposition. By adopting a story-like format, these scenarios describe the application of project management topics to practice.

Each part of the scenario delves into the perspectives of the client and the engineering team, providing a holistic view of the project dynamics. From the client's viewpoint, the narrative explores differing priorities and objectives, including evaluations of how success is quantified and how project objectives are communicated. The engineers' narratives offer a contrasting experience of a novice engineer and a seasoned project manager, highlighting how early choices and interactions among stakeholders can lead to a cascade of ethical and legal challenges as the project evolves.

By following the project from its initial planning stages to a final retrospective, the scenarios serve as a pedagogical tool and a nuanced exploration of a professional journey in the engineering world.

Appendix A, Part 1 of the Millbrook Logistics Park scenario, commences with an examination from the client's viewpoint on the initial stages of pre-project planning and the origin of the project. This introductory segment provides an overview of the key stakeholders involved, delineating their roles and perspectives on the project's opportunity. The client, TerraHaven, in collaboration with the engineering firm ApexTech and the legal firm Meadow Law Partners, engage in preliminary discussions to conceptualize the project scope. These initial planning sessions are instrumental in laying the groundwork for the 10-part narrative detailing the development of the Millbrook Logistics Park.

CHAPTER 2

Origins and the Initiation of Projects

Where do projects come from?

CHAPTER OUTLINE

Management Essentials for Civil Engineers: A Practical Guide to Business, Communication, Ethics, and Risk, First Edition. Cody A. Pennetti, C. Kat Grimsley, and Brian M. Grindall.

This chapter begins by examining the origins of a project and then provides the fundamentals of the procurement process. After client needs are identified, the procurement process often begins with an announcement that a client is seeking interest from qualified professionals to perform the project scope of work with a request for proposal (RFP). The method of advertisement varies dramatically between public and private clients. The procurement process continues as the client evaluates proposals based on firm qualifications, staff expertise, schedule, price, and other factors. Toward the end of a procurement process, a contract is issued to hire the successful firm; however, the procurement processes will continue throughout the project's duration as work is monitored or new services are necessary.

2.1 Introduction

Long before an assignment arrives at an engineer's desk, the client has identified a *need*. Understanding client's needs, goals, and motivations before a project's inception can help engineers achieve project success more effectively. Indeed, a project's objective and measure of success is often about more than the technical solution. For example, a road project is less about the technical design of a road and more about improving the accessibility, safety, and mobility of a community for economic benefits. Similarly, a stormwater management project is not about the detailed design of the system but is instead about protecting the natural water systems, promoting community resilience, and establishing natural habitats. This context is important when considering how best to communicate the need for a project (the client's perspective) and how to provide engineering consulting services.

There are substantial differences between the goals and needs of private-sector clients compared to public sector clients. Understanding these variances is crucial, as they shape the client's objectives, introduce unique constraints, and define distinct measures of success. These factors, in turn, influence an engineer's strategy for securing new business and successfully meeting project demands. For instance, the financial conditions of a project can vary greatly; private-sector clients may be driven by the returns on private equity investments, while public sector clients often depend on allocated public funds.

Such financial frameworks have a direct bearing on each client's risk tolerance and prioritization. Clients will invariably seek engineers who demonstrate an awareness of sector-specific conditions and can align their approach accordingly. Incorporating this understanding into the engineering process is not just about adapting to different financial models; it's about recognizing the broader implications these models have for project management, risk assessment, and overall client relationships. Table 2.1 outlines some of the clients' significant categories and contrasts of potential needs and goals.

Clients will invariably seek engineers who demonstrate an awareness of sector-specific conditions and can align their approach accordingly.

TABLE 2.1 Potential objectives of a client

Type of Client	Examples of Clients	Client Goals
Private	For-profit developers	Return on investment
Quasi-public	Hospitals, universities	Long-term vision to serve stakeholders
Public	Department of Transportation, school system, courts	Solving structural or societal problems and generating the capacity to provide services to citizens
Mission-driven	Nonprofits such as affordable housing developers	Providing facilities that align with or support the mission, often the end-user

2.2 Procurement

As defined by the Project Management Institute (PMI), *procurement* is the process of purchasing or acquiring the external services or products necessary to complete a project. There are two perspectives of procurement: that of the buyer and that of the seller. The buyer, or *client,* is typically a private-sector commercial developer or a public-sector agency with a development-related need. The seller, or service provider, is the technical consulting firm (engineer, architect, etc.) that will provide the necessary services. As buyers, clients will procure services from external entities (e.g., an engineering firm) to prepare the designs, plans, geotechnical reports, traffic studies, and permits for a wide range of civil infrastructure and real estate development projects. Engineering firms will acquire new work by submitting qualifications, proposals, and pricing to the client to be hired to perform the desired work.

When a client's project encompasses a variety of services, such as architectural design, site planning, geotechnical analysis, and field surveying, it often necessitates a collaborative effort from multiple specialized teams. Typically, one lead firm – referred to as the prime consultant – secures the contract with the client and subsequently engages other firms, known as subconsultants, to perform specific tasks. For instance, a site-civil engineering firm might win the main contract and partner with a geotechnical engineering firm, a surveying company, and others to fulfill the project requirements. These collaborative efforts are termed *teaming arrangements,* and the consortium of these firms establishes what is known as the consultant design team. Further details on such arrangements can be found in Chapter 4.

To perform shared project work, the procurement process requires formal agreements that protect the buyers and sellers involved. Sometimes, a client may already have contracts for certain services, creating a forced teaming arrangement wherein a newly hired firm must work with an existing firm. For example, a private developer may prefer a specialty landscape studio, which requires the civil engineering team to coordinate with that consultant through the client. Similarly, a public agency may have an existing basic ordering agreement (task order, on-call contract, or master service agreement) for services such as surveying and mapping, and the civil design will rely on information provided by that team.

Several options exist for how work can be advertised to procure services from engineers, architects, and other consultants. The public and private sectors have very different procurement processes; each client will also have unique characteristics to consider. Further, clients evaluate multiple proposals, creating a competitive procurement process. To fully understand the goals of a project and possibly improve chances of being hired, the best practice is to understand the origin of the work and the client's motivation, which could include seeking a return on investment, improving transportation operations, solving community issues, maintaining existing services, or in response to a host of other needs.

Solicitation announcements for new projects, commonly called a request for proposal (RFP), will include project requirements on scope, schedule, qualifications, and other content. Occasionally, a request for a letter of interest (LOI) is published before an RFP, or a request for qualifications (RFQ) must be answered before an RFP is provided. Preliminary steps before issuing a RFP usually arise in situations where the project scope is sensitive, or when the RFP process necessitates considerable coordination with potential teams. For instance, a developer may opt to select an engineering firm confidentially if they prefer not to publicly reveal their interest in a site before finalizing their building plans. By doing so, they can maintain discretion until they are prepared to disclose further project details.

Most RFPs include a section titled Objectives, which is written by the client to communicate the purpose of the project. The project objective is often a brief statement on the first pages of the RFP. Still, it is perhaps the most crucial information provided as it describes the buyer's definition of project needs and success criteria. When possible, a direct conversation with the client will improve the team's understanding of the project objectives, which could otherwise be misinterpreted with written content (A-2 PURPOSE/OBJECTIVE). The County has issued this Request for Proposals (RFP) to solicit Proposals from interested and qualified firms to perform conceptual designs and develop construction plans for an enhanced pedestrian facility across Aspen Avenue, between Swamp Fox Road and Anchor Street. Improvement may include street lighting, signage (wayfinding and vehicular and pedestrian warning), ADA-compliant ramps, enhanced pedestrian signal timing, modification to existing traffic signal layout, relocation of existing transit stop(s), and a laddered, textured, or raised crosswalk. The goal of this project is to improve pedestrian access and safety to the Aspen Avenue metrorail station from the neighboring area.

Several anticipated design solutions (ramps, signal timing, relocation of transit stops, etc.) are listed that describe the possible technical approach. However, the technical work is successful only if the project's objective is satisfied. The paragraph ends with a statement about the project goal: ". . .improve pedestrian access and safety to the Aspen Avenue metrorail station from neighboring areas." Understanding the client's success criteria enables engineers to tailor solutions that directly meet these goals. For instance, rather than only presenting designs for traffic signals and crosswalks, an engineer could highlight past projects that have demonstrably enhanced pedestrian safety under similar conditions. This targeted approach is more likely to resonate with the client's primary objectives and will likely be met with appreciation for the engineer's insight into their fundamental needs.

This targeted approach is more likely to resonate with the client's primary objectives and will likely be met with appreciation for the engineer's insight into their fundamental needs.

From the client, a clear objective is critical for a successful procurement process and informs the scope of work (SOW). The SOW, as detailed in the RFP, lays the groundwork for project planning, enabling the engineering team to devise a suitable schedule, budget, and collaboration strategy to meet the project's demands. It's important to note that while the scope outlines the requirements, it does not prescribe the methodology; crafting the execution strategy falls to the engineering firm. Often, the initial SOW needs further clarification as it may present goals in broad strokes. For instance, the SOW may define the desired parameters of a new retail center but leave it to the bidding firm to consider applicable constraints, refine the project parameters, and propose an actionable approach. Chapter 5 includes more information about establishing and managing scope. The following section highlights key procurement differences between the public and private sectors.

2.2.1 Distinction between Public and Private Procurement

Several types of clients undertake development projects based on their unique needs and goals. Just as these clients and their objectives differ, so are their approaches to hiring the project development team. Arguably, the most significant distinction in procurement processes is between the public and private sectors, as summarized in Table 2.2. Public procurement processes are typically more resource intensive, requiring substantially more time and money when compared to the informal process that private-sector clients most often use. While not required, private developers may elect to

TABLE 2.2 Differences between the public and private sectors in procurement processes

Private Sector	Public Sector
• Opportunity may not be advertised.	• Opportunities are required to be publicly posted.
• Invitation to bid on work is selective.	• Bidding is open to the public.
• May be relationship-dependent.	• System of checks and balances is designed to avoid personal favoritism in awards.
• SOW may be incomplete and/or developed collaboratively.	• SOW will be predefined as part of a procurement process.

have a formal process that shares some characteristics with public processes in terms of advertisements, selection criteria, and notice of award.

Public Sector

Public-sector clients (federal, state, and local government entities) typically initiate projects to support citizen needs or facilitate government administration. Such projects range from constructing infrastructure, such as new roads or utilities, to developing new buildings such as courthouses, government offices, or schools. Before the project is advertised as an RFP, a public agency will work through five pre-project steps:

1. Establish the need and objectives.
2. Publicly communicate and refine the objectives to establish a scope.
3. Estimate project costs to allocate funds.
4. Estimate the schedule.
5. Establish the RFP.

This process may take years to complete before an RFP is issued, often requiring internal proposals and negotiations. It is important to note that the public sector will almost always own or have long-term rights to the land it seeks to develop. However, the land-owning department may differ from the division or bureau soliciting a development team. To move forward with a development project, public-sector clients at all levels are subject to laws and regulations that guide their operations. For public entities, purchasing items or contracting for professional services is achieved through a process known as acquisition (or procurement).

Specific regulations contain detailed definitions, bidding procedures, and contracting obligations (among other requirements) that the procurement process must follow for each acquisition or project. Commercial real estate development involves a prescribed bidding and contracting process to identify and hire development services firms, such as engineers, architects, or construction contractors. Examples of such regulations at the federal level are the Federal Acquisition Regulations (FAR) Part 36 Construction and Architect-Engineer Contracts and the Brooks Act (Public Law 92-582) for design services, although other regulations apply. Similar regulation also exists at the state level, for example, in the Virginia Public Procurement Act (Code of Virginia § 2.2-4300 et seq), which, among other things, specifies how project bidding and negotiations take place, defines design-build contracts and other terms, and dictates limits on architectural and engineering contracts.

Often, a public entity will provide information on the requisite procurement methods so that internal staff and bidders (the engineering firms proposing services) understand the formal processes and constraints. The purpose of this section is not to provide extensive detail on the web of public regulation but rather to highlight how these constraints influence the procurement process. Engineering firms or design teams that regularly work with public-sector clients must develop a specialized understanding of the procurement process and the regulations that guide them to win and complete public development projects. Further specifics of the public-sector procurement process are discussed in future sections throughout this chapter.

Private Sector

In contrast to their public-sector counterparts, private-sector developers often make project awards in a far less structured manner. This is partly due to the wide variety of size and sophistication of private developers and the fact that private-sector actors are not subject to the same level of regulatory oversight as their public counterparts. Indeed, even the language used to describe the process differs; for example, private-sector clients generally do not use a formal *procurement process* or even the term *procurement* when referring to hiring their development team members. Instead, private clients can directly hire their preferred team members, often based on existing professional relationships and informal price negotiations regarding the proposed work.

Public-sector clients typically begin their procurement process only after they (a) own, or have a long-term ground lease on, the land to be developed and (b) have a defined public need driving the project; however, private developers encounter a much broader range of starting conditions for a project. A private developer may own property with no specific plan for its development or, conversely, may have an interest in a site but no development right when they decide to undertake a project. Either scenario can create challenges for the engineering team.

Private-sector clients will often conduct a *site search*, looking at numerous properties to find land suitable for a future development project. Depending on the market and the proposed type of project (retail, office, multifamily, mixed-use, etc.), a private developer may need to evaluate several sites before negotiating to buy a particular parcel. During the site search, the private client may ask their engineering colleagues to give an initial opinion about potential site layouts. This work may be unpaid or provided at a nominal rate as part of the engineer's business development activity. Although providing a "test fit" site plan may increase the chances that the engineer will be hired once the client contracts for the land, it also carries risks. Critical boundary, soil, easement, and other information are typically unknown during this early stage. The engineer must make assumptions about site conditions that could later prove incorrect during the site study phase, known as the *due diligence* stage, which occurs after the client has contracted for the site but before the actual purchase.

Rather than go through a formal process, private clients who already own a site might verbally communicate a high-level need and rely on their engineering partners to help develop the scope, schedule, and price necessary to complete the undefined project. While this offers more flexibility to both parties, it can introduce considerable confusion, particularly if the client does not have a clear idea about the project goals. In this case, the engineer effectively becomes responsible for working collaboratively with the client to identify the underlying needs and define an appropriate SOW. The lack of a shared agreement on what work is included versus implied in the SOW is one of the challenges that can arise. These challenges may be further exacerbated because clients' backgrounds and technical knowledge vary greatly. Unlike a public organization with a dedicated procurement office and technical staff, a private client may have a background in business administration or could be an entrepreneurial landowner with little or no knowledge of the development process. This knowledge gap often leads to disparities between how clients describe their needs and what they need for professional services, as lightheartedly represented in Figure 2.1. In many cases, this leads to assumptions about what work is done by the client or the engineer (such as permitting) or confusion about the scale of the project, costs, and schedules. In any scenario without a detailed RFP, detailed communication between the engineer and client is critical to develop a SOW that clearly defines all project needs and the responsibilities of each party. More information about the development of proposals is provided in Chapter 3.

Note that a private-sector developer may choose to operate more like the public sector and establish similar formal hiring processes. This is typical of larger firms working on complex, high-value projects. These clients may voluntarily decide to follow similar steps as those seen in public procurement processes or may have internal policies that direct their operations, such as for a publicly traded private-sector firm or a real estate investment trust (REIT). Embracing a formal bidding process establishes a fair review of qualifications and maintains competitive pricing from multiple technical services firms. A formal procurement process can also serve as evidence of evaluation procedures and results that can be shared with stakeholders.

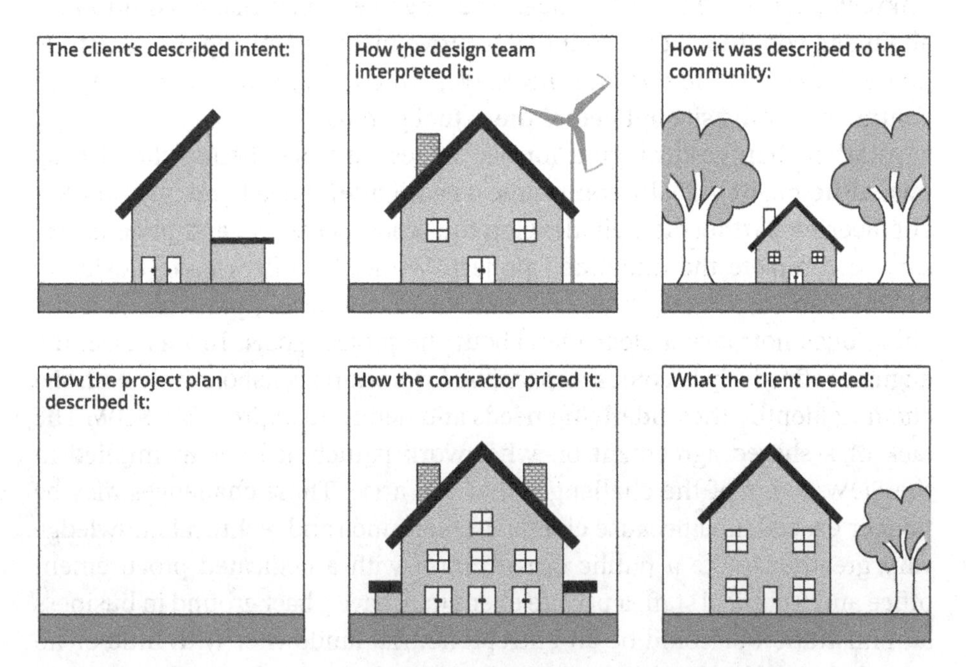

FIGURE 2.1 The importance of communicating and managing expectations.

2.2.2 Factors that Influence Procurement

Multiple conditions influence a project and the procurement process, often called *enterprise environmental factors* (EEF), as defined by PMI. The EEFs can come from internal or external sources and represent factors beyond the control of the project team – whether from the client's or the engineer's side. Aside from the client's objectives and requirements, these EEFs will influence the project scope, schedule, budget, and other factors. Several EEFs and examples are provided below:

Multiple conditions influence a project and the procurement process, often called enterprise environmental factors.

- *Marketplace conditions.* Material prices, availability, and resource/workforce capability may influence or constrain the project during design and construction. Significant changes in marketplace conditions for office, retail, and housing may require flexibility in the design process or schedules. Many development projects span several years from concept to completion, which means the project must adapt to the varying conditions over time. The longer the project time frame, the greater the risk that a change in market forces will occur and affect the project.
- *Existing and new products.* In the context of the built environment, other regional projects and existing conditions will influence the project. For example, newly created transportation infrastructure may offer new site access opportunities; competing retail activity may challenge leasing agreements or assumptions; unforeseen site conditions, such as wetlands or soils, may reduce the buildable site area. These factors must be considered as they arise, typically during the later project procurement phases.
- *Legal requirements.* Legal requirements influence the project scope and methodology and can extend beyond local zoning codes and private easements or covenants to also include environmental regulations, health and safety standards, accessibility requirements under laws like the Americans with Disabilities Act (ADA), and even international standards if the project spans multiple countries. Additionally, legal requirements might also involve labor laws, intellectual property rights, and contractual obligations. The project team must

understand and comply with these requirements to avoid legal complications that can lead to delays, increased costs, project cancellation or legal liability.

- *Unique local requirements.* Local social norms, community perspectives, and regional comprehensive plans can have a significant influence on the likelihood of project success. Engineers should note that definitions of some code terminology will vary across city lines, and each locality has different processes, staff members, community perceptions, and nuances for securing permits.
- *Financial systems.* The financial systems that act as EEFs encompass the economic models and fiscal policies that affect project funding, cash flow, investment return expectations, and financial reporting requirements. This includes understanding the tax implications, availability of credit, interest rates, inflation rates, and the financial health of stakeholders. For publicly funded projects, this might also involve adhering to government budget cycles and procurement guidelines, or for private projects, navigating the investment strategies and expectations of stakeholders.
- *Political climate.* Changes in political leadership or policies can impact project priorities, funding, and public support. For example, a change in administration may alter infrastructure funding priorities or environmental policies. This may influence the risk appetite of the stakeholders and can dictate the pace and approach of the project.
- *Technological advancements.* The pace of technological change can affect the design process, construction techniques, and operational efficiencies, potentially offering new solutions or rendering planned approaches obsolete. For example, a new hospital project may be influenced by new medical equipment, which may prompt design changes.

Project management in civil infrastructure and real estate development requires a keen understanding of technical and organizational dynamics as well as a comprehensive awareness of these EEFs. Civil engineering projects are inherently linked with the geographical and community contexts in which they are situated, both influencing and being influenced by the communities and spaces they occupy. Project success hinges on understanding and harmonizing with many interconnected factors that arise from their geographic and societal context. By acknowledging and preparing for these factors, engineers can better navigate the complexities of procurement and project execution, ensuring that they remain adaptable and responsive to the changing landscape in which they operate.

2.2.3 Project Delivery Methods

Project delivery methods and the associated teaming arrangements will influence procurement processes. The most common delivery methods are design-bid-build and design-build. Traditional *design-bid-build (DBB)* projects are structured around an initial contract with a design team (or teams) that performs design work for the project; these same project design deliverables are subsequently used to establish a construction scope, which is then used to begin a new procurement process to hire a contractor for project construction. With *design-build (DB)* projects, the RFP solicits a team of designers and contractors working together. The choice of delivery method will influence the proposal requirements, evaluation criteria, agreements, and other specifics. For example, it is common to see a DB use an evaluation method that places a significant weight on the total project price, almost entirely governed by construction costs. Alternatively, DBB often prioritizes technical qualifications in the design phase and then the lowest bid in the construction phase. More information on the contractual arrangements of different delivery methods is included in Chapter 4, but the typical relationship between DBB and DB projects is shown in Figure 2.2.

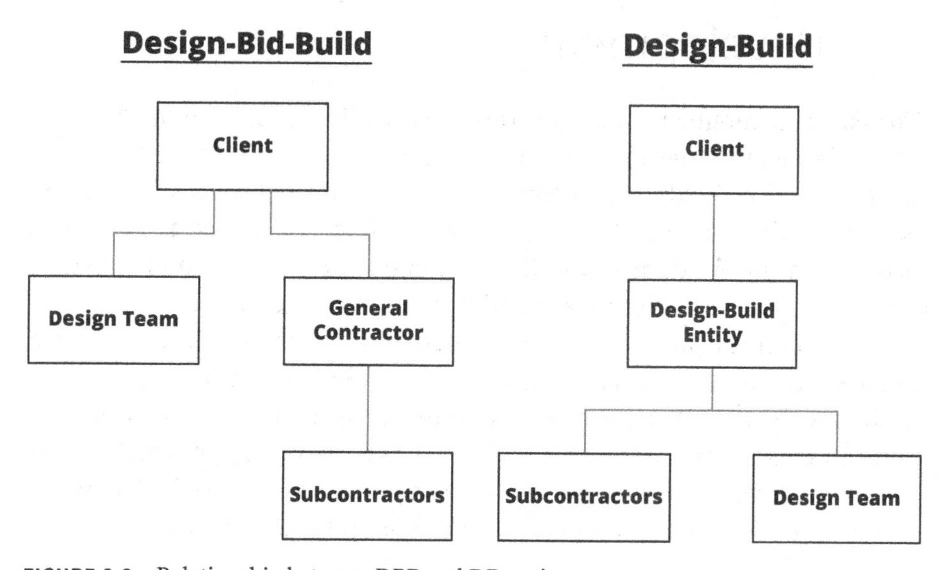

FIGURE 2.2 Relationship between DBB and DB projects.

In addition to a project-specific delivery method (DB and DBB), other procurement formats cater to different project types. A typical example is an RFP that establishes a basic ordering agreement (BOA), task order contract, master services agreement (MSA), or similar structure that seeks a qualified team that can be assigned multiple projects in a given time frame. In federal procurements for engineering services, these contracts are typically referred to by the acronym IDIQ, which refers to a contract of indefinite delivery, indefinite quantity. These types of RFPs seek to establish an on-call arrangement with a seller, preventing the need to issue an RFP for every project the client requires. Under these arrangements, the client reviews the qualifications of the engineering firm ahead of each new project and often has an agreed billing rate in place. The scope listed in the RFP includes a range of possible services instead of a project-specific scope. When the client requires a task, a scope will be developed and assigned to the selected engineering firm, design team, or larger team; these are often referred to as *task orders* in public-sector projects. For example, a contract term might be established for one to five years, requesting a pool of potential services and an expectation of the size of multiple projects to be assigned in that time frame. With these contract types, the selection process is expected to include multiple qualified firms.

2.3 Overview of the Procurement Development

The exact procurement strategy varies by each client (and type of client) but generally includes the following three stages: (1) prepare the procurement notice, (2) advertise the opportunity, (3) review bids and contract the award. Before an RFP is released, the needs must be defined, the project requirements determined, and funds made available to commit. This can be a protracted process involving several divisions of a particular entity or entities.

With a private project, the procurement process could be just a conversation that leads to a contract. In this case, the three stages would be informal, such as a client confirming internal financing structures, a casual meeting with an engineering firm, and reviewing one (or more) proposals. Alternatively, a private project may choose a more formal approach. The public sector follows a prescribed process and must publicly advertise a project, review qualifications and proposals, and announce the intent to award. Public-sector processes are often published in a procurement manual, adding transparency.

Next, RFPs are published or distributed. The RFP must be announced publicly for a public entity, allowing any qualified individual or organization to bid on the project. Depending on the agency and the project, the public sector may host public pre-bid meetings where any interested party can learn more about the project and ask for clarification. Publicly posted RFPs will frequently have a mechanism for public clarification and submission of questions and may include amendments to the RFP. These amendments answer questions or clarify the project objective or scope publicly or could modify project parameters.

Finally, the client will review proposals, evaluate them, and then decide to award a team and enter into negotiations for the contract. Public agencies can use several different bid structures for their procurement, each with its own evaluation process. Depending on the project and bid format used, a public entity may award one contract based on the proposal or may require several steps, such as ranking respondents, interviewing top candidates, and then negotiating.

2.3.1 Procurement Process

The procurement framework itself can be incredibly complex, particularly if a public-sector project triggers multiple levels of regulation due to joint project participation, oversight, and/or funding sources. A more detailed description of the formal public procurement process is outlined in this section. The following list provide more detail on several of these processes.

The procurement framework itself can be incredibly complex.

1. *Proof of concept and high-level cost estimate.* A proof of concept is often developed to establish the baseline scope of work. From this concept, a design and construction cost estimate is developed to determine financing requirements. A cost estimate informs the project budget; however, the RFP does not always advertise this information. The budget may be included to help inform technical consultants regarding the expected level of effort, or the budget may be evaluated later, along with bids from qualified candidates.

2. *Prepare the RFP.* The client establishes the SOW to identify the project requirements based on the client's needs and objectives. This should be specific and clear to mitigate interpretation issues in later project phases. The SOW and other parameters (cost, schedule, quality, etc.) are included in the RFP.
3. *Advertise the RFP.* This may be done publicly with a procurement system (typical of government agencies) or sent to a select group of bidders (typical of private-sector clients). The advertisement of the RFP should include a proposal submission deadline, the required content of the proposal, and deadlines with processes for questions and answers.
4. *Proposal evaluation.* After the proposal submission deadline, the client reviews each qualified proposal. In some cases, this includes a formal scoring system and ranking by a panel of experts; in other cases, the evaluation might be more qualitative. Proposals are evaluated across multiple criteria with varying weights, often including qualifications, team locations, organization size, cost, prior experience, team availability, resources, and other factors.
5. *Short lists and interviews.* A possible intermediate step before selection is creating a short list and interviewing the best-qualified bidders. This occurs typically with a smaller, select group of bidders, rather than the entire initial pool of submissions, when the client has determined they want to invest more time in hearing from select candidates before making the selection(s).
6. *Selection.* After the evaluation (and possible interviews), the client selects one or more technical consultants they wish to work with. Typically, selection notification is made through written correspondence, but a phone call may be used in certain circumstances or for private-sector projects. This step signifies the intent for a client to procure services from a selected engineering firms or design team.
7. *Negotiations and award.* Selecting a qualified engineering firm is one of the last steps before awarding a contract. After a selection, the client and chosen engineer or design team must negotiate terms and prices before a formal contract is awarded and the notice to proceed (NTP) is issued. There is no project until there is a contract and the project cannot start until the NTP.

When responding to RFPs with a project proposal, it is critical to follow the submission format and content precisely as defined. The proposal submission format is often prescribed in detail with an RFP, requesting certain sections, formats, content, and forms. For example, most RFPs require content such as organizational charts and team member resumes. Some requirements are as specific as establishing a minimum font size and maximum page count. If information is missing or does not meet the requirement, the proposal might be deemed noncompliant and would not be accepted for review. Another critical element is proposal deadlines – missing a deadline by even a minute will likely result in an immediate rejection.

2.4 Procurement Schedules

The schedule of the procurement process will vary, but in many cases of civil infrastructure and land development projects, there are just a few months or weeks from RFP issuance until the proposal deadline. A sample public-sector procurement schedule might resemble the example in Table 2.3. With only a few weeks to organize a team and prepare a proposal, many engineering firms will strategize before RFP issuance. By understanding the origin of a project, a design firm can create a competitive advantage by *pre-positioning* for an expected RFP. This should involve building relationships with the client, identifying possible team members, and attending meetings

By understanding the origin of a project, a design firm can create a competitive advantage by pre-positioning.

TABLE 2.3 Sample public-sector procurement schedule

Event	Date
RFP Issuance	June 01
Nonmandatory Proposal Site Visit	June 10 at 2:00 pm
Deadline for Questions	June 15
City Issues Responses to Questions via Amendment	June 25
Proposal Due Date and Time	July 15 at 4:00 pm
Short List & Interviews	August 15
Contract Award	August 30

or conferences where a client discusses future work or similar activities. For example, a public agency may hold several planning and community outreach meetings to discuss project scope and alternatives before the RFP is finalized and advertised. For the private sector, pre-positioning might involve *business development* efforts, where the seller establishes or grows the professional relationship with the client. The ability to pre-position can make all the difference in an effective response to an RFP because the engineer will better understand the context of the work and the project objectives; more information on proposal development is provided in Chapter 3.

The issuance of the RFP signifies the date that the RFP is publicly posted, allowing an engineering team to review RFP documents. Public agency RFPs are commonly advertised through a procurement website, ensuring open access for all interested parties. To provide more context about the project requirements and objectives, many schedules include time for a client meeting, which might last a few hours and may be mandatory or nonmandatory. These meetings often include a presentation by the client and are held onsite so prospective engineering partners can see existing site conditions.

Proposal discussions and questions are carefully administered. For example, questions may require a formal submission from the engineer and a public response from the client to ensure transparency and equity in the public process. Again, this question-and-answer process intends to provide a fair opportunity for all bidders by ensuring all parties have access to the same information. The client will issue an RFP amendment with a list of responses as questions are received. It is common to see reformatted questions to consolidate inquiries from bidders and to mitigate subversive misdirection (e.g., an offeror asking about rock blasting regulations when there is no rock onsite). Amendments may also be issued if unforeseen external environmental factors or new constraints arise during the procurement process or if a question prompts the need for a detailed response and clarification. As amendments are issued, the engineer must manage and address new information during proposal preparation. For this reason, in many cases, there is a requirement to include an acknowledgment of receipt of RFP amendments when submitting the proposal.

As previously mentioned, the procurement process holds a strict deadline for submission, and late proposals are rejected as a matter of this process. This practice is another measure intended to provide a fair opportunity for all bidders. In some cases, the procurement schedule may face delays, postponement of selection, or cancelation with (or without) a new posting of a new RFP. External environmental factors may force the client to reevaluate priorities or secure new funding based on bidder pricing. Sometimes, the client may not receive enough proposals to make a selection. As with other procurement processes, notification of delay or cancellation is typically advertised publicly. Unlike the engineer's obligation to submit a proposal according to a strict RFP deadline, procurement schedule changes are at the unilateral discretion of the client.

Within a few days or weeks of the RFP submission deadline, there is often direct correspondence to bidders that they have either been (1) selected, (2) not selected, or (3) shortlisted based on their submitted proposal package. A status of *shortlisted* means that the larger pool of initial candidates has been reduced and the engineer is included in the smaller, remaining pool of the highest-ranked submissions. In this case, the client requires more information before making a final selection. Additional information may be in the form of requesting new details or scheduling an interview between the client and engineer with an open discussion about the project approach, team qualifications, proposed schedule, or other project-related items. Eventually, a selection is made from the top shortlisted candidate(s), and the process moves into a contract negotiation and award phase.

The total number of steps in the procurement process may vary, and projects can include more or fewer steps than those shown in Table 2.3. The format can reference phases, such as a single-phase process that makes a selection from the responses to an RFP, or a multiphase process that begins with a request for qualifications (RFQ), shortlists the candidates, issues an RFP to the candidates, and then makes a selection from the proposals. For example, a larger and more complex project may include precursor activities to review qualifications (e.g., letter of interest, (LOI), statement of qualifications (SOQ), or expression of interest (EOI)) before the distribution of the RFP.

The project size may dictate the procurement process, where less complex projects are generally more direct with the procurement process. Still, additional steps may be seen if a project is deemed to have sensitive information as part of the RFP scope, which may be the case for public safety facilities, government buildings, or similar projects. By including a pre-qualification phase, the client limits the dissemination of potentially sensitive project information while providing an opportunity for all qualified candidates.

The sample schedule shown in Table 2.3 serves as the outward-facing representation of the process associated with a procurement. However, within the client's organization, ongoing internal processes maintain schedules not shared with engineering bidders. Given the complex procurement process, it is expected to see a dedicated staff member, typically called a procurement or contracting officer (CO), assigned to each RFP and involved in the procurement process as the client's official point of contact. In addition to the CO, a client might have internal selection committees and committee chairs, with oversight from contract and financial officers, reviewing submitted documents. Before the issuance of the RFP, the client will finalize a procurement strategy and obtain internal approval of the process, scope, budget, and schedule. Evaluation scores are documented and shared as determinations are made for selection or *regret letters* issued to each engineer-bidder. During this preparatory phase, the client's timeline is established to maintain project momentum, providing timely answers and selections to each bidding firm.

2.5 Overview of the RFP

Consistent formatting and content of an RFP ensures that the client receives consistent responses. A government agency or a private organization may have a procurement manual that dictates the contents of an RFP. An RFP could be concise, asking only for evidence of qualifications, resumes, and a price, or may include more complex content outlined over dozens of pages. Within an RFP, there are often specific instructions about formatting or a list of content or questions the client seeks answers to. As each RFP is unique, the project team must review all RFP content in detail to ensure a valid proposal is made. Even with the same client, RFP requirements can vary across projects. An example table of contents for an RFP is shown in Figure 2.3.

> The project team must review all RFP content in detail to ensure a valid proposal is made.

Request for Proposal

1. Purpose & Intent
2. Background
3. Scope of Services
4. Proposal Requirements
5. Solicitation Requirements
6. Evaluation & Award
7. Additional Terms

FIGURE 2.3 Sample RFP table of contents.

This sample table of contents is just one format of an RFP that highlights the major categories:

1. Purpose and Intent
   - Generally, the introduction content includes notes about the client with a narrative about high-level expectations. For example, a county seeks to acquire engineering services for one (or more) firms to complete a scope of services.
2. Background
   - This content describes the origin of the project or purpose of the RFP, providing some context about why the scope of services is being requested.
3. Scope
   - This defines the work to be performed, often broken into various subcategories. For example, the scope might reference the project size, the services requested (structural, transportation, environmental, etc.), the expected schedule, the project deliverables, and other work.
4. Proposal Requirements
   - The expected format of the proposal is defined, which typically requires information such as a cover letter, understanding of the scope, firm qualifications, team member resumes, and other administrative forms. Other directions might include proposal formatting and submission requirements, such as minimum font size, maximum pages, and submission deadlines.
5. Solicitation Criteria
   - This documents the RFP process, referencing an intent for competition, the possibility of addendums, the potential to reject noncompliant proposals, and similar information.
6. Evaluation Criteria
   - This includes the qualitative or quantitative way proposals will be evaluated, scored, and then ranked for selection (See section 2.6 for more details).
7. Contract Terms
   - These are often included by the proposing team so the engineer understands their commitments. The engineering firm's attorney typically reviews these terms before developing a proposal.

A significant challenge can arise when a client does not know exactly what should be included in the project parameters or RFP. This can lead to unclear requirements in the RFP and the need for extensive subsequent amendments or clarifications. In many cases, a client may have a clear idea of the project objective but have challenges communicating the technical scope of work. Just as engineering firms may struggle to communicate their qualifications in a proposal, clients often face challenges with publishing an RFP. Given the complex scope, large teams, and long RFP development process, sometimes an RFP includes conflicting requirements, unreasonable schedule expectations, missing information, and ambiguous requirements. For this reason, many RFPs provide an opportunity to submit requests for information (RFI) via formal written channels. The questions are then answered and shared with all offerors, typically with clarifying addendums to the RFP. The process may be less formal with private clients, where discussions can lead to a mutual understanding of the project intent and scope (then documented in the proposal of services).

A client may have a clear idea of the project objective but have challenges communicating the technical scope of work.

2.5.1 Diversity and Inclusion in Procurement

Many clients recognize the benefit of working with a diverse team and understand the obligation to mitigate past discrimination in the infrastructure design industry. For example, the U.S. Department of Transportation manages the Disadvantaged Business Enterprise (DBE) program, which supports women and minority-owned businesses. As stated by the U.S. Department of Transportation, "The primary remedial goal and objective of the DBE program is to level the playing field by providing small businesses owned and controlled by socially and economically disadvantaged individuals a fair opportunity to compete for federally funded transportation contracts." In federal public-sector procurement, many RFPs mandate that certain types of small businesses be included in the proposed design team and that a certain amount of the project work and funding be allocated to the small business team member.

Projects benefit from diverse perspectives. Civil engineering projects cater to various users, and DBE participation improves how solutions and designs are developed. Many DBEs offer specialty services based on unique core values. For example, a DBE may specialize in sustainable design infrastructure with a rich portfolio of prior projects not seen with a more prominent general consulting firm. Local DBEs can also foster better engagement with the community, understanding historical context, local operations, community values, etc. These partnerships can generally lead to more competitive pricing, innovative solutions, and an opportunity for teams to distinguish themselves. Even without a formal requirement to include a DBE, some proposals can incentivize participation. For example, some scoring points may be provided if the offeror involved a DBE, or there could be a qualitative statement that DBE participation is favored.

Civil engineering projects cater to various users, and DBE participation improves how solutions and designs are developed.

2.6 Overview of RFP Evaluation Criteria

An RFP will typically include information that allows the engineer to understand how a selection will be made. Sometimes, this might only be a qualitative statement about the selection method (e.g., the most qualified firm will be selected based on the information provided in the proposal). A standard evaluation tool used in public-sector selection processes is a scoring mechanic that evaluates multiple criteria. This evaluation method often includes a total possible score wherein various criteria are weighted with possible points totaling a maximum value (e.g., 100 points). Each firm's score is then used to rank the engineering bidders for short-listing or selection. The criteria and weight of the evaluation metrics will vary across clients and potentially with different projects. This is often referred to as multicriteria decision analysis (MCDA). A sample of evaluation criteria for an example project is shown in Table 2.4, where each criterion would reference an RFP requirement.

In the sample shown in Table 2.4, the list of assigned points provides evidence of how the client rates the importance of each criterion. In this example, the most weight is assigned to project understanding and qualifications (Criteria #4), with similar importance placed on team or organizational experience (Criteria #2). This is important to consider because it communicates how the client prioritizes each criteria, which can influence how the bidder (engineering firm) crafts a proposal. For example, if a client lists a criteria of "sustainability approach," it would indicate that this is important to the client and that the engineer should weave in sustainability

TABLE 2.4 Sample project evaluation criteria

#	Criteria	Points
1	General Information (R1–R4)	5
2	Qualifications and Experience of Proposed Team Members (R5)	25
3	Experience with Similar Project Types (R6)	20
4	Project Understanding and Technical Approach (R7)	35
5	Risk Management Approach (R8)	10
6	Administrative Requirements (R9)	5
	Total	**100**

concepts across other proposal sections, perhaps describing the engineering firm's internal sustainability operations and demonstrated success. It is also common for RFPs to request other qualification criteria, such as geographic proximity to the project or client testimonials. Some criteria, such as providing general information or assessing administrative requirements, are used to determine if a proposal submission is *responsive*, meaning that the bidder has taken the time to follow the client's instructions and adequately address all required information. The omission of critical information or RFP administrative items could disqualify the proposal.

Many RFPs will request budgets as additional criteria, sometimes evaluated separately from the technical qualifications. As an example of a calculated selection process that considers both cost and qualifications, suppose a city is seeking proposals for a new public parking garage advertised as a design-build project. The RFP states that the proposals will be evaluated with a 25% weight on the technical engineering approach (such as the criteria shown in Table 2.4) and a 75% weight on the submitted price. The evaluation of technical scores of submissions from firms A, B, and C are shown in Table 2.5.

TABLE 2.5 Evaluation of technical scores

Firm	Technical Score	Weighted Score	Price	Price Score	Weighted Score	Final Score
A	81		$49.8M			
B	70		$45.3M			
C	79		$47.7M			

Next, the pricing scores are calculated. The scoring for pricing is such that the lowest price receives 100 points, and each competing offer is scored as 100 × (lowest/offer) to determine a numeric score, as shown in Table 2.6. For example, Firm B holds the lowest price at $45.3M, so they receive 100 points. Firm A submitted a price of $49.8M, so they receive a score of 100 × (45.3/49.8) for 91 points. The weights of 25/75 for technical price are used to compute the final combined score to select the winning proposal, also shown in Table 2.6.

TABLE 2.6 Calculating combined score to select the winning proposal

Firm	Technical Score	Weighted Score	Price	Price Score	Weighted Score	Final Score
A	81	20.25	$49.8M	91	68.25	**88.5**
B	70	17.50	$45.3M	100	75.00	**92.5**
C	79	19.75	$47.7M	95	71.75	**91.5**

When evaluation criteria include both cost and qualifications, the technical approach can carry considerable financial weight. For example, if Firm A has received a technical score of 100, their combined score would have exceeded the low bid score. This means an excellent technical proposal would have overcome the $4.5M construction cost difference.

When evaluation criteria include both cost and qualifications, the technical approach can carry considerable financial weight.

In addition to the MCDA selection method, many other standard evaluation methods are commonly seen in practice; additional information and context for each selection method are provided below:

1. *Least cost*. With this selection method, a qualified design team is selected based on price alone. The caveat is that a prequalification process often verifies that a bidder meets the minimum technical requirements for the SOW. Sealed bids are sent to the client by an established date, and the client then evaluates all prices to identify the lowest qualified bid. This approach is best used for projects with a very well-defined SOW, such as routine design and construction, or if the project delivery method

is design-build, wherein the design-build team is accountable for project costs.

2. *Qualifications only.* The selection is made solely on the qualifications of the engineer or design team. In these cases, a budget and schedule have been predefined and possibly advertised with the RFP so the firm understands the project constraints. This method is common for specialty projects, such as advanced engineering and research work, where the details of project scope are challenging to define, and the client and engineer must negotiate a fair price that is within budget – or the scope is refined to accommodate cost constraints.
3. *Qualifications-based.* This qualifications-based selection format can use MCDA when evaluating the proposal, and then prices (or staff rates) are negotiated after a selection. The client may have a preestablished price that is disclosed or reserved and is looking for an engineer who is qualified and meets the established budget. Many government agencies will follow this methodology, seeking to select the most qualified firm but recognizing that price and budget will influence the ability to select the desired technical consultant.
4. *Qualifications and cost-based.* In this selection format, cost is part of the MCDA process, with varying weights (e.g., 75% price and 25% qualifications) assigned to the selection criteria. This way, technical qualifications are grouped with one weight, and cost is grouped as another. This selection process seeks to balance qualifications with the cost of a bidder. The scoring of each submitted price is often evaluated as a function of the lowest received bid (or the budget), and all other prices are ranked accordingly. This selection method is common on design-build contracts where construction cost is a significant factor, but the qualifications and technical approach can influence the selection.
5. *Sole source.* A sole source selection means that a client has identified only one qualified technical consultant to perform the work; this firm is effectively, automatically selected and then negotiates costs with the client. This is typical for specialized projects where no other entity is (reasonably) capable of performing the work based on expert knowledge of the project, unique manufacturing capabilities, or similar specialized experience.
6. *Fixed budget.* This is the preferred selection format for projects where the available budget may be the governing factor for selecting bidders

and refining the project scope. In this case, the budget is communicated to potential bidders, and elements of scope, or the quantity of units, are driven by the available budget. For example, an engineering research project may have a limited budget and request the bidder to determine the number of case studies that can be published with the project.

Legal conditions or operational procedures may govern the choice of evaluation method and are likely influenced by the type of project work. For example, selecting a team to design a new highway interchange could use both quality- and cost-based methods, considering the various interchange design options and the design/construction cost of the proposed options. However, a routine infrastructure maintenance project, such as designing and repairing sidewalk curb ramps, might be best served with selection using the least cost option. As a final example, an owner looking for on-call services with multiple tasks may decide on a qualifications-only contract and negotiate the scope and cost for each task order.

Total Evaluated Price

Other evaluation methods exist, often tailored to a project or market based on the client's needs. For example, some government projects may use a value-adjusted total evaluated price method, which has a unique perspective on how value-added services can influence the cost factor of a proposal. These adjustments could include a monetary value assigned for a bidder's approach to early delivery, the use of local vendors, minimizing traffic disruption, exceeding minimum environmental regulations, reducing operational costs, or other criteria. This approach is similar to qualification and cost-based methods but quantifies the monetary value of the proposed methodology. In this way, an engineer may have a higher upfront cost, but the *evaluated price* is competitive because the methodology or scope has added value to the project. Figure 2.4 shows how various consultants (#1–5) might have an adjusted *total evaluated price* if a proposal meets certain value thresholds.

With the example in Figure 2.4, some contractors (#1 and #5) have not met technical qualification thresholds or have exceeded the allowable budget. The other candidates (#2, #3, and #4) are in the competitive range. Bidder #4 received the highest technical score and initially had the highest cost; however, if the proposal includes criteria that meet certain value thresholds (e.g., a design that allows for lower operating costs), the *total evaluated price* is improved, and #4 becomes the best option for selection.

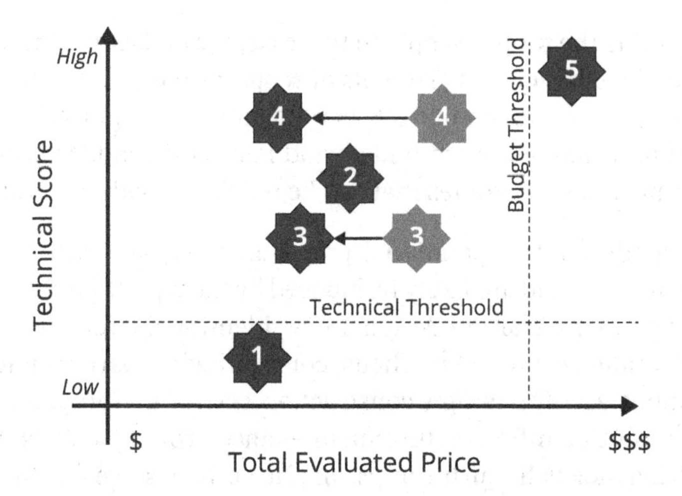

FIGURE 2.4 Comparing technical score against total evaluated price.

2.6.1 Shortlisting and Interviews

After evaluating a written proposal, the client may select a few candidates and ask for more information to advance the selection process. This is referred to as being *short listed*, which indicates that a bidder has ranked well among the larger pool of candidates, but a final selection has not been made. For example, an RFP open to the public might receive dozens of proposals from interested design firms, but only the top three might be shortlisted.

Following the notification of being shortlisted, the next step often includes an interview, wherein the client asks for a formal presentation by the proposal team describing their qualifications and project approach. This allows the client to ask questions directly to the project team. Similar to the requirements established for the proposal submission, an interview will include parameters such as information to be presented, the maximum duration of the presentation, the maximum number of participants, questions to expect, or similar information. This interview process will often include additional qualitative evaluation criteria.

2.7 Awards and Agreements

Once the client has reviewed and evaluated all compliant bids, a winning firm will be selected for project award. The selection process and methodology can vary, with a more formal system established for public-sector agencies or private firms choosing to use an established selection process. While it is rewarding to be notified of selection for a project, engineering team leads must remember that the award does not mean the team has a contract for the work. Similarly, as rewarding as it is for the client to have selected a qualified team, some steps must be taken before a contract is established and work can begin. Both parties must understand that the selection or an intent to award differs from a formal agreement. Selecting an engineer or design team only signifies a client's interest in working with the selected firm. Still, contracts must be formalized before the project is awarded. Both engineers and clients can encounter billing and liability problems if work begins before a contract is signed. For public-sector clients, work that begins before an official notice to proceed (NTP) is an *unauthorized commitment*. Technical work on a project should only begin after a formal binding agreement for the design team to provide a defined SOW has been signed by both parties and, in the case of a public-sector client, an NTP has been issued.

The award does not mean the team has a contract for the work.

Legal aspects of procurement are critical – the terms and conditions established during the procurement phase will be a reference throughout the project and if (or when) conflicts arise. The terms may stipulate the timing of payments, adherence to the schedule, ownership of intellectual property, and other factors that must be considered. This section only serves as an introduction to some of the legal aspects, with more information provided in Chapter 4, which includes content such as:

- The necessity of agreements before work begins.
- Mutual understanding of the contract terms and conditions.
- Review and negotiations of contract terms and conditions.
- Subconsultant agreements between design team partners.

Procurement structures will vary – the engineers may directly contract with the client, they may have a contract with a prime consultant (who has a contract with a client), or there could be a joint venture (JV) where multiple parties are operating as a single entity for the client, or other similar arrangements. The engineer may also procure services, as they seek to satisfy project scope requirements by teaming up with other partners (e.g., a civil firm might select as survey firm as a subconsultant). This inherent complexity means the procurement process occurs long before the technical work can begin and can take several months or years (longer durations are typically associated with public clients). The procurement process, documentation, and agreement will serve as the foundational documents of a project that define the scope and schedule of the work.

2.8 Procurement Ethics

There are numerous ethical considerations involved in the procurement process. The following discussions provide examples of points to consider but are far from the only ethical scenarios in procurement. When focusing on the procurement process, clients and engineers should apply the basic ethical principles outlined in Chapter 1 and refer to their firm and industry codes of conduct for guidance when necessary.

2.8.1 Business Development and Gifts

A standard part of a firm's operations typically involves building client relationships, which can benefit the engineer and the client. For example, teams or organizations familiar with each other will benefit from improved project communication and shared systems or operating practices developed over time. Similarly, an engineer may better serve a well-known client whose needs, preferences, and motivations are familiar to the engineer through repeated discussions or project partnerships. In exchange, engineering firms anticipate earning future business due to ongoing client relationships.

The firm's effort to build and maintain these client relationships is *business development*. Business development can happen organically through networking at professional industry events, communication and marketing efforts, or client-specific interactions. Many of these interactions involve a monetary cost to the engineering firm. This might include personnel time, monetary support for industry events, or strategic

Business development can happen organically through networking at professional industry events, communication and marketing efforts, or client-specific interactions.

research and development investments that align with client needs. This cost may be supported by a dedicated business development team or managed budgets that monitor the cost and return of the investments. Examples of common business development expenses include:

- *Preliminary site assessments.* Prior to project announcements, engineers may provide prospective clients with complimentary (or low-cost) preliminary assessments of potential sites, investing a few hours or days in research and communication to facilitate early decision-making.
- *Competitive pricing strategies.* To secure contracts, firms might offer financial concessions, accepting lower profit margins to cultivate client relationships, respond to competitive pressures, or in anticipation of future projects that may stem from an initial engagement.
- *Relationship-building gestures.* Informal gestures, such as bringing snacks to meetings or sending holiday gift baskets, serve to strengthen rapport and goodwill with clients on a more personal level.
- *Branding and team-building gifts.* Items like company or project-branded merchandise could be given to new subcontractors as a token of partnership and team unity.
- *Celebratory events.* The engineering team lead may sponsor dinners to celebrate project milestones with the team, consultants, and clients, reinforcing team cohesion and acknowledging shared achievements.
- *Client entertainment and reciprocity.* Firms may extend invitations to clients for industry events, covering ticket costs as a gesture of appreciation and networking, while also welcoming similar gestures from clients or project partners in kind.
- *Encouraging client feedback.* During the project, and at the end, inviting clients to provide informal (open dialog) or formal (survey forms) feedback on experiences and satisfaction will demonstrate the engineering firm values this information.

All the examples above are common practice, and the ethical implications of these and other examples of business development "gifts" are highly subjective. A gift's timing and cost may influence how it is perceived ethically.

A gift's timing and cost may influence how it is perceived ethically.

When considering the timing of client gifts, context is crucial. A gift may be seen as a normal token of appreciation when given after a project award or during its execution. However, if presented just before a new project announcement, it might raise concerns about undue influence, whether this perception is by competitors or the client itself. The situation becomes more complex when the timing of a gift coincides with the bidding process by chance or appears to be strategic. Budget constraints or event scheduling may also dictate when gifts are given. Additionally, in a large firm managing concurrent projects for a single client, gift-giving for one project may unintentionally coincide with the procurement phase of another, inviting misinterpretation.

Regarding the cost of gifts, the ethical evaluation is not always straightforward. It might be an oversimplification to deem inexpensive gifts acceptable. The context of the project's scale is significant; what might be considered costly in one scenario could be relatively minor in relation to a large project's overall budget. Therefore, the proportionality of the gift's value compared to the project cost can also be a determining factor in assessing its appropriateness.

Another consideration is gifts between prime contractors and their subcontractors. For example, if gifts are exchanged in the context of one project, but the prime subsequently recommends the subcontractor to the client, the potential connection between the two events may create an ethical question. Suppose a civil engineering firm is regularly a subconsultant for an architecture firm that works on public school projects. The civil design scope is relatively small for a new public school project, so the client primarily evaluates the architecture firm and trusts them to select the right civil subconsultant. If the architecture firm regularly receives gifts from the civil engineering firm, then it could be perceived as an ethical issue – is the architect making the best choice for the client or favoring the civil engineering firm that provides the most gifts?

There are generally no restrictions on business development practices in the private sector, making it incredibly important for engineers to consider their actions' perceptions and ethical implications. Public-sector entities have policy controls prohibiting their employees from receiving gifts altogether or being limited by a low dollar value. These policies are intended to prevent undue influence in project awards, conflicts of interest, and any perception of favoritism. Because these policies exist, drawing boundaries around ethical behavior with public-sector clients is relatively more straightforward. For example, bringing snacks to meetings is likely acceptable, but offers of event tickets, dinners, gift cards, etc., should be avoided. However, even with these restrictive policies, there can still be ambiguity. If an engineer hosts a team dinner, a public-sector client may want to join and insist that doing so is acceptable if they are not the formal project point of contact, or they choose to participate in a group event but pay for themselves. A public-sector client may believe that accepting a gift is permissible if the incremental cost of the item or ticket is marginal. In these and other scenarios, the engineer faces an ethical quandary about whether it is the firm's job to understand and enforce public policy on government clients.

In Appendix A, Part 2 of the Millbrook Logistics Park, the narrative progresses through the development of a Request for Proposal (RFP) by the client (TerraHaven) for the upcoming project. Part 2 outlines the process of RFP preparation from the client's perspective, emphasizing the challenges encountered in establishing a detailed technical scope of work. Furthermore, it documents informal interactions between the client and a prospective engineering firm (ApexTech) that raise potential ethical issues. These discussions are critical in understanding the ethical considerations necessary to ensure transparency and fairness in procurement activities within civil engineering projects. The exposition of these processes and challenges provides insight into the complexities of procurement management.

CHAPTER 3

Project Pursuit Processes

If the team can't communicate that they're qualified, maybe they aren't.

CHAPTER OUTLINE

Management Essentials for Civil Engineers: A Practical Guide to Business, Communication, Ethics, and Risk, First Edition. Cody A. Pennetti, C. Kat Grimsley, and Brian M. Grindall.

This chapter explores how the engineering team evaluates a request for proposal (RFP) and prepares a response based on the project needs, as introduced in Chapter 2. The first step in a pursuit is deciding whether to invest the time to submit a proposal for work. This is followed by establishing the project team through formal agreements before working through the steps to develop the content and submit a proposal.

3.1 Introduction

Writing a proposal (responding to an RFP) should not be taken lightly – it is a significant effort that requires substantial organizational resources. The submitted proposal represents the organization's services and quality of work, which means the package needs to be refined and vetted to demonstrate technical and managerial competence. Before starting a proposal, deciding if the pursuit is worth the investment is important. Preparing a proposal often requires a team of professionals to spend days, weeks, or months writing and coordinating content. In many cases, the team of professionals writing the proposal are technical and management professionals working on other active projects, which means the proposal work can represent a significant additional effort. While they are supported by marketing and communications team members who help write, edit, and prepare media for the proposal, there is a an indirect cost to the organization for the time spent on proposals. To ensure the effort is justifiable, many organizations have a go/no-go decision before starting a proposal, as Section 3.2 of this chapter outlines.

As defined in Chapter 2, the procurement process can be time-intensive and rapid. The process includes short deadlines for submissions, typically followed by interviews, and an often-tedious contract negotiation phase – all before the firm is officially hired and the engineering work can begin. Whether the process follows a formal style seen with public projects or an informal set of prompts with a private client, the content preparation for a proposal follows similar steps.

Early pursuit efforts by the engineering firm will focus on pre-positioning for the RFP, building relationships, and developing an understanding of the project objectives. As the scope of work is refined, project teams are established so all necessary services can be provided through one or more firms.

After an RFP is issued, the team will take ample time to review all requirements, the scope of work, selection criteria, schedules, and contract terms before deciding to submit a proposal.

The development of the proposal often moves through internal review phases, validating whether the content addresses all requested items and that the format and quality meet corporate standards. After proposal submission, additional efforts may be required to prepare for interviews or discuss details of the contract terms, price, schedule, or other information not initially required during the selection process. However, the team cannot move through these advanced proposal stages until undertaking the first critical step: a decision to pursue the project.

3.2 Go/No-Go Decisions

Even if the project's technical scope is intriguing to the engineering team, an organization often requires formal go/no-go decision-making before an engineering firm invests the resources to pursue a project. This process evaluates the feasibility of success, both of winning the work and of undertaking the project itself, by considering the project scope details, the client relationship, the client's reputation, the contract terms, the organization's available resources, and other factors. At the highest level, an early decision could be made based on criteria such as those defined by PMI:

1. Unfavorable legal terms and conditions
   - The contract will dictate many of a project's operating, payment, and liability conditions. An engineering firm may not pursue a project if there are excessive liability provisions, indemnification without limitation, hefty insurance requirements, lengthy payment terms, and others that could deter the firm's decision to submit a proposal. Many organizations will require internal counsel to review contracts to identify these risks before investing resources in the pursuit.
2. Client relationship
   - There are two key factors concerning an evaluation of the client relationship. First is determining what past work (if any) the engineering firm has performed for the client before and evaluating if

that prior project was mutually successful. The second pertains to determining whether a conflict of interest exists with the client, which can occur, for example, if the engineering firm already works with a competitor.

3. Lack of qualifications
    - Professional ethics require an engineering firm to exercise honesty when acknowledging the limitations of service. Suppose the engineering firm has little to no experience providing the requested services and no ability to partner with a qualified team. In that case, it is unlikely that the pursuit will be successful. It is important to distinguish between a firm's institutional qualifications and the qualifications of the individuals who will be performing the work.
4. Conflicts with existing or future work
    - Similar to a potential conflict of interest with a client, there may be concerns with the project scope of work if it conflicts with the firm's other project commitments. Many organizations have a diverse project portfolio, serving public and private clients. This may result in a conflict wherein the engineering firm has obligations on both sides of the project as public agencies negotiate requirements with a private entity. For example, an engineering firm may serve a public transportation agency on a highway project, seeking cooperation from a property owner adjacent to the roadway, which the engineering firm also represents.
5. Lack of resources
    - An engineering firm will not win every pursuit, meaning many proposals are developed simultaneously with the same key staff leveraging the same resources. Early pursuit decisions must recognize the resource limitations of the organization and feel confident that the available staff can reasonably serve the project. In other words, if the proposal team cannot submit a compelling response, it may make better strategic sense for the team to pass on the bid and dedicate their time elsewhere to ensure the firm's greater success. Note that this may mean the firm declines to bid on a project even when they have the technical competency to complete the work. This can be an uncomfortable outcome if it is perceived as "losing business" instead of as a strategic decision.

6. Financial risk
   - As noted in Chapters 2 and 7, not all project financial terms are the same. The schedule of payments, the insurance requirements, the liability, and other factors may carry an unattractive risk to the engineering firm. For example, a project may restrict payment until long-range milestones are met, which burdens the firm's cash flow. In such cases, the firm may determine that pursuing the bid is unjustified even if the team can provide the required services.
7. Lack of project knowledge
   - Without an effective pre-positioning tactic, an engineering firm likely does not have the time and resources to understand the project objectives when the RFP is published. A proposal should not serve as an introduction to the client. A lack of understanding of the project can result in a misdirected pursuit package (even if the content is strong technically) and a poor first impression if the proposal strategy is misaligned with the client's goals.
8. Significant resources required for the pursuit
   - Just as a lack of technical resources may be a factor in the pursuit decision, the high demand for resources or time to prepare the proposal must also be considered. Even if the engineering firm is well-positioned to win the project, the proposal must communicate the engineering firm's qualifications effectively. Without the necessary resources, the quality and content of the proposal will suffer, as will the firm's reputation.
9. High-risk project type or client
   - Not all RFPs lead to a project, which makes the time and cost invested in preparing the response a sunk cost for the firm. Even if the team wins the bid, the client may decide to delay or postpone a project for multiple reasons; funding sources, client reputation, and external factors may influence the project initiation. The engineering firm should only pursue work if there is a high degree of confidence that the project will eventually progress. Some organizations may choose not to work with certain types of clients because they don't feel they can provide the necessary level of service.

10. Values or goals don't align with the organization

- The culture and values of an organization will often guide initial decisions for a pursuit. The opportunity may be passed if a project (or client) does not align with the engineering firm's values. Differences in political or socioeconomic alignments can drive this, or if the engineering firm determines that a project is not a worthwhile use of resources.

An engineering firm must evaluate whether a potential project aligns with its strategic initiatives.

Generally, an engineering firm must evaluate whether a potential project aligns with its strategic initiatives. This assessment considers multiple criteria because pursuing and executing a project will depend on more than firm profitability. An organization will consider both threats and opportunities in the project work. Will this project allow the organization to grow, and is the organization prepared to grow? Will this project damage or strengthen the relationship with other core clients? These questions and the criteria listed in this section inform an organization's go/no-go decision process. In addition to go/no-go qualitative assessments, engineering firms use several quantitative methods to determine the practicality of investing in a pursuit, explained in the next section.

3.2.1 Pursuit Decision Process

The go/no-go decision process is a firm's first action before pursuing new work. Sometimes, the project criteria may warrant an immediate decision to pass on the opportunity. For example, a large engineering firm may not be positioned to serve a small client that has a limited project budget (i.e., the firm has established a minimum contract value threshold to warrant a pursuit). Similarly, suppose the last project with a client was not profitable or otherwise beneficial (or enjoyable) for the engineering firm – that experience will influence the decision to engage with that client again.

Typically, organizations employ a formal process that utilizes both qualitative and quantitative metrics to assess opportunities. In this process, criteria like those detailed in Section 3.2 are systematically evaluated. Many go/no-go evaluation criteria consider the client and project conditions, questioning whether funding is secure or if there are conflicts with the client or project scope. This decision process also considers criteria such as client relationships, project knowledge, assumptions about competitiveness, and pre-positioning efforts. Financial information, such as the client's expected budget and the firm's estimation of project costs, will inform whether the project can be profitable.

The qualitative aspect of this approach is discussion-oriented, acknowledging the value of expert insights in facilitating swift decision-making. In some cases, participation by key team members can quickly reach a consensus based on expert knowledge. However, to ensure structure and consistency in these evaluations, a qualitative review should incorporate elements like talking points or a questionnaire. Such tools not only guide the team through the review process but also create a documented record of the decisions made. The documentation is crucial as the engineering firm evaluates future pursuits and seeks to benefit from prior evaluations.

For a quantitative approach, the criteria are evaluated with metrics indicating the self-assessed numeric score. Multiple criteria can be established by the engineering firm, with weights assigned to each condition based on the perceived importance. An engineering firm should have a well-documented scoring mechanic to ensure consistency. Table 3.1 provides an example of a go/no-go decision matrix that an engineering firm might use, as referenced from the criteria identified in Section 3.2.

Scoring is often completed through the pursuit team's self-evaluation or informed by other internal participants who can remain honest and objective about the team's capabilities and the project parameters. The process includes scoring metrics, but even one criterion could identify a critical reason not to pursue the project work—for example, unfavorable legal terms or misalignment of values. The evaluation should be well-grounded, often critiqued by someone in the organization who can provide a fair assessment, maybe because they operate under a different business unit and have no vested interest in the opportunity. Even if the project team is passionate about the scope of work, artificially gaming these scores will only waste organizational resources if there is no perceived value to the firm or little chance of success.

Even if the project team is passionate about the scope of work, artificially gaming these scores will only waste organizational resources if there is no perceived value to the firm or little chance of success.

TABLE 3.1 Go/no-go decision matrix

#	Criteria	Score Range	Scale	
			Low	**High**
1	Legal Terms and Conditions	0–10	Unfavorable legal terms.	The contract uses the firm's legal terms and conditions.
2	Client Relationship	0–10	Never worked with the client.	Regular client with a positive experience.
3	Qualifications Alignment	0–10	The firm does not currently have most of the required expertise.	The firm has expertise in all required domains.
4	Potential for Conflict of Interest	0–10	Known conflicts, such as current work with client's competitors.	No known or anticipated conflicts of interest.
5	Resource Availability	0–10	Current resources (staff) are already overcommitted.	Adequate resources (staff) are available for the project work.
6	Financial Viability	0–10	Questionable funding or budget could result in financial loss.	The budget is lucrative and could yield high profits.
7	Project Knowledge and Pre-Positioning	0–10	The bidder was unaware of the project before the RFP issuance.	Strong pre-positioning and prior project research and context awareness of project objectives.
8	Resource Availability for Proposal Development	0–10	The proposal and marketing team is overcommitted, or RFP deadlines are too short.	Adequate resources are available to develop the proposal.
9	Risk Profile	0-10	High risks are identified (technical, financial, legal, reputation).	Minimal risks were identified; there was confidence in the risk response techniques.
10	Value Alignment	0-10	The project may have an ethical or strategic misalignment with the engineering firm.	Project objectives and values align with the firm.

This scoring system shares similarities to the RFP evaluation criteria identified in Chapter 2 (Section 2.6), but this scoring method is for internal use by the engineering firm (the bidder, or seller of services). The engineering firm should align the go/no-go scoring metrics based on the RFP evaluation criteria. For example, if geographic proximity to project work is a scored criterion with the RFP, it should be considered with the go/no-go decision.

The decision-scoring system uses numeric values but primarily relies on subjective and qualitative evaluation to determine the score. Still, the evaluation process provides a record of the result, identifying the factors that influenced decisions. This record could be helpful when evaluating retrospective analysis based on the outcome of the pursuit (a win or loss). This is important data even when the engineering firm decides not to pursue a project, as it informs recommended actions to pre-position for the next opportunity. For example, the next go/no-go decision process might score better if the team can strengthen the client relationship, improve the resources available, and gain additional knowledge about the project background.

Chapter 7 provides more information on project cost management, but an organization must also consider cost conditions during the initial phases of work. The intent is to understand if the project budget can accommodate the proposed scope of work, which informs the pursuit budget and expected revenue.

3.2.2 Cost of Sales

The resources required to develop a proposal (and hopefully win the pursuit) are only one function in the *cost of sales*. Before the proposal is developed, an engineering firm often invests other time and resources to pre-position for the RFP. These efforts include strategic planning meetings, meeting time with the client, marketing, business development, time and costs to establishing teaming partners, and other similar efforts. All costs are associated with the pursuit and should be tracked appropriately to enable an organization to compare the cost of sales for different projects and clients. Developing a proposal is often the most significant cost of the pursuit, but it's essential that all time and costs associated with pre-positioning are also tracked. In this way, an organization can better understand the pursuit costs and the realized benefits.

Preparing a proposal package, especially with design-build projects that require upfront design development, requires significant time and resources that can be budgeted based on the anticipated project revenue. For example, an organization may only pursue a project if the profitability justifies the pursuit's expense and risk. As a basic example, an organization may determine the pursuit budget (internal time and resource allocation) as a function of the probability of award and the net fee.

The probability of winning may be measured qualitatively (low to high) or through a quantitative metric established by the engineering firm. For example, if the firm is an incumbent and the client survey has been favorable, the probability of winning the new project could be near 70%. Conversely, if the engineering firm's personnel feels they have the technical capabilities to meet the project scope but have no direct relationship with the client team and expect to face a lot of competition, then the probability could be considered 35% or less. Although the probability value is quantitative, it is not inferred as a statistical probability but a metric used to inform the pursuit process, as shown in this example:

Net fee:	$75,000
Probability of winning:	35%
Weighted revenue (Net fee × Probability)	$26,250
Pursuit budget (*X*% of Weighted revenue)	$2,650

In the given example, the net fee earned by the engineering firm is determined by the proportion of work it undertakes from the total project fee (gross fee). It's important to distinguish the net fee from the gross fee. Each collaborating partner or department involved in the project is allocated only a portion of this gross fee; therefore, each team will evaluate its cost of sales based on its portion of revenue. For instance, if a new water treatment plant has an estimated design gross fee of $8,000,000, a civil engineering firm might estimate its net fee as $600,000 representing its portion of the site design effort.

The probability of winning is based on pre-position evaluation but can only be estimated by the project team. A fraction of the weighted revenue (e.g., 1%) is then used to guide the pursuit budget. The pursuit budget is an allowance the engineering firm uses to determine how much time (cost) the firm should spend on proposal development. In rare cases, a client may reimburse the engineering firm for some of the pursuit expenses incurred. However, this is typically only with teams that have been prequalified and are directed by the client to develop preliminary designs and cost estimates for large infrastructure projects.

Uncategorized business development time will obscure the return-on-investment calculations. For example, if an engineering firm uses a general method to track business development time, it would be impossible to distinguish which clients require more (or less) business development effort. Ideally, an engineering firm should have a method to track time spent for various clients or market segments to determine the cost and benefit of business development strategies. An engineering firm will monitor sales costs and consider ways to minimize them as they learn which pursuit strategies are most effective.

An engineering firm should have a method to track time spent for various clients or market segments to determine the cost and benefit of business development strategies.

3.2.3 Intangible Business Value

The direct financial costs and expected revenue serve as quantitative metrics of value; however, even with uncertain financials, an engineering firm may submit a proposal if it identifies business value. Specifically, the *intangible* business value is more difficult to determine using quantitative metrics. Instead, qualitative metrics and strategic planning are required to determine if a positive intangible business value will be gained from the project.

A project is initiated because the client has identified a need or an objective that requires a change. Similarly, the engineering firm may accept a project because it benefits the firm. For example, an engineering firm could recognize a new growth market and choose to work on the project at a financial loss to build skills and qualifications for future work or establish a new relationship. Or the firm may wish to support a nonprofit organization and recognize that the project financials should not be measured with standard profit and loss metrics. These investment strategies, aligned with organizational values, are more challenging to recognize and justify

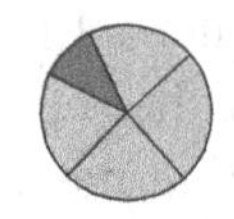

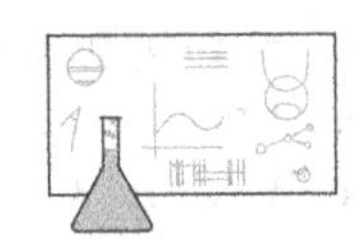

Brand Building
High-profile project to enhance reputation

Diversification
New market, service area, or industry to diversify firm portfolio

Innovation
Opportunity to work with new tech or solve unique challenges

Social Responsibility
Positive social, environmental, or community impacts

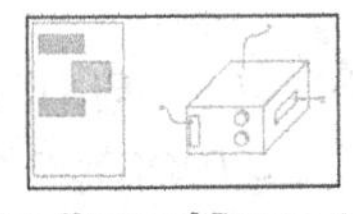

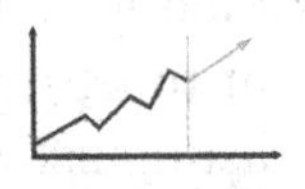

Opportunity to to develop new processes, tools or solutions

Market Position
Demonstrate expertise and leadership to become the go-to firm

Opens a gateway to future projects or long-term contracts

FIGURE 3.1 Examples of nontangible business values

> These investment strategies are more challenging to recognize and justify but can also provide the most rewarding work with new growth opportunities.

compared to quantitative go/no-go metrics – but these efforts can also provide the most rewarding work with new growth opportunities. The PMI provides a few examples of intangible business value elements, such as goodwill, brand recognition, trademarks, strategic alignment, or reputation. Figure 3.1 provides examples of other nontangible business values that would influence an engineering firm's decision to pursue a project.

Just as the go/no-go criteria (listed in Section 3.2) include several conditions that may deter an organization from submitting a proposal (e.g., lack of qualifications), some factors may encourage a proposal submission. For example, the engineering firm might determine the project's expected financial performance will not meet target revenue goals; however, the project could strengthen the firm's portfolio and benefit future pursuits that lead to larger, more profitable projects. For this reason, the firm would move forward to realize strategic objectives. These conditions will often require additional scrutiny, recognizing that direct project revenue and profitability would be unfavorable. Still, an organization may choose to invest in a pursuit because:

- Client connections or relationships could grow.
- Qualifications can be improved for future work.

- There is a teaming opportunity that strengthens engineering firm's relationships.
- The organization supports charitable, environmental, or other project values.
- The project has high visibility and improves brand recognition and reputation.

Even if direct project revenue is unfavorable, an engineering firm may invest in the project and client by spending the resources required to win the pursuit because it recognizes an intangible business value.

3.3 Team Formation

The design team for a project will have a prime consultant (or prime contractor) who has established an agreement with the client (buyer). The prime is often responsible for leading, developing, and submitting proposals. The organizational hierarchy of design teams will vary by project type. Still, one organization generally operates as the prime, and subconsultants (or subcontractors) work on the project through the prime agreement. Subconsultants (or subcontractors) are often engaged to support the requisite services and therefore create a project team comprised of multiple companies. This practice is common when one design firm does not offer all necessary services or if the teaming arrangement strengthens the qualifications because of prior project experience. Sometimes, there are limitations on how much work can be assigned to a subconsultant. For example, a public agency may require that the prime contractor hold at least 51% of the scope and fee. Other project conditions may require that a prescribed amount of work be held by a DBE (disadvantaged business enterprise), often in a range of 10–20%.

The staff from subconsultants can be intermixed in the organization chart, operating next to the prime team members. This is most common when the prime seeks to expand resource availability, offering additional staff to support project work by operating in tandem. Alternatively, the subconsultant may serve as the sole team member that provides one service, operating almost independently from the rest of the project team. For example, a local professional surveying firm may be hired to support the field survey services for the prime consultant. There could be several layers and strategies for the teaming arrangement on a land development project. If a project requires a significant amount of vertical construction, it is common for an architectural firm to lead the project and pursuit. The architect can connect with discipline leads to provide the necessary services, such as mechanical, electrical, structural, and site-civil engineering services. Still, the subconsultants may require services from their subconsultants, such that a site-civil firm may engage a geotechnical, environmental, or surveying firm to support the site-civil scope of work. Additional layers beyond the first subconsultant are referred to as *tiers*, such that the geotechnical firm working for the site-civil firm is considered a second-tier subconsultant.

Teaming arrangements are often made based on professional relationships and organizational qualifications. For some projects, there is value in a message that all requisite services can be provided in-house, which means subconsultants do not need to be engaged. This is common when the engineering firm advertises the efficiency of operations or the institutional knowledge of the organization and the project team. Alternatively, the prime consultant should recognize the benefits of diversifying the project team, seeking to broaden the team's credentials or provide unique perspectives to the project.

Teaming arrangements are ideally determined before the RFP advertisement as the team pre-positions for the expected work. The pursuit strategies, methodology, and the choice to submit a proposal are often confidential and proprietary. Before forming the project team, the prime and subconsultants may require nondisclosure agreements or other teaming agreements to formalize the arrangement.

Teaming arrangements are ideally determined before the RFP advertisement.

3.3.1 Teaming Agreements

During the pursuit phase, a team of consultants from different organizations may elect to establish teaming agreements to strengthen and protect professional relationships. The other purpose of teaming agreements is to plan the project's scope, schedule, and financial responsibilities. In the procurement phase, it is common that the scope, schedule, and cost factors are not finalized. As the project proceeds to award and execution, there should be a clear determination of how work is shared and split across all teaming partners. This determination is essential for work split across different organizations and also important for work shared between different departments in the same organization, which may have disparate resource availability and financial constraints. The types of legally enforceable teaming agreements often consist of contracts where the various participants can act (and be expected to act) per the terms and conditions of the agreement and ultimately be held liable if they fail to act accordingly.

There should be a clear determination of how work is shared and split across all teaming partners.

Many pursuits will require preparing sensitive information that an organization wishes to protect. This could include a project approach that uses trade secrets, company financial information, or institutional and client knowledge that improves the chance of success. Establishing a project team does not imply exclusivity – a subconsultant may choose to partner with multiple prime consultants (unless the teaming agreement prohibits this condition). In addition to protections established from formal agreements, teaming ethics must be considered.

Many ethical scenarios apply to teaming arrangements because of at least two competing imperatives (although there are certainly others): on the one hand, a project manager has an obligation to leverage competitive advantages to win new business for the firm; however, the engineer also needs to protect the reputation of the firm (and individuals in the team) to win future business. An engineer – or firm – that gains a reputation for unethical teaming practices is unlikely to be invited to join future teams, which can jeopardize future earnings. Thus, it is essential to consider actions from an ethical perspective to balance short-term project-specific gains with long-term results.

As mentioned, without a formal agreement, a teaming arrangement may not be exclusive nor have any enforceable confidentiality. This means an engineer could work with multiple prospective partners to learn as much as possible about an opportunity and then choose to team with the most strategic partner or even participate in bidding as part of multiple teams. In this scenario, the engineer may be able to take project information from one potential partner and use it as part of a different team or even for future projects. While this may be legal in the absence of any confidentiality agreement, the professional ethics of such behavior are less clear and highly dependent on context and interpretation. For example, a firm's pricing, process innovations, and strategy (among other things) are often kept confidential. Yet, such competitive elements may be shared with a project manager while exploring bid strategies and potential teaming arrangements. In this scenario, the engineer could decide not to use this potentially sensitive information. However, the engineer could also decide that using or sharing the information is permissible because the partner firm did not require a nondisclosure agreement and, therefore, did not truly consider the information to be protected or confidential. In this latter interpretation, the engineer may determine that the teaming partner bears responsibility for the decision to protect (or not to protect) its own intellectual property, and its failure to do so is not the engineer's responsibility. Note, of course, that there is no guarantee that the partner firm will agree with this interpretation.

A less clear example might occur if the potential teaming partners have stated that they do not intend to share certain project elements. Later, the engineer is accidentally included in an email (or other communication) that contains confidential information. In this scenario, the intent to protect information has been expressed, and the engineer knows that the information has been shared in error. While there may not be any formal confidentiality agreement in place, the engineer has reason to believe that the teaming partner expects the information to be kept confidential, and any onward usage or sharing would not be appreciated.

Engineers should note that ethical decisions are not necessarily binary; in other words, in these scenarios, the engineer's choices are not limited to using or not using the potentially confidential information. Examples of other options include checking internally at the engineer's firm to see if there is an existing precedent or policy on how best to proceed, asking the partner firm if the information can be used, informing the partner firm that the information will be used to provide fair notice and allow the other firm time to make internal changes or updates, or working with the partner-firm to establish an agreement retroactively that might include certain usage rights for the engineer's firm.

3.4 Proposal Development

After the initial go/no-go decisions and teaming arrangements are made, the proposal preparations begin. Note that the go/no-go is likely to be revisited throughout the proposal development phase if RFP amendments are issued, teaming arrangements change, or other external environmental factors influence the project's value and chance for success. This section will focus on the process of developing a proposal, which will vary by organization but generally follows an industry standard of practice. This section is not intended to guide strategic proposal development but introduces some best practices.

Both technical staff and a marketing and communications team often prepare the proposal package. If a company is structured with a marketing and communications team, these team members will often serve as proposal managers, working with technical staff to establish review dates, look over the RFP together, develop strategies, and review the formatting and delivery requirements. The technical staff will serve as the primary authors of the proposal, writing the detailed approach for the work and defining the corporate and individual qualifications. Ideally, the proposal's authors are the same staff members involved in the project; however, resource limitations often prohibit this – or an organization may have defined roles for technical proposal writers.

An initial step for proposal development is carefully reviewing the RFP formatting requirements. These requirements often restrict the total page count or font size, requiring the team to consider the best information to include. Even without a specified page count, proposals should be concise. Many RFPs are prescriptive in the sections required for the proposal, often with a short description of what information is required in each section. The format and terminology often vary by client, and the proposer may need to interpret RFP requirements (or submit questions) to validate that the correct information has been provided.

The proposal development process and formatting requirements are often seen in both public and private project types, although public projects tend to be more stringent in formatting requirements. Many proposals offer flexibility in a free-form proposal style, where each organization may have its own branding and format. In some cases, public RFPs require specific forms to be filled out to ensure consistency in formatting and content. Unfortunately, this can be so strict as to disqualify a proposal based on the font size of page numbers. The content in this section is generally applicable to both public and private proposal development, with crucial distinctions identified throughout.

3.4.1 Proposal Color-Coded Review Process

Preparing a proposal often requires market research and discovery efforts, identifying the essential project criteria or goals while considering the best strategic approach for the pursuit. As the team is formed, the pursuit's schedule and financial conditions are determined through internal negotiations to provide each team member with the appropriate resources while meeting project requirements. The iterative development process for the proposal will often refer to a color-coded nomenclature for each phase of development and review. The format and inclusion of each phase will vary by organization and project. Still, it generally follows the order of blue, pink, red, gold, and a final 'white-glove' review before submitting the proposal. Each phase represents a different level of development and review. Reviews are commonly performed by staff not directly involved in the proposal development, so an objective opinion can be provided because the RFP content may be subject to interpretation. The effort and deliverables associated with each phase will be referenced in this section, and a summary of each phase is shown in Figure 3.2.

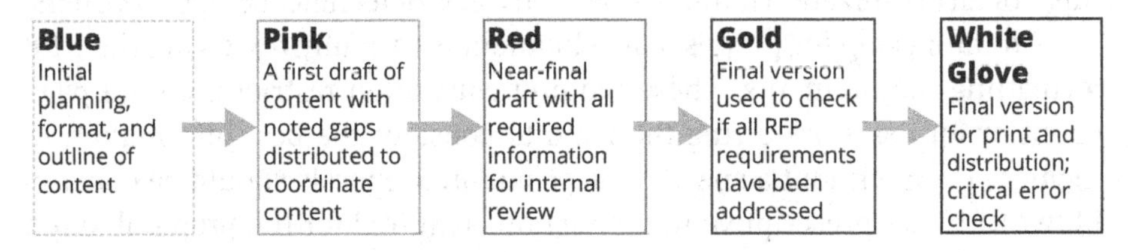

FIGURE 3.2 Color-coded review process

The color-coded review process establishes checkpoints for proposal development. The content and format requirements for each phase may be detailed or informally understood. With standard practice, the proposal authors and internal reviewers have a mutual understanding of the content of each draft. For example, the blue draft is a starting point that creates the framework section of the proposal and includes formatting as required per the RFP, maybe with some initial content for personnel, sample projects, and similar qualification content. This provides the proposal development team an opportunity to evaluate how the qualifications serve the project and begin to think more strategically about the *win themes* (a firm's unique characteristics and innovative approach). The pink draft includes enough content for a clear message that covers most of the RFP requirements with the personnel listed, representative projects selected, and draft text about the project approach. By the time a red draft is published, all content should be nearly finalized, and the review is a final assessment to validate the RFP requirements are met. This includes formatting reviews, such as checking that figures are labeled, grammar is correct, spelling is correct, and only minor edits might be returned to the proposal authors. The gold draft provides one last check before prints, often used to that all RFP requirements have been met and formatting is correct (page numbers, dates, signatures, etc.). The final document used for delivery is called the white-glove version.

3.4.2 Proposal Planning

During the go/no-go decision process, an organization will evaluate the project scope requirements and assess how best to demonstrate their qualifications. Section 3.2.1 provides a reference to this assessment. Early proposal planning tends to focus on identifying the right team members, organizational structure, representative work, and how to author an approach that demonstrates project understanding.

Many proposals will communicate qualifications regarding both the organization's and project personnel's qualifications. Qualifications of an organization could include company size, geography, industry ranking, prior contracts with the client, corporate-wide completed projects, and similar statistics of corporate success. These corporate metrics indicate the scale and quality of resources available to the project team; however, even with an impressive corporate portfolio, the work performed for the project is done by individuals who must demonstrate their qualifications. For example, when submitting a proposal to design a new university dormitory, the organization could have a portfolio of dozens of projects, but the proposed team will need to demonstrate knowledge of that work as well. In this case, many proposals will match sample projects to those listed on individual resumes to communicate that the project team has the right experience.

Even with an impressive corporate portfolio, the work performed for the project is done by individuals who must demonstrate their qualifications.

Identifying the right staff and organizational structure of a team depends on at least (1) qualifications for the specific type of work, (2) the relationship between the engineer and client, and (3) the availability of the staff. Without even one of these, the proposal (and project) will face challenges. There are often several options for choosing the project team arrangement.

For example, maybe a staff member is well-qualified to hold several roles but doesn't have the time or resources available to hold a large role on the project. Similarly, a team member may have only moderate experience to show on a resume but has a very strong relationship with the client (and proposal review panel) that will show favorably.

Staff resumes are often tailored for each pursuit to highlight the most relevant projects and match project experience to the requested services. For example, an RFP for developing a new public school would have team member resumes that include prior school projects, public projects, higher education projects, and projects in the same geographic location. Ideally, the referenced projects are recent, and the team member has a similar role. Some RFPs request that past example projects be no older than a specified number of years to demonstrate relevance. For brevity of the proposal, many RFPs require only key staff member resumes to be included. Most team members will have a long portfolio of projects that can be referenced, but the referenced projects are selected based on relevance to the pursued project's scope.

3.5 Proposal Format

The format of the proposal should follow the exact format of the request for proposal. In Chapter 2 (Section 2.5), an overview of the typical RFP is provided with an example table of contents – a common inclusion is the *Proposal Requirements* section, which will often provide the required format for the proposal submitted by the consultant. The requirements section is a frequent reference during proposal development to validate that formatting complies with requirements. Even minor deviations from the requirements (too many pages, wrong font size, missing content) could disqualify the proposal.

Minor deviations from the requirements could disqualify the proposal.

As an example, a general RFP could include the following requirements:

1. Maximum file size and page count
2. Minimum font size
3. Paper size format and numbering requirement
4. Copies of licenses and agreements
5. Content to be self-contained

The last note, "content to be self-contained," might be overlooked based on the engineer's organization and staff's expectations of brand awareness. For example, the names of the largest engineering firms may be well-known to the industry, or there could be local awareness about a consultant's services and project experience; however, the reviewers of the proposal may not have the context or may be instructed to only evaluate the proposal by what has been included in the document. Hyperlinks to company websites, staff publications, and other broadly published media material are typically ignored. These practices are more familiar with public agencies or large private organizations, so any inquiry or challenge to the selection process can reference the standardized format. Only the content shown in the proposal and defined sections will be evaluated.

3.5.1 Communication in Proposals

Proposals in various formats, including RFP responses and proposal presentations, are important business development tools used by clients to select project teams. They represent an opportunity for an engineering firm to communicate its competencies, promote the unique value of its brand or project approach, and otherwise make a positive first impression on the client's proposal review team. As such, proposal writing is as much a type of business and marketing communication as it is technical communication. Similarly, pursuits that include oral presentations require the project team to consider messaging, medium, and content delivery by the project team. However, proposal work is a non-billable endeavor that does not always result in the engineering firm being hired. It is often an activity added to an engineer's busy schedule of completing ongoing work for other projects. Thus, it may be tempting to rush through the proposal writing process in order to reduce costs; however, achieving successful results requires an investment of the engineering firm's time and attention.

It is important that proposals do not contain spelling or typographical errors, grammar mistakes, or informal language. These mistakes can negatively influence the proposal reviewer's opinion about the firm's competency and attention to detail. A client reviewing a proposal is unlikely to award a project to an engineering firm that lacks sufficient quality control to spell correctly in

A client reviewing a proposal is unlikely to award a project to an engineering firm that lacks sufficient quality control to spell correctly in proposal documents.

proposal documents. Proposal authors should also note that some content is made public, and the project team should exercise caution regarding submitted content. As proposal writers, many engineers will focus on technical accuracy, but this is only part of how a project team must demonstrate credibility. This technical writing should be balanced with communicating the understanding of project objectives, values, and measures of success.

Another common mistake in proposal writing is the overreliance on promotional phrases or catchy "buzzwords" that do not demonstrate the firm's track record or how the team plans to approach the proposed project. While some promotional posturing is required, the proposal should provide clients with a clear outline of the firm's experience on similar past projects and the relevant details related to the proposed project. Facts and evidence are more convincing than generalizations. For example, any firm could describe itself as cutting edge, always innovative, a client advocate, best in class, highly efficient, and capable of delivering exceptional quality. However, these are baseless claims and do not provide factual information to a proposal reviewer. Instead, qualifications are best represented by listing examples of innovative projects, evidence of client advocacy, listing project awards, and presenting metrics that show adherence to cost and schedule benchmarks, along with other relevant performance indicators

3.5.2 Proposal Organization

The required format and organization of the proposal should be referred to during the proposal development. Each client may require or expect a different organization of proposal content or apply terminology differently. For example, a client might ask for the firm's demonstrated experience, which could reference staff experience, corporate experience, prior work with the client, or similar others. In this case, there is an essential distinction in understanding if the client is interested in the role and experiences of the proposed team members on a similar project. Alternatively, the client might be interested in understanding the consultant's institutional knowledge and the history of the company, portfolio of projects, total qualified staff, or other corporate information. For this reason, most RFPs will include a list of required information and dictate the table of contents with descriptions of the requisite information.

A proposal generally has these elements:

1. Cover
   - Common requirements include identifying project name, consultant contact information, submitted date, and other identifying information.
2. Summary Letter
   - This could be a cover letter, often a one-page summary of the team's interest and high-level themes about the consultant's approach. Alternatively, this might be a client-issued form that includes schedule and pricing summary information.
3. Table of Contents
   - A list of included content, often used to verify all sections of the RFP have been included in the proposal.
4. Project Approach
   - The format of this section is highly variable by project types and clients but requires the project team to demonstrate their understanding of the work and describe an approach to achieve the project's objectives. Specific and measurable written content is expected, and generalizations on an approach are often discouraged. In some cases, listing the project staff associated with specific tasks and deliverables is encouraged.
5. Qualifications
   - This section often requires one or more of the following contents: (a) qualifications of the proposed staff with resumes; (b) project sheets that showcase similar completed projects completed by the staff and/or the organization; and/or (c) metrics on the submitting firms' size, sales volume, geographic locations, or similar others.
6. Timeline
   - This section may require a preliminary schedule, demonstrating that the submitting firm understands the project timeline and milestones. The RFP may also request information on staff's other obligations to verify the proposed team has time to work on the project (if selected).
7. Budget
   - This would be excluded from a qualification-based proposal. Still, when a proposal is evaluated by price (or needed to validate that the cost is within the client's budget), this section will often require information on staff hours, billing rates, and supplemental costs (travel, printing, and others).

8. Diversity and Inclusion Plan

- Often a requirement of the RFP, this section should cover how the submitting firm has implemented diversity and inclusion concepts in the teaming arrangement and proposed project approach. This could include information about DBE partners and references to a project approach that considers these concepts.

9. Other Documents

- Other documents required for proposal compliance, including liability statements, license certificates, and others.

Of these typical proposal sections, the Project Approach and Qualifications will often make up most of the proposal pages and are heavily weighted in the evaluation criteria with the most points. The expectation is that each task identified in the RFP will have a direct response. If the RFP includes five disparate tasks, the proposal must reference each of the five tasks directly with a clear description of how the work will be performed. Similarly, if the RFP requires a list of previously completed projects, the proposal should provide the project information and the team member's involvement and role. Any listed projects and the personnel are ideally referenced in other proposal sections (such as the proposed approach or evidence of innovative practices).

Qualifications of the submitting firm and the staff are the other focus of the proposal evaluation. Resumes are often limited to one page and are typically required to document the staff's qualifications. The summary qualifications of team personnel are tailored to their appropriate role for the project. Suppose an engineer is listed in the role of quality manager; then the listed experiences would reference projects where they held the same or similar roles. Some RFPs may include requirements for project roles that must be validated for staff resumes. For example, the project manager might be required to have a certain number of years of work experience, be licensed as a professional engineer (for the project location), be certified as a project management professional (PMP), and have worked on similar projects in recent years. These conditions require that staff resumes are formatted for each pursuit, describing how the staff's prior work is related to the current project. Figure 3.3 provides a sample layout for a proposal resume for a key team member (note that this is formatted differently from traditional job application resumes).

Summary qualifications of team personnel are tailored to their appropriate role for the project.

[Name], [project role]
[introduction of team member and description of their role for the project]
[summary]
• [years of exp.]
• [education]
• [awards]
• [publications]
• [affiliations]
[Example Project 1]
[narrative detailing the team member's experience and role on this project with details about how it applies to the current pursuit]
[Example Project 2]
[narrative detailing the team member's experience and role on this project with details about how it applies to the current pursuit]
[Example Project 3]
[narrative detailing the team member's experience and role on this project with details about how it applies to the current pursuit]

FIGURE 3.3 Sample layout for a proposal resume

With clients from private entities, the proposal format may not be prescriptive, and the submitting firm must choose the appropriate content. The selected format could be similar to the one outlined in this section, providing a cover letter, approach, personnel resumes, and other qualifications.

Proposal Style and Design

The graphic design, format, illustrations, font, and other characteristics are communication factors, and a proposal will often serve as the first impression of the engineering firm's qualifications. Many organizations have invested in marketing and branding guidelines identifying how proposals (and other similar marketing material) should be constructed.

Many organizations have invested in marketing and branding guidelines identifying how proposals should be constructed.

These guides will include information about tone, maybe with key terms or phrases that they feel will portray the organization as professional, casual, innovative, or modern.

Branding, colors, logos, and format are carefully controlled, restricting proposals from using custom corporate colors or project-specific logos. To promote consistency, an organization will have a custom color palette that coordinates their standard graphics, such as icons, colored fonts, and technical charts. Project and personnel photos are organized, often with document management systems, to enable proposal teams to find relevant project photos or team headshots quickly. These resources, when available, enable efficient proposal production and ensure consistency across an organization. More information on corporate branding and communication is included in Chapter 9 (Section 9.3).

3.5.3 Standard Format Proposals

In certain public projects, particularly those at the federal level, the client (usually a government agency) prescribes a specific format and content for proposals. This standardized approach enables reviewers to assess proposals efficiently and consistently, streamlining the procurement process. It also promotes transparency and fairness by providing a uniform submission structure. Proposal reviewers can easily locate essential information, although this format limits the use of creative design elements like graphics, colors, fonts, and other stylistic features. This way, the standardization shifts the reviewers' focus to the written content and the qualifications presented.

For instance, the US General Services Administration (GSA) utilizes Standard Form 330 (SF330) to evaluate the qualifications of architects and engineers. This form systematically organizes essential information like company details, contact information, resumes of key staff, and examples of past projects. The SF330, while resembling a general proposal's table of contents, is a predefined, fillable document with specific requirements and limited space for additional content. It mandates concise descriptions for each project listed in a personnel's resume, detailing the individual's role, project completion dates, and services provided. Project examples must also include a project description and a list of team members, showcasing the team's consistency and qualifications.

The relationship between the project team and their qualifications is further illustrated through a matrix. This matrix links personnel names to their significant roles in sample projects, as shown in Figure 3.4.

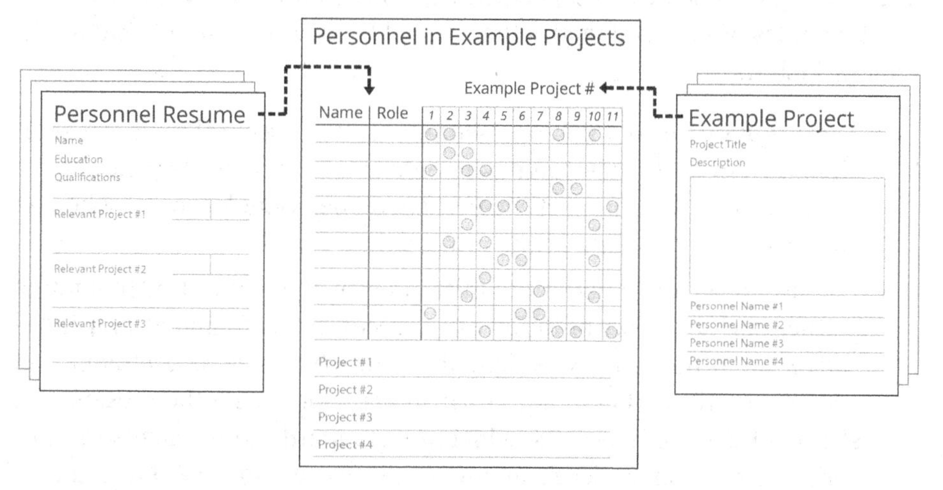

FIGURE 3.4 Matrix of project team members, representative projects, and qualifications

During the client's review of submitted proposals, this personnel-project matrix (similar to the style shown in the center of Figure 3.4) is used to assess the alignment of team members with the project's requirements. It is a visual tool to check if team members have previously collaborated and what their respective roles are in those projects. The aim of standard forms like SF330 is to uphold ethical practices and communicate the project team's capabilities.

> The ethical concern arises when information is conveyed in a manner that suggests the personnel or engineering firm possesses capabilities beyond their actual expertise.

3.5.4 Ethics in Proposal Writing

Engineers represent their firms when preparing content for a proposal or other client presentations. They are tasked with presenting themselves and their team to win new business. However, overly enthusiastic efforts in this regard can create an ethical dilemma if the proposal content is manipulated to falsely influence the client's hiring decision. The ethical concern arises when information is conveyed in a manner that suggests the personnel or engineering firm possesses capabilities beyond their actual expertise.

Examples of proposal tactics that should be examined for ethical implications include:

- The proposal makes misrepresentations, such as inflating the individual, team, or firm accomplishments and experience beyond reality. For example, listing project experiences on personnel resumes who had only trivial involvement in the project.
- The use of technical jargon to misrepresent the team's understanding of the work or describe prior experiences as value-added and innovative when the work was actually a jurisdictional requirement and used standard methodology. For instance, portraying the erosion and sediment control measures, which are standard requirements in most civil engineering projects, as an innovative environmental management strategy. This could involve using elaborate technical language to describe these basic compliance measures, thereby misleading the client into thinking that these standard practices are exceptional or unique contributions by the team.
- The proposal overpromises (either deliberately or through implication) the qualifications and services of the project team. For example, including key personnel with strong resumes without intention to involve them in the project work.

Engineering firms submit proposals to win work, so proposals are drafted to be as compelling as possible. However, these ethical issues should be monitored and can arise during proposal development. For example, is there a clear distinction between the firm's portfolio of projects and the team member's experience? Did the team member have a meaningful role in a reference project?

Personnel resumes that include a list of projects should explicitly describe the personnel's role and responsibility. Similarly, representations about other qualifications, experience, and knowledge can be presented generally as that of the firm versus that of the specific project team members. To illustrate the difference, consider the following line used in a project proposal:

"Our project team has 45 years of combined experience, including PMP and PE designations."

As written, the representation can describe either the inexperienced top-heavy Team A or the experienced Team B, as outlined in Table 3.2. Suppose the proposal does not provide the professional specifics for each team member and their assigned hours dedicated to the project. In that case, a client may think they are hiring a more experienced team than the one being proposed. Public-sector RFPs will ask for details about each team member and their role in the project to make the most informed decision; unless otherwise instructed, there are no requirements to provide specifics in private-sector proposals.

However, the ethical implications of best describing a team can be unclear. Suppose Team A is a highly driven and effective team. In that case, a client may not realize their capabilities by judging individual qualifications alone, so a more general description may be appropriate to communicate the team's overall capability. Conversely, if Team B has never worked together, focusing on their experience could be construed as misleading the client to believe that they are better able to deliver work simply as a result of their number of years in practice; however, longevity does not necessarily correlate with either competence or capability. Ultimately, during the proposal stage, the personalities and abilities of proposed team members may only be known internally to the engineering firm, which must balance representing these honestly and accurately with the need to win new business.

TABLE 3.2 **Comparison of combined experience composition**

Team A			Team B		
Name	**Experience**	**Project time**	**Name**	**Experience**	**Project time**
Gary	40 years, PMP, PE	15%	Nolan	20 years, PMP, PE	100%
Misha	2 years	100%	Javier	15 years, PMP, PE	75%
Hector	2 years	100%	Adela	10 years, PMP, PE	75%
Luca	1 year	100%			

3.6 Post-Submission of the Proposal

The submission of the proposal is a major milestone accomplished by a significant amount of planning and coordination; however, there are several more steps between the proposal submission and the contract award. As noted in Chapter 2, the proposal document may serve as one component of the client's evaluation. In many cases, the client will notify any bidders that they have (or have not) been selected or shortlisted. If a bidder is not selected to move forward, the bidder may request a debrief.

There are several steps between the proposal submission and the contract award.

The various post-submission processes are defined in this section. Interviews and presentations are a common post-submission requirement. These may be formal steps laid out in the RFP, with a defined process, format, and evaluation criteria (which is common with public clients). Conversely, there may be an informal meeting to review and discuss the proposal content (standard with private clients).

3.6.1 Proposal Interviews

Like the proposal document, the interview format, content, and duration are often prescribed in advance, and all consultants must adhere to the requirements. This might include the maximum number of consultant attendees, the allowable time for presentation and questions, and provisions for providing supplemental material. Similar to page limitations assigned to proposal documents, the interviews may have a strict limit on presentation time and time allotted for questions.

The interviews are organized by the various client personnel interested in an open dialog with the consultants working on the project. This provides an additional level of evaluation for the client who is interested in direct interaction with the project team. This allows the client to evaluate other skills, such as verbal communication and gauge the team's knowledge with impromptu questions. The client may also reveal some additional project context in this setting, discuss potential scenarios, or seek demonstrations of the teams' approaches.

The preparation of interview material often follows a similar color-coded review process (blue, red, gold, white glove, as noted in Section 3.4.1), which includes an internal review of the presentation materials and the communication of the interviewees. Like an internal proposal review, the processes often involve staff that have not worked on the proposal content so they can provide a clear perspective on the interview material. The review process may also include mock interviews, as the engineering team anticipates questions that might come up from the client during an open format discussion.

3.6.2 Debrief

A proposal *debrief* is a crucial process for gathering information for the project team and the engineering firm. The debrief may be optional and only be provided by the client if the engineering team has requested it. The debrief format can vary, often restricted to communicating only through written questions or review of any documented feedback provided by the review panel. Alternatively, it could be a less formal conversation where the bidder is prepared with their own questions to the client.

The debrief process validates the engineering firm's approach (if the team was selected), or it could inform the engineering firm on how to improve their proposal for the next opportunity. If a scoring system was used in proposal evaluation, the scores are typically shared with the engineering firm, including their ranking with competitors. The questions that the bidder may ask include:

- Why was the team selected (or not selected)?
- What were the team's perceived strengths and weaknesses?
- Was the pricing competitive?
- Was any of the content unclear, or did anything seem irrelevant?
- Was the format and language of the proposal well-received?

Many discussion topics can be communicated between the client and engineering firm during the debrief, and the engineering firm should record these results for internal use. The debrief information is then used for future pursuits, either with the client or similar project topics, informing future go/no-go decisions. For example, if the client noted a weakness in a skill set with a prior proposal, the project team would identify how they could improve their skills and be more critical in their internal evaluation of a go/no-go decision and the proposal review.

The debrief notes should be tied to the proposal document, which categorizes the proposal as a strong (or weak) example for the written content and project team. With the significant resources required to develop a proposal, an organization must understand which prior proposal content is best reused (or perhaps rewritten) for subsequent pursuits. For example, if the risk management and quality control plan was praised by a client, then the engineering firm knows it should write similar content in future proposals. During the debrief, it is common to discuss upcoming RFPs or hear general advice from the client on what should be improved or what qualifications are important with the following opportunities.

3.7 Proposal to Contract

A proposal is not intended to serve as a legally binding agreement; rather, a proposal customarily serves as a presentation of the project team's qualifications, approach to the work, and interest in undertaking a specific project. Given the preliminary nature of proposals, many details associated with the project may be unknown or insufficient and, therefore, inadequate for contract language. However, elements of a proposal can often serve as the basis of a contract. For example, a proposal may identify the project deliverables and outline specific milestones across a project schedule. The final contract documents will often draw upon these specific proposals by reference or incorporation.

For private-sector clients, a request for a proposal may start informally, perhaps from a phone conversation or a casual meeting where the project is introduced. In such scenarios, there is no formal RFP with a predefined scope. Instead, the client may focus more on the project's broader objectives, goals, and constraints. In such scenarios, the content of the proposal will largely be crafted by the engineering firm proposing the services. As a result, a proposal may have less emphasis on qualifications and instead focus on the approach, schedule, and price.

Deemphasizing team qualifications is common when the client has approached an engineering firm based on the team's reputation or professional relationship. However, if multiple engineering firms are invited to submit proposals, then the client might solicit proposals emphasizing specific information about key staff and their availability – all of which will facilitate a comparative formal evaluation or qualitative review by the client. While a proposal may start as an informal pitch for services, it can very quickly evolve into a contract that details the project scope, approach, and team. The fluidity of negotiations in the private-sector significantly elevates the importance of the proposal's content, particularly concerning the proposed scope and any assumptions and exclusions.

As discussions and negotiations about the project advance, elements of the initial proposal discussion and written content will gradually get integrated into a document intended to serve as a legally binding contract. In some instances, a proposal is deliberately structured with the potential to serve as a contract (note that *proposal-contract* is not a recognized legal term). The transformation from a proposal to a contract happens as essential contract elements are fulfilled, including consideration (usually financial), mutual agreement on the scope of work, terms and conditions, and formal acceptance documentation, like a signature block. More details on contract formation are described in Chapter 4.

Once a proposal meets all the requirements for contract formation, it becomes a legally binding document – which would permit the parties to enforce it as a matter of law. The evolution from proposal to contract demands high precision and clarity in defining the scope of work, project timelines, payment terms, and other contractual obligations. For this reason, many engineering firms have a structured proposal and contract format, and the document is carefully reviewed by internal legal counsel before it is delivered to a client. Ambiguities or oversights can lead to misunderstandings or disputes once the proposal content is contractual. Therefore, the proposal must present a viable, well-thought-out plan with comprehensive details to prevent misinterpretation. The language should align with best practices in scope definition and project planning, as discussed in Chapters 5 and 6, rather than focusing on marketing the team's qualifications and general approach.

In Appendix A, Part 3 of the Millbrook Logistics Park, the narrative focuses on ApexTech, the engineering firm, following their receipt of the Request for Proposal (RFP) from the developer, TerraHaven. This segment details ApexTech's initiation of a formal project pursuit, outlining its methodology for developing a competitive proposal. The firm's strategy is described as ApexTech evaluates its qualifications, shortcomings, and the potential for collaboration with strategic partners. Additionally, this section elaborates on the rigorous process involved in proposal preparation, including the iterative drafting of content, the acquisition of critical project information, and the estimation of project costs. This examination highlights the complexities and strategic considerations inherent in the proposal development phase of civil engineering projects.

CHAPTER 4

Contractual Frameworks and Liability

The information contained in this book should not be construed as legal advice.

CHAPTER OUTLINE

Management Essentials for Civil Engineers: A Practical Guide to Business, Communication, Ethics, and Risk, First Edition. Cody A. Pennetti, C. Kat Grimsley, and Brian M. Grindall.

This chapter describes the core concepts of contracts, including the evolution from agreements to contracts. Various project delivery systems are defined with reference to teaming arrangements and contract enforcement. Areas of law and sources of liability are also described, which extends to professional licensing and ethics and methods to limit liability. This chapter connects the legal aspects of engineering with other chapters.

4.1 Introduction

Engineers must understand how agreements and contracts establish project parameters, including payment terms, work scope, schedules, and liability. Agreements are generally encouraged in society. Society benefits from the combined effort when separate parties collaborate to achieve outcomes. For example, a multilane interstate highway cannot be built by a single engineer but can be built when many project team members and stakeholders work together. As important as agreements are to the fabric of a well-functioning society, agreements are also ubiquitous. With an increasingly interconnected world, the capacity for individuals to enter into and satisfy agreements occurs on a 24-hour basis.

The weight and authority attributed to an agreement in society will vary from agreement to agreement. Many agreements arise informally each day – such as scheduling a meeting at a work site to speak with a colleague or agreeing to drop off a package at a client's office. Failure to satisfy such an agreement won't likely be the subject of a lawsuit where a court would enforce the agreement. On the other hand, some agreements are more formal – for example, agreeing to sell a piece of property on a specific day of the month or executing a construction contract and proceeding with work in response to a notice to proceed. Failure to satisfy agreements such as failing to deliver the purchase price by the appointed time on the appointed date or failing to proceed with work in a timely manner, will be enforced by society with the authority of courts. Figure 4.1 provides a graphic relationship of agreements and contracts.

An agreement creating obligations that are enforceable at law is a contract.

An agreement creating obligations that are enforceable at law is a *contract*. Many legal requirements must be satisfied for an agreement to be considered a contract. Failure

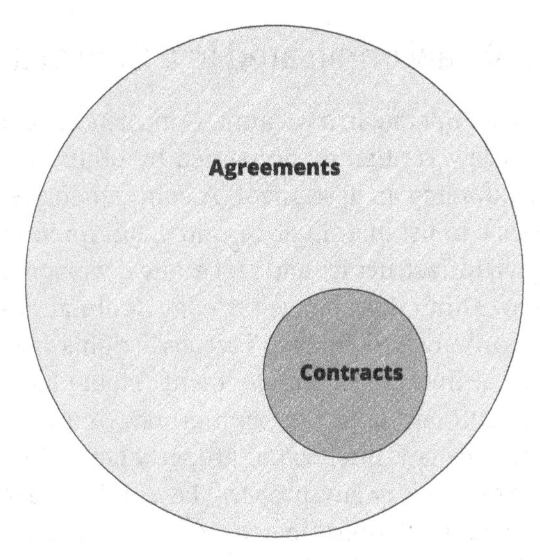

FIGURE 4.1 A contract is an agreement that establishes legally enforceable obligations.

to satisfy the appropriate legal requirements will result in the agreement not being enforceable in court. Whether an agreement arises to the level of a contract is a legal question. The law of contracts or "contract law" is a body of law relating to consensual agreements formed by two or more parties to perform an act or refrain from performing an act. Contract law is one of the most complex areas of law in the United States. A complete and exhaustive discussion of contract law is beyond the scope of this book. Rather, this chapter explores the distinction between a nonbinding agreement and a legally binding agreement and focuses the significance of the distinction from a business perspective.

4.2 Agreements and Contracts

An agreement creating obligations that are enforceable at law is a contract. In order for an agreement to be considered a contract, a number of legal requirements need to be satisfied. Failure to satisfy the appropriate legal requirements will result in the agreement not being enforceable in a court. Generally, agreements become contracts through the processes of offer and acceptance, mutual consent, and the exchange of consideration.

4.2.1 Areas of Law Implicated in a Contract

Contract law does not operate in a vacuum. Contracts arise in different contexts, and contract law is often accompanied by other areas of law when establishing and enforcing an agreement. A comprehensive list of all areas of law is not possible to list in a single resource, but important areas of law associated with civil infrastructure and real estate development may include administrative law, admiralty or maritime law, animal law, aviation law, bankruptcy law, banking and finance law, civil rights law, constitutional law, corporate law, criminal law, employment (labor) law, environmental law, family law, health law, land use (zoning) law, immigration law, international law, insurance law, intellectual property law, military law, natural resources law, personal injury law, property law, and tax law.

For example, a lease relationship between a landlord and tenant where a guest of the tenant suffered bodily harm while inside the premises may involve *contract law* (governing the lease document) as well as *property law* (a body of law relating to the nature of ownership of the property) and *tort law* (a body of law relating to the breach of a duty that the law imposes on parties). Each contractual relationship may be subject to additional areas of law depending on the context of the legal relationship and factual events. Civil engineering projects include a range of activities and relationships associated with contract law that vary by project demands, such as insurance agreements, easement agreements, site access agreements, project contracts, and employment contracts.

4.2.2 Evolution from Agreement to Contract

In practice, agreements often evolve from a statement of intent (such as a letter of intent or request for proposals) toward a legally binding contract (such as a purchase and sale agreement or a general contract). Starting with merely an idea, agreements evolve from a proposal to collaborate to a legally binding contract that can be enforced by a court. For example, an owner desiring to engage an engineer might issue a request for proposal (RFP) followed by the

Starting with merely an idea, agreements evolve from a proposal to collaborate to a legally binding contract that can be enforced by a court.

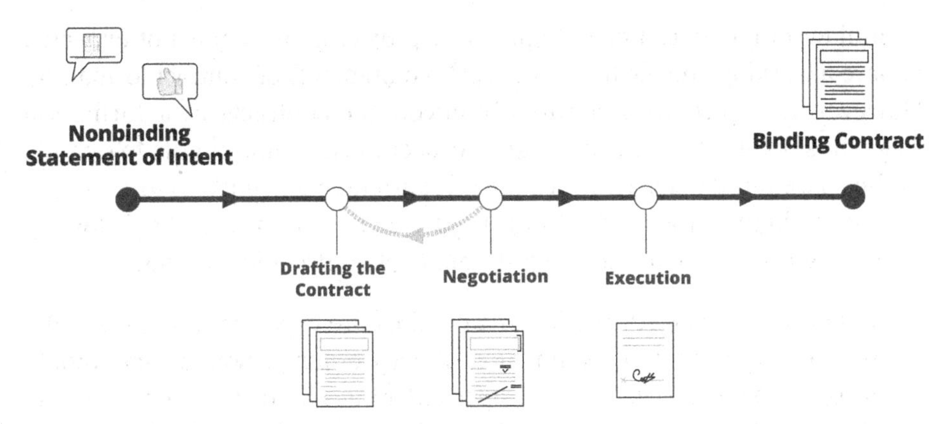

FIGURE 4.2 Evolution from agreement to a binding contract.

engineer proposing a scope of work and terms for performance, followed by a formal but nonbinding proposal being approved by the engineer and the owner followed by a legally binding design services contract between the engineer and the owner. When the owner requests the proposal or the engineer issues a response to the request, an agreement is not likely to exist as a legal matter at this point. The engineer and the client may further discuss the owner's proposal and the engineer's response through back-and-forth communications – which may increasingly evolve from diverging proposals and counter-proposals to mutually agreeable terms and conditions. This evolution from agreement (nonbinding statement of intent) to a binding contract is shown schematically in Figure 4.2.

Ultimately, the exchange may evolve to the point where the proposal hardens into an agreement representing a mutual understanding of the scope of work and terms for performance. A legally binding contract may be created upon satisfaction of the legal requirements for contract formation. However, the discussion may subside in many cases before an agreement arises to qualify as a legally binding contract.

4.2.3 Contract Formation

The law of contracts identifies several legal requirements that must be satisfied for an agreement to constitute a legally binding contract. These requirements generally derive from common law, and the number and organization of the requirements can be the subject of debate among legal scholars.

A nonlawyer (such as a civil engineer or project manager) is not expected to have a working knowledge of the legal requirements of contract formation. However, for those who commonly encounter contracts at a formation stage, it is helpful to understand and respect these requirements to operate in the space more efficiently. The legal requirements needed to form a contract vary slightly from jurisdiction to jurisdiction; however, the following list represents a few common requirements of contract formation:

1. *Competent parties.* All parties to a contract are expected to maintain the power to read and understand the contract. Certain persons are commonly recognized not to have full capacity to enter into contracts – such as minors (individuals who have not yet attained the age of 18 years old), individuals with cognitive limitations (individuals who have mental incapacities) and persons under the influence of drugs or alcohol. Further, corporations, partnerships or other entities are expected to establish the appropriate authority to enter into a contract. For example, an engineering firm often establishes a contract value threshold based on title (e.g., $50,000 for an associate, $500,000 for vice presidents) to identify which employees have authority to execute what contracts on behalf of the engineering firm.
2. *Legal subject matter.* An agreement formed for an illegal purpose (such as to perform a tort or crime or to violate a public policy) cannot be enforced as a legally binding contract. For example, a contract for work that includes a budget for kickbacks or bribes to the local government might be considered unenforceable.
3. *Consideration.* An agreement requires the parties to exchange one obligation for another, which can be (a) something of value (money or service) in exchange for a promise or (b) a promise in exchange for a promise. For example, engineering services are typically compensated with an agreed price, which would reflect consideration in the form of money in exchange for a promise to perform services. Alternatively, an engineering firm might agree to perform services on a pro-bono basis, which might be unenforceable for lack of consideration.
4. *Mutual asset or mutual understanding.* The parties to the agreement must objectively reach an agreement on all the material items in the contract. Perhaps one of the most difficult factors to determine, the concept of mutual assent or mutual understanding is an essential requirement and is discussed in more detail in this chapter. This is especially

challenging with engineering projects where the scope of work can be technically complex to describe and understand by a nontechnical audience. A well-written scope and requirements will mitigate this challenge (see Chapter 5).

5. *Delivery*. The parties to the agreement must deliver the offer and acceptance that evidences the mutual asset or mutual understanding. For example, there is a distinction between a proposal, the award selection, and an executed contract. After the proposal, many projects transition to an "intent to award" to distinguish between an offer and acceptance and a binding contract. Proposal, award, and acceptance varies from project to project based on the terms and conditions of the proposal and award process (with further variability based on whether the project is public or private).

This is a highly generalized list of contract requirements. Applying the above requirements to an agreement to determine whether a formal contract has been created and is enforceable is a legal exercise that requires the assistance of a lawyer.

4.2.4 Mutual Assent or Mutual Understanding

From a business perspective, the general requirement that the parties achieve mutual assent is perhaps the most critical contract requirement as well as the most difficult contract requirement to prove. Mutual assent, mutual understanding or a *meeting of the minds* essentially requires the objective determination that the parties reached an agreement on all material items in the contract. A determination is usually a promise involving an offer and acceptance.

> The general requirement that the parties achieve mutual assent is perhaps the most critical contract requirement as well as the most difficult contract requirement to prove.

In formal terms, a person making the offer is called the *offeror* and the person receiving the offer is called the *offeree*. In order for an offer to be effective in forming a legally binding contract, the offer must meet certain legal requirements including:

- The offer must be communicated to a specific offeree (e.g., an engineering firm),

- The offer must be a serious offer, and
- The offer must be definite and certain enough to be accepted by the offeree (i.e., can't be merely an invitation for offers such as a newspaper advertisement or a request for proposals).

Typically, once an offer is made, a legally binding contract is not yet in existence. Rather, the offer needs to be accepted before the offer is terminated by the offeror or the offeree (such as an offer terminated by a certain date if an acceptance is not made). If an offer is accepted, then mutual assent has occurred. Evaluating whether an offer is accepted often involves consideration of factors such as the intent of the parties, the timeliness and manner of acceptance, and whether acceptance was conditional.

Determining whether mutual assent occurs and whether there is a meeting of the minds between the parties to an agreement is often a facts and circumstances analysis turning on the specific language of the contract and/or the conduct of the parties in performing the contract requirements. When evaluating whether a meeting of the minds exists between two contract parties, the first source of consideration is the written contract. For this reason, contract requirements should adhere to best practices of being unambiguous, nonconflicting, achievable, and measurable (as noted in Chapter 5):

- Unambiguous
  - Ambiguity in contractual terms can lead to misunderstandings and disputes. A typical example is the term "substantial completion." Without a clear definition, parties may disagree on when the work is considered substantially complete, impacting key project milestones like scheduled deliverables or payments. To avoid this, such terms should be explicitly defined, perhaps by specifying the criteria and metrics of completed work that constitutes "substantial completion."
- Nonconflicting
  - Conflicts arise when the same terms are used differently within a contract, or when there's conflicting information. A typical example is the inconsistency in defining time in project schedules. If one section of the contract refers to schedule durations in business days (excluding weekends and holidays) and another section in calendar

days, it could lead to significant misinterpretations about deadlines and milestones. Clearly defining how time is measured throughout the contract is essential to ensure mutual understanding.

- Achievable
  - Setting unachievable requirements in a contract can lead to unrealistic expectations and disputes. For instance, expecting an engineering plan to be delivered without any errors is impractical. Engineering services are generally performed with good-faith efforts and aim to deliver a high-quality product, but perfection is not always feasible. Contracts should set reasonable standards for deliverables, acknowledging the possibility of minor revisions or corrections.
- Measurable
  - Ensuring contract terms are measurable is crucial for evaluating performance and compliance. For example, if a contract includes a requirement to minimize earthwork costs, this should be accompanied by specific, measurable standards. A requirement for an (achievable) amount of cut or fill expressed by a volume measurement (e.g., cubic yards). This clarity allows both parties to objectively assess whether the material meets the contractual requirements. Measurable terms facilitate straightforward evaluation and reduce the likelihood of disputes over subjective interpretations of contract language.

Where the written contract may not be sufficient to confirm whether mutual assent occurred regarding a material term, then the inquiry may turn to the contract parties' intentions. The client's objectives could be misaligned and request unnecessary requirements that are not needed to achieve their objectives. This case could occur when communication between different stakeholders have pre-conceived notions. For example, a developer may be asked to furnish a simple boundary survey to satisfy a real estate transaction. In this case, if the developer requests surveying services without stipulating the type of survey needed, then the bidder may inadvertently price and scope work to provide surveying services for topographic, planimetric, zoning information, flood zones, easements, and other superfluous work.

Consequently, the engineer's role in negotiating the contract and then performing the contract will serve as a significant factor in determining whether a meeting of the minds occurred on a particular issue. In the event contract parties dispute the terms of a contract (i.e., no meeting of the minds), then the lawyers representing each of the contract parties in such a dispute will generally investigate the parties' efforts during the contract formation stage (including the conduct of the parties in negotiating the contract). Consequently, the nonlawyers must understand the significance of sufficiently capturing intentions in the written contract and then performing the contract in a manner consistent with intentions.

4.2.5 Contract Enforcement

Whether an agreement was formed and whether an agreement has been broken is a paramount consideration in all contract relationships. Contract enforcement is a separate body of law providing parties to a contract the legal tools to have a contract enforced or a breach of contract remedied. A *default* or *breach of contract* occurs when one or more parties to a contract fail to perform a material part of the contract. Several different remedies may be available to the aggrieved or nondefaulting party when the other party defaults or breaches a contract. These remedies can be judicial in nature (a remedy requiring a lawsuit) or nonjudicial in nature (a remedy stated in the contract which represents an attempt to avoid a lawsuit). Given that the complex and comprehensive nature of the law of contract enforcement is beyond the scope of this text (or any one text for that matter), legal counsel should be sought immediately once a dispute concerning a contract arises.

Whether an agreement was formed and whether an agreement has been broken is a paramount consideration in all contract relationships.

4.3 Construction and Design Contracts

The process of designing and constructing a civil infrastructure or real estate development project requires multiple parties to perform a variety of services (e.g., architects, structural engineers, geotechnical engineers, mechanical engineers, landscape architects, land planners, surveyors). When negotiating and performing contracts for construction and design services, the multidisciplinary relationships can be referred to broadly as the construction and design enterprise or design team. For example, a new retail center will need multiple design disciplines, which can have a range of hierarchical contract structures. Still, the collective group is considered the design team by the client. The various constituents of a construction and design enterprise will inevitably rely on other constituents to perform in a timely manner. Sometimes reliance is based on a contract with the other constituent, and sometimes it will not.

4.3.1 Contract Stakeholders

The contracting process associated with construction and design enterprise can involve a variety of common participants or stakeholders:

1. *Client* – a person or entity which owns the project or has authority over the project sufficient to direct the work (i.e., the infrastructure operator or a contract purchaser of a property).
2. *Design professional/team* – a person or entity responsible for designing the project, which often includes a variety of professionals such as architects, engineers and other design specialties.
3. *Subcontracted design professional* – a person or entity the design professional may engage to perform a specific portion of the design services work.
4. *Construction manager* – a person or entity responsible for planning, advising or supervising functions of the general contractor, design professionals, the owner and the jurisdiction. A construction manager can be engaged to serve an owner, a lender, a local jurisdiction, a tenant or other significant stakeholder.
5. *Contractor or general contractor* – a person or entity employed directly by the owner of the project to construct or coordinate the construction of the project in accordance with the plans, specifications, contract documents, and applicable laws, rules, and regulations. The general contractor has the prime responsibility for construction of the project compared to a subcontractor responsible for a portion of the overall project.

6. *Subcontractor or trade contractor* – a person or entity engaged by the general contractor to perform specific portion of the construction services work.
7. *Supplier* – a person or entity supplying materials or equipment for the project but does not perform labor at the project.
8. *Design-builder* – a person or entity employed by the owner to both design and construct the project (often an entity serving in a combined role of architect/engineer and general contractor).
9. *Inspecting architect, engineer or consultant* – a person or entity engaged by an owner, lender or jurisdiction to (a) independently review and verify the plans and specifications, and (b) monitor the design and/or construction process in order to confirm the progress of construction conforms to the plans, specifications, contract documents, and applicable laws, rules, and regulations.
10. *Construction lender* – a person or entity providing loan proceeds to pay for the labor or materials associated with acquiring, developing, and constructing the project.
11. *Other significant stakeholder* – several other significant stakeholders can affect the larger contract relationship such as tenants, local jurisdictions, and insurance companies.

These stakeholders will influence other project operations, as described with project risk management (Chapter 8) and stakeholder relationship management (Chapter 10).

4.3.2 Delivery Systems

Construction contracts and design services contracts can be structured in a variety of ways to reflect the roles construction contractors and design professionals perform in delivering the design and construction services on a project. Often, the delivery systems will be organized considering two phases: a *design phase* where the project is designed and permitted, and a *construction phase* where the project is constructed or built. The extent to which each of these phases occurs sequentially or concurrently is a critical factor to consider when selecting a delivery system. In addition, issues of timing, quality, cost, risk allocation, and compensation will influence the extent to which an owner selects and formats its relationship with the construction contractor and design professional. In addressing these issues, the ultimate structure of delivery systems will uniquely reflect each project.

Design-Bid-Build Contract Delivery

Design and construction phases are separated and sequential in traditional design-bid-build project delivery methods. The owner independently engages the design team and the construction contractors. Design and construction services occur sequentially, with completion of the design occurring prior to the selection of the construction contractor. Typically, the owner first establishes a relationship with the design professionals and the general contractor is engaged separately later. The general contractor is responsible for establishing the relationship with the subcontractors, sub-subcontractors and material providers. The contractual relationship for design-bid-build project delivery methods is shown in Figure 4.3.

The design professional will be responsible for designing the project during the design phases, and later would be responsible for construction administration. The general contractor will be responsible for performing

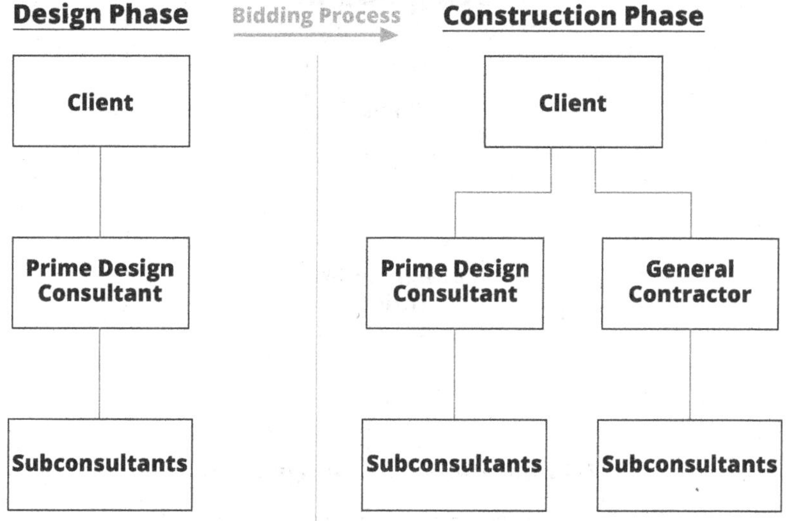

FIGURE 4.3 Contractual relationship for design-bid-build project delivery.

the construction services and obtaining the materials while maintaining control over the means, methods, sequences, techniques, and procedures of the construction services.

The design-bid-build method is often established through multiple contracting documents: (1) a separate design services contract for the design services with the architect or engineer, and (2) a separate construction services contract with the general contractor. Construction services contracts are typically broken into two documents: a foundational document confirming the business terms and supporting document containing the more detailed terms and conditions (or "general conditions") applicable to the project.

Design-Build Contract Delivery

With the design-build project delivery method, the design and construction phases are integrated with responsibility for design services and construction services established by a single party – the *design-builder*. The client engages the design-builder which consists of the construction contractor, and the design professional. Distinct from the design-bid-build method but similar to the construction manager method, design and construction services occur concurrently during the design phase. The contractual relationship for design-build project delivery methods is illustrated in Figure 4.4.

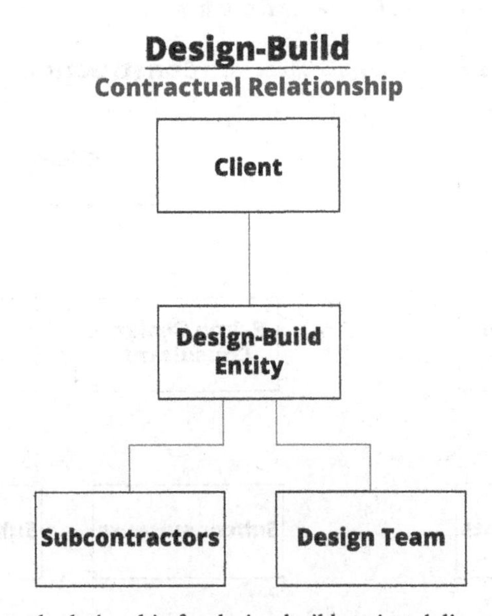

FIGURE 4.4 Contractual relationship for design-build project delivery.

Another distinction is that the design-builder will be responsible for designing and constructing the project. The design-build method employs contracting documents that unify the design and construction functions.

A unique characteristic of design-build projects is the inclusion of RFP design documents, which usually represent 30% design as prepared by the owner and a different design entity. These contract design documents are used by the bidding contractor and the design team to price construction and final engineering. Modifications to the RFP design documents will require contract modifications and often a change in the project cost for new work, deleted work, and redesign efforts. This characteristic of design-build means that the engineer often operates through the contractor, so the owner has less authority to instruct or communicate with the design team. While this condition limits an agile and adaptive design condition, there are benefits realized by the owner. In general, the owner has a single point of responsibility (the design-builder, consisting of the contractor-engineering team) and should, therefore, expect fewer design-related change orders. Faster delivery and reduced construction costs (because of the upfront competitive bidding) are other benefits associated with design-build contract delivery.

Construction Manager Contract Delivery

In the construction manager project delivery method (or construction manager at risk, CMAR), the design work and construction work are separated by contract but overlap in progression with early involvement by a construction manager. The client independently engages the design team and a construction manager. Distinct from the design-bid-build method but like the design-build method, design services and construction services occur concurrently during the design phase. The design professional will be responsible for designing the project; however, the construction manager advises the owner and design professionals regarding design considerations and often administers the construction activities. This involvement informs the design team and owner regarding considerations for costs, schedule, and methodology. This contractual relationship is shown in Figure 4.5.

The construction manager method may include similar contracting documents as those used in the design-bid-build method along with additional documents establishing the construction manager's role. The client will determine the level of involvement of the construction manager, ranging from operating as a heavily involved owner's representative

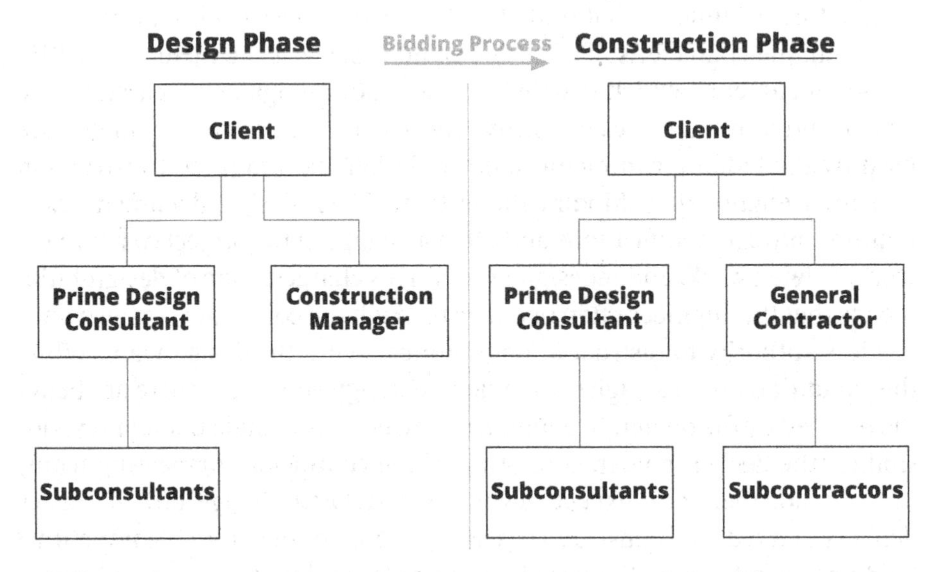

FIGURE 4.5 Contractual relationship for construction manager project delivery.

or providing guidance for the team. Following completion of the design phase, a general contractor will be engaged who will be responsible for building or constructing the project, which may or may not be the original construction manager.

Variations

Many projects will deploy delivery systems that vary from and combine attributes of the design-bid-build, the design-build, and the construction manager project delivery methods. Common variations include the turn-key method, the multiple prime method, and the fast-track method. Ultimately, the process of identifying the appropriate delivery system for any project is not a prescriptive exercise where a particular method can be bluntly applied to project dynamics. For some projects, the appropriate method may be immediate and obvious. For other projects, the appropriate method may reflect a variation of the common methods or a highly customized approach. For all projects, care and consideration should be taken to assure that the ultimate delivery method adopted for the project best addresses issues of the timing, quality, cost, risk allocation, and compensation.

4.3.3 Contract Tiers

Construction contracts and design services contracts are often tiered to create a hierarchy of relationships among stakeholders providing construction services and professional services. The service provider may perform some of the services sought by the owner, or the service provider may contract with another provider to perform the services. For example, the owner has a contract with the civil engineer, and the civil engineering firm has a contract with a geotechnical engineering firm. In most cases, the service provider who originally agreed to perform the services will ultimately be responsible for successful completion.

The service provider who originally agreed to perform the services will ultimately be responsible for successful completion.

Contract tiers can operate vertically where a primary service requires several layers of subconsultant or subcontractor services. For instance, a new commercial office building may begin with a contract between the owner and an architect – in turn, the architect enters into subcontracts with other design firms (site-civil engineering, structural engineering, etc.), which may in turn have sub-subcontracts with other supporting design firms. Alternatively, the tiers can operate horizontally where the prime service provider performs most of the labor. For instance, a transportation project may have one prime civil engineering design firm contracted with the owner with in-house design team for most design efforts, with maybe a few specialty services such as archeology or irrigation design. An example of vertical and horizontal contract tiers is shown in Figure 4.6.

The extent to which the tiers are more predominantly vertical or horizontal in nature depends on the unique demands of a project and the unique competencies of the service providers for both design and construction services.

Prime Contract

A prime contract is the primary document between the owner and the service provider – serving as the first tier of contract that often governs all the subcontracts, sub-subcontracts and material provider contracts. For construction services, a general contract or the design-build agreement often serves as the prime contract where the owner engages the general contractor – which

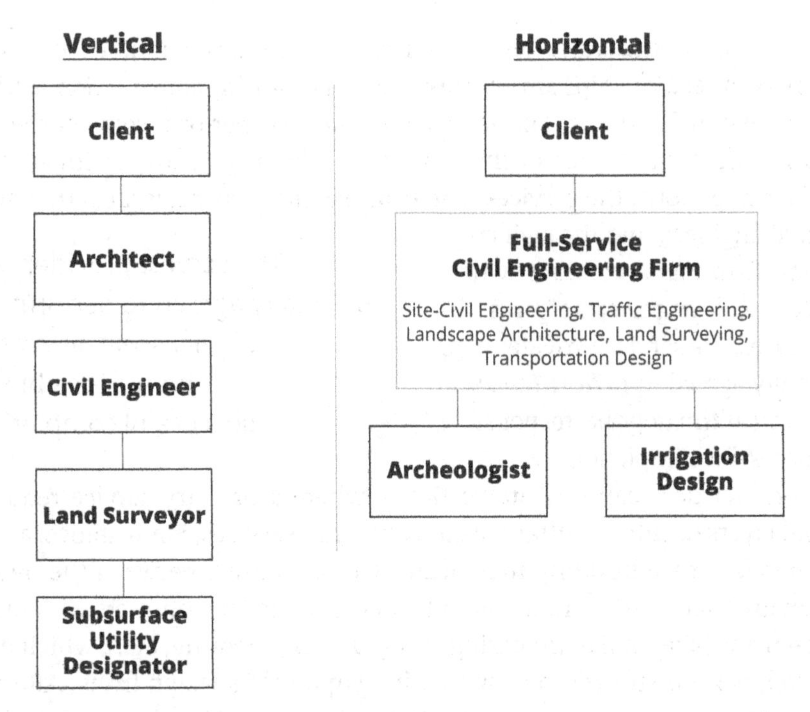

FIGURE 4.6 Vertical and horizontal contract tiers.

then is responsible for engaging other parties needed to perform construction services for the project (such as subcontractors and material providers). Similarly, an architect agreement, design-build agreement, or other type of design services agreement often serves as the prime contract where the owner engages an architect or engineer – which then is responsible for engaging other parties needed to perform design services for the project (such as specialized design professionals or consultants). Projects typically have a single prime contract for construction services and one or more prime contracts for design services; however, depending on the nature of the delivery system deployed for a project, more than one prime contract may be required, or both the construction services and the design services are combined in a single prime contract (such as the design-build method).

Subcontracts

A subcontract is a subsidiary document between the primary service provider and another service provider – serving as the second tier of contract.

Some subcontracts can reflect a significant portion of work (e.g., a land planning agency as the prime and civil engineers as a sub) while other subcontracts can reflect a (financially) minor (but still critical) portion of work (e.g., architect prime and a field survey sub). Further expanding the construction and design enterprise, subcontracts can be formed in further tiers below the initial subcontract with the prime contract. For example, an architect may engage a civil engineering subconsultant, and the civil engineer then engages a traffic engineering firm. The role of the subcontract within the larger negotiations of a prime contract is often addressed in a formulaic or boilerplate manner – whether the prime contract will refer to anticipated subcontracts generally and in broad terms without too much distinction among the various subcontracts that may ultimately be entered into at the project. For example, some requests for proposals (RFP) only require confirmation from the prime that all requisite services can be performed, while other RFPs may require a list of all subcontractors, the qualifications of each, and the designated role for each. No matter how the allocation of work occurs, it is critical that the parties coordinate the subcontracting effort to assure timely completion of the work. In many cases, the project manager will need to communicate and coordinate contract requirements across the teams to validate that all tiers understand their responsibilities and have the requisite information to deliver.

Material or Supply Contract

Material or supplies may be obtained under a variety of circumstances within the construction and design enterprise. The general contractor, a subcontractor, the owner, or another stakeholder (such as a tenant or other end-user) may be responsible for obtaining certain materials. For example, a site plan may include only generic site design information necessary for regulatory approval, but tenants could influence the required materials. Suppose a retail pad site is depicted in the engineering plans, the developer and the prospective tenant would need to determine if material changes are necessary to accommodate the intended use and establish ownership and compensation for material changes. These could include site generator pads, truck loading docks, eating areas, site furnishings, utility connections, and others. No matter how the material is obtained, it is critical that the parties coordinate the process to assure timely completion of the work.

4.4 Considerations for Construction and Design Services Contracts

Every design services and construction services contract must adequately describe the relationship including explanation of the roles, the scope of work, the schedule, the compensation, the costs, and other significant elements of the relationship. Here are eight common issues to consider when forming a design services contract:

1. *What is the engineer supposed to design?* A design services agreement needs to adequately describe the project envisioned by the owner and the designer's or engineer's role within the project which is commonly referred to as the scope of work (SOW). The challenge is in how the scope is communicated, which includes how the client describes the project and how the engineer interprets the scope. More detail is included with Chapter 5.
2. *What duties will be included and excluded within the scope of the designer's or engineer's work?* Separate from identifying the scope of work, a design services agreement should identify the specific duties of the team members as well as identify work that may be excluded from the scope of work. Also, it is important to identify the owner's responsibilities and the general contractor's responsibilities to the extent that they relate to the engineer's scope of work. For example, an owner may be obligated to furnish surveys, geotechnical information, delineations, or performance specifications prepared by other consultants or a general contractor may be responsible for designing specific elements of the project. More detail is included with Chapter 5.
3. *What standard of care will the designer or engineer be required to satisfy?* The standard of care that must be exercised by a designer or engineer while performing duties under a design services contract may be more important than performing the services at all. The law often requires a designer or engineer to perform to the level of a reasonably qualified professional in the same discipline practicing in the same location. However, this standard of care may be the subject of considerable negotiation in a design services contract. The ultimate standard of care that must be used by a designer or engineer can be the subject of controversy

and litigation. The standard of care that must be exercised by a designer or engineer while performing duties under a design services contract. In some cases, the approach is more important than the deliverable. For example, there should be no expectation or assurance that the engineering plans are free of error. However, was the work performed and delivered in a way that meets a reasonable standard of care (which could eventually be determined by arbitration, a judge, or jury)?

4. *When will the design or engineering services be performed?* The schedule during which the designer or engineer is to perform the duties within the scope of service is a critical component of any design services contract. The design services schedule often impacts several other components of the project beyond the design services contract. For example, lease agreements with prospective tenants could be established early based on an engineer's scheduled design completion date and the subsequent permitting and construction schedules. To avoid misunderstandings that could undermine the entire project, a design services contract should set forth a relatively detailed schedule with relevant milestones. Similar to mutually understanding scope, the schedule is influenced by complex processes that should be communicated between stakeholders. For example, the schedule to receive site plan approval is likely different from the schedule to achieve a land disturbance permit (which might be four to six weeks later).
5. *How will the designer or engineer be compensated, and when will payment be made?* Compensation of a designer or engineer can be calculated and paid in a variety of means and methods such as fixed-price, time and materials percentage of costs of constructing the project, incentive or bonus structures, contingency arrangements, or a combination of the foregoing. How much will be paid and when payment is to be made are critical provisions of any design services contract. More information is included in Chapter 7.
6. *Will the designer or engineer be required to satisfy any cost constraints?* Designers and engineers are often expected to provide cost estimates or projections associated with portions of a project's overall budget. As a default standard, designers and engineers should expect to advise the owner of reasonably anticipated project costs with the design scope of work. However, the specific obligations of a designer or engineer to provide timely and accurate cost projections may be the subject of

considerable negotiation in a design services contract. For example, does the owner expect detailed cost estimates with some assurance, which could be perceived as beyond the scope of work associated with design services?

7. *How will disputes be resolved among the designer or engineer and the owner, general contractor, lender and other design or engineering professionals?* Given that disputes are unpleasant, costly, and generally avoided, the dispute resolution provisions are critical components of a design services contract. The design services and construction industry have particularly evolved with sophisticated approaches to dispute resolution, ranging from litigation, executive negotiation, mediation, arbitration, or a combination of these. Section 4.5 describes these types of dispute resolution in more detail.
8. *How may the design services contract be terminated and what happens upon termination?* The owner, the designer, engineer, and other stakeholders of the project often make significant commitments to a project in reliance on the anticipated performance of a design services contract. A lender may agree to proceed with a construction loan upon satisfaction that an owner has engaged a particular engineer to perform certain work necessary for the project's success. An engineer may hire additional staff in reliance upon a specific design services contract with the owner. However, under certain circumstances, an owner, designer, engineer, or other significant stakeholder (such as a lender or a tenant of the owner) may want to terminate a design services contract. Under what circumstances a design services contract may be terminated by an owner (such as termination for cause or termination for convenience) and the resulting consequences of termination (such as reimbursement of costs or payment of compensation), a designer, engineer, or lender may be the subject of considerable negotiation in a design services contract.

Negotiating vs. Bidding

Whether a project can be built at an affordable price within an acceptable period often dictates the fate of a project. Making this determination occurs through a process where the stakeholders (namely the owners, contractors, and engineers) either discuss or bid on the cost and schedule components. Whether and to what extent stakeholders negotiate or bid on material terms and conditions of construction contracts or design services contracts can be influenced by several stakeholders including the owner, a lender, a jurisdiction, or an owner's tenant.

In a negotiated process, the parties arrive at agreement on material terms (such as cost of work, manner and timing of payment and construction schedule) and then convert the agreement to a contract. During the negotiation process, the parties will often agree on the form of contract to utilize as the basis of the negotiated construction services of design services contract.

In a bidding process, an owner will select a contractor or engineer when a critical term (which is often the variable of cost) as the primary factor of engagement. The bidding process often commences with an advertisement for bids setting forth information intended to notify potential contractors of potential work on the proposed project. Subsequently, an owner often issues a "bid package" that contains all of the relevant information needed by the bidding contractors to prepare and submit a reliable bid. The bidding process concludes with a services contract – which may or may not be the subject of negotiations. Notably, public contracts are often non-negotiable where the owner presents a contract document that is intended to be accepted by the winning bidder without any revision to the legal provisions. In theory, bidding is an efficient method of engagement where multiple interested service providers have access to the project requirements and are then given the opportunity to compete to offer the best possible price (or value) for the owner. However, bidding is also susceptible to vulnerabilities where subjective factors such as quality and competency are not objectively comparable between competing bidders (i.e., the low bid may not be particularly competent).

Public contracts are often non-negotiable where the owner presents a contract document that is intended to be accepted by the winning bidder without any revision to the legal provisions.

Contract Forms and Templates

There are generally no legally mandated forms or templates of contracts for construction services or design services. For convenience, trade groups or industry associations have promulgated a variety of forms and templates for use in the construction and design context. Many of these forms and templates are widely recognized as accepted standards in the construction and design services industry for architects, contractors, engineers, attorneys, owners, and other stakeholders involved in civil infrastructure and real estate development projects. Some commonly utilized industry-accepted forms and templates include American Institute of Architects (AIA) Contract forms, Design-Build Institute of America (DBIA) forms and Engineers Joint Contract Documents Committee (EJCDC) forms. Often, these forms and templates are available in a wide array of formats to reflect the various contexts commonly encountered in various projects.

Industry-accepted forms and templates serve as a baseline document that are to be negotiated between the parties. Sometimes, the negotiations are reduced to simply filling in blanks within the form or template. Other times, the negotiations result in significant modifications to the form or template language. If industry-accepted forms and templates are to be modified, the parties should pay particular attention to the specific modifications as well as the impact the modifications have to the underlying form or template (including unintended consequences). Retaining a lawyer to negotiate such revisions is often necessary to ensure the desired outcome is reached. Whether following the language in the template or extensively negotiating the provisions, the parties should ensure that the final agreed-upon document accurately reflects their intentions.

In addition to industry-accepted forms and templates, many stakeholders may utilize end-user contract forms for construction services and design services prescribed by the stakeholder. Some owners contract for such a high volume of work that the owner will require contractors and service providers to adopt the owner's prescribed form. An example stakeholder in this context is government agencies, which require bidding and negotiations to adopt forms and templates unique to each government agency. This is common for term contracts (basic ordering agreements, BOA or master service agreements MSA) as described in Chapter 2.

4.5 Construction and Design Services Contract Provisions

The "best contracts" define with precision and avoid ambiguity – in detailed written statements, clearly and completely expressing all aspects of the parties' intentions concerning the project. The contract should be comprehensive because disputes often arise when one or more of the parties to a contract seeks to take advantage of an ambiguity.

Construction and design services contracts typically consist of more than a single document and include the following components:

1. *Business terms*. The business terms of the contract can be located across several documents but are often concentrated in a primary document, including a description of the parties, the project description, the schedule of when the work must be commenced and completed, compensation, and the conditions under which work can be suspended or the contract terminated.
2. *General and supplemental conditions*. The general conditions elaborate on the business terms by allocating legal responsibilities between the parties – especially the owner, contractor, and designer. The general conditions also describe the role of significant stakeholders beyond the contract parties – such as the lender, inspector, or subcontractors. The general conditions are often standardized and applicable from project to project (i.e., boilerplate); therefore, general conditions can sometimes be modified or supplemented with special conditions.
3. *Plans and specifications*. The plans and specifications consist of the architectural and engineering components of the project identifying and describing what is to be constructed. Plans are the design drawings that depict or illustrate the project and its components. Specifications are the narrative descriptions of the manner of work and the materials being incorporated into the project.
4. *Submittals*. Submittals elaborate or complete elements of the plans and specifications that may not have been finalized at the time of bidding or the time when work commenced under the contract. Submittals can take the form of shop drawings, narrative descriptions, mock-ups or samples, or analysis from test results.
5. *Schedule of values*. The schedule of values segregates the performance elements of the project into readily identifiable components. These are often developed by cost estimators working with the project team members. Stakeholders can refer to the schedule of values when allocating costs or making payments associated with the performance of each component of work.

6. *Modifications (change orders).* Contract modifications or amendments arise after the contract is executed. Modifications can be unilaterally initiated by one of the contract parties (a *change directive*) or agreed to mutually by the contract parties (a *change order*).

Beyond the contract components, construction and design contracts commonly contain detailed provisions concerning the following significant legal considerations:

1. *Indemnification.* A legal relationship between two or more parties providing for compensation where one of the parties is harmed or suffers a loss. Indemnification relationships can be formed by contract or by operation of law. Common contract indemnification relationships include protection against loss, liability, or damage arising out of or in connection with (a) any cause, (b) any cause except negligence or conduct of the other party, (c) any cause including negligence or conduct of the other party, or (d) errors or omissions of one party.
2. *Duties of the parties.* Legal duties of the parties to a contract can be expressed in the contract documents or imposed on the parties by operation of law. Duties commonly imposed by the owner include disclosing significant information to the service providers, making decisions promptly, avoiding interference with the work, and assuring separate service providers perform their respective work. Duties commonly imposed on a service provider include building the project in conformance with the plans and specifications (in the case of a construction service provider), designing the project in accordance with applicable laws (in the case of a design services provider), promptly identifying errors or omissions in the construction documents, performing the work according to a particular standard of care, and identifying and promptly notifying the owner of anticipated cost overruns in the budget or delays in the schedule. Many contracts involve mutual duties that may be imposed on both parties symmetrically (such as duty of good faith and fair dealing) or asymmetrically (such as a prime consultant agreeing to a requested schedule acceleration without coordinating with subconsultants that would also need to meet the accelerated schedule).
3. *Compensation.* Common compensation models include fixed-price method, time and materials compensation, and unit or segment pricing. In addition, compensation can include incentives where the owner may share in the economic benefits associated with a contract

being performed ahead of schedule or below the anticipated cost. Regardless of the compensation model employed in a contract, the compensation provisions reflect significant legal consideration – with an emphasis on the timing, conditions, and calculations of compensation payments made during the contract relationship. More information is included in Chapter 7.

4. *Subcontracting*. Construction and design service providers will engage directly with an owner to perform a broad category of work – some of which they can perform directly but some of which they will need assistance from others to perform indirectly for the owner. Subcontracting describes the experience where a service provider engages another service provider to perform a specific portion of the scope of work described in the prime contract. The contract will often contain detailed provisions governing whether, under what conditions and to what extent a service provider can subcontract work to others. Typically, subcontractors (or subconsultants) are hired by, report to and are paid by the primary design services provider or the primary construction service provider (the general contractor). Because the primary service provider is responsible for assuring the work on the project occurs in accordance with the contract, the law often imposes a duty to coordinate and supervise the work of subcontractors. Customarily, parties expect the terms and conditions of the prime contract to be reflected in or flow through to the subcontracts to assure continuity, consistency, and reliability across the construction and design enterprise.
5. *Bonds and performance guarantees*. Construction and design service providers are often required to provide assurances or bonds from third-party sureties that certain components of the work are completed. Typically, the third-party surety (often, an insurance company) assures that a specific act will be performed under certain conditions. Bonds can arise in a few contexts such as bid bonds (assuring that contractor will enter into a contract following a winning bid), performance bonds (assuring that contractor will perform the work described), a payment bond (assuring that the material and labor will be paid in full without liens being filed on the project), or release bonds (assuring that a lien will be released). Some jurisdictions require bond agreements informed by the civil engineer's design as part of the permitting process (e.g., bonding public infrastructure elements of the project based on quantities calculated from design drawings). The availability of bonds to address certain risks varies significantly from project to project and from time to time as the construction industry evolves through economic cycles.

6. *Dispute resolution.* A dispute arising out of a contract can be costly and can compound problems on the larger project. For example, an architect's (prime consultant) omission of certain design requirements could cause rework for subconsultants, which could impact cost, schedule, and quality of their work. As a result, establishing a method by which disputes can be quickly identified and then resolved is a significant consideration in construction and design contracts. Methods for dispute resolution range from mandatory one-on-one negotiation between responsible officers of the contract parties (also known as *executive negotiation*), delegation to a professional (such as an architect), mediation by a third-party professional, binding arbitration, and litigation. The range of dispute resolutions methods can be employed across the entire contract in various ways.
7. *Statutory rights or obligations.* Many jurisdictions impose statutory rights or obligations on the contract parties such as limiting the scope of indemnification, restricting the waiver of future mechanics lien rights, or agreeing to be obligated perpetually into the future.

Ultimately, the process of negotiating and enforcing a construction or design contract involves legal considerations that inherently require the advice and counsel of a lawyer who is familiar with the laws, rules, and regulations relating to the contract. Whether the lawyer leads negotiations as the primary negotiator for one of the parties or serves as an advisor to a party behind the scenes will reflect a number of factors such as contract form, competencies of the parties, and nature or scope of the project. In any case, the benefit of legal counsel will enhance the process and promote a successful outcome.

4.6 Liability

Every person and organization may be exposed to legal liability at any time. Legal liability may be defined in several ways. From a layman's perspective, *liability* can be understood as the state of being legally responsible for something. According to the *Blacks Law Dictionary* (the predominant dictionary relied on by lawyers and judges), liability is defined as "The quality or state of being legally obligated or responsible subject to both civil and criminal liability." A liability is essentially a legally enforceable obligation – or an obligation that can be enforced as a matter of law by a court order or by act of government. Legally enforceable obligations can arise in many contexts. For example, where a client engages an engineering firm:

Every person and organization may be exposed to legal liability at any time.

- the contract for services can represent a legally enforceable obligation to perform on the part of the engineer and to pay on the part of the owner;
- the insurance policy obtained by the engineer pursuant to the contract for services may represent a legally enforceable obligation on the part of the insurance company to pay for certain claims arising under the insurance policy;
- a property survey that is stamped and certified by a surveyor may represent a legally enforceable obligation that the survey has been prepared pursuant to a specific set of standards;
- and the employment agreement between the engineering firm and its employees represents a legally enforceable obligation on the part of the firm to compensate the employee and to perform on the part of the employee.

Whether arising from a voluntary agreement (such as a contract for services) or resulting from a duty imposed by society (such as a survey prepared pursuant to an industry standard) legally enforceable obligations are ubiquitous.

The concept of legal liability should be distinguished from social liability (behavior or conduct that poses a threat of embarrassment) or from the concept of liability commonly used in the accounting context (money owed to buy an asset).

4.6.1 Role of Liability in Society

Legal liability serves an important function in society – providing for an avenue by which legal culpability (or blame) can be reconciled with a set of facts and, in many (but not all) cases, a responsible party can be compelled to remedy a harm they have caused. Failure on the part of an owner to compensate an engineering firm for services rendered pursuant to a contract for services can give rise to an award of damages (i.e., payment of dollar amounts) by a court to the engineering firm. A legally enforceable obligation represents a serious commitment on the part of society to assure that the obligation is in fact performed. The failure to perform an obligation that is enforceable at law can result in the mobilization of a significant number of societal resources to remedy the failure – by requiring financial payment (such as an award of damages by a court requiring a client to compensate an engineering firm pursuant to a contract for services as well as reimburse the engineering firm for legal costs) or by restricting a liberty right (such as a decision by a court to send an individual to prison or prohibit a company from participating in an area of business).

Legal liability also reflects an attempt in society to balance two competing policies. On one hand, legal liability is a necessary attribute of an open system consisting of innumerable discrete relationships encouraged to freely negotiate their path in the world. For example, a client may want to solicit non-binding bids for services on a project from as many qualified engineering firms as possible without running the risk of being obligated to compensate each and every bidding firm simply for submitting a bid. On the other hand, unfettered liability can discourage persons from taking risks that society desires to otherwise encourage. For example, an engineering firm that is awarded a bid needs legal assurances that it will be compensated in the future for services being performed today.

4.6.2 Areas of Law and Sources of Liability

Law does not operate in a vacuum. Rather, law functions in the context of human affairs and is often organized accordingly. Liability follows the same organization and can arise under a broad array of areas of law ranging from administrative law, admiralty or maritime law, animal law, aviation law, bankruptcy law, banking and finance law, civil rights law, constitutional law, corporate law, education law, entertainment law, employment law, environmental and natural resources law, family law, health law, immigration law, international law, intellectual property law, military law, tax law, etc. *ad nauseum*. The context in which an interaction occurs can often involve the application of multiple areas of law – such that the liability imposed on individuals can arise from many different areas. For example, where an engineering firm is

Law functions in the context of human affairs and is often organized accordingly.

accused of providing defective design or planning services, the client may sue the engineering firm and the individual engineers who worked on the project for breach of contract under contract law as well as professional negligence under tort law; the insurance firm issuing the professional services insurance coverage may have claims arising under contract law or insurance law against the client, the engineering firm, and the individual engineers. The state or local jurisdiction that relied on the services and plans may have claims against the engineering firm arising under professional licensure rules and regulations; and the individual engineers may have claims against the engineering firm for indemnification arising under contract law or corporate law.

In addition to the different areas of law, law derives from various sources – generally including statutes, common law, or contract law.

Statutes and Regulations

Statutory law consists of laws written and enacted by the legislative branch of the government. Legislative bodies (such as the U.S. Congress, state legislatures, or local boards of supervisors) enact statutes or ordinances affecting civil or commercial affairs as well as criminal conduct. Statutes can sometimes provide for civil liabilities (i.e., a fine, injunction, or other nonpenal consequence) as well as criminal liabilities (i.e., incarceration, fines, restrictions, etc.). Statutes are often augmented by rules and regulations that expand upon the statutes – supplementing the statutory requirements. For example, a state may pass laws relating to the management of stormwater, and state agencies and/or local governments may pass rules and regulations expanding on the laws by providing direction, guidance, and technical relevance to the laws.

Common Law

Common law consists of a body of unwritten laws based on historical legal precedents or decisions established by the courts. Common law evolves over time based on opinions and interpretations by judicial authorities and public juries. Common law can inspire statutes as well as motivate parties to a contract to agree to rules different than the common law. For example, common law may require that a property owner protect and hold harmless visitors to a property – such as an environmental consultant visiting a property to conduct tests – where the property owner and the environmental consultant may agree to a different standard of care within the services agreement providing that the environmental consultant will not be given such protections by the property owner.

Contracts

As noted earlier in this chapter, a contract is a legally binding agreement that sets forth each party's responsibility to each other. A contract is negotiated by the parties and provides for obligations as well as remedies when one of the parties to the contract fails to perform the obligation. A breach or default of a contractual obligation can result in remedy enforceable at law. Remedies can be specified in the contract document or imposed by a court at law or equity. For example, an engineering firm may agree with a client in a contract for services to perform certain services by a deadline and, failure to meet that deadline, may result in the engineering firm having to pay the client a very specific amount of money (i.e., liquidated damages, or LDs).

4.6.3 Professional Licensing and Ethics

Legal liability is distinguishable from professional licensing requirements or professional ethics considerations. Many professions – including engineering – maintain a code of ethics prescribing behavior consistent with a common set of values within the profession – such as the National Society of Professional Engineers (NSPE) Code of Ethics for Engineers.

In some professional settings, a code of ethics may be voluntary and represent a purely aspirational set of values – violation of which will have no meaningful consequence. For example, a local engineering community organization may have self-imposed ethics emphasizing practices in resilient and sustainable design; however, failure to adhere to these guidelines would not result in formal disciplinary action because they are not legally binding or tied to professional licensure (even if these guidelines are highly valued among the members).

In other professional settings, a code of ethics may be mandatory and impose professional obligations – violation of which may threaten the professional license needed to participate in a professional capacity, such as the NSPE Code of Ethics is mandatory for professional engineers. Failing to disclose a conflict of interest, as required by code, could lead to investigations and potentially result in suspending or revoking an engineer's professional license.

Such professional ethics and licensing ramifications oftentimes involve conduct associated with legal liability but can operate independently of any legal action. For example, a professional engineer who neglects to follow proper safety codes, resulting in a structural failure, could face legal liability for negligence. This liability exists regardless of the engineer's adherence to a professional code of ethics. The legal action is based on the breach of duty in the standard of care to the public, a legal standard independent of the professional code of ethics.

4.6.4 Limiting Liability

The specter of legal liability motivates many businesses and individuals to structure their relationships and interactions with society deliberately, and in doing so, liability can be reduced or avoided entirely. Limiting liability can manifest in many ways, including conduct, contract, ownership, and insurance.

By Conduct (Act or Omission)

Many businesses and individuals can reduce or avoid legal liability by taking certain actions. Many engineering firms impose mandatory training concerning standards of professional care – which can prevent errors and enhance insurance defenses where errors occur. In addition, legal liability can be reduced or avoided by deliberately choosing to avoid certain actions. Some engineering firms may avoid participating in certain services entirely where the conduct needed to provide such services is inherently risky, such as the delineation of hazardous substances following an environmental disaster.

By Contract

Businesses and individuals can often reduce or avoid liability by deliberately anticipating obligations and accounting for obligations. Many contracts for services specifically describe extraordinary circumstances beyond the reasonable control of either of the parties – often referred to as *force majeure* – and elaborate on the respective rights and obligations of each of the parties upon the occurrence of a force majeure event. With the benefit of forethought and a legally binding agreement, parties can "contract around" or "negotiate away" certain liabilities. For example, engineering firms may negotiate provisions in a contract for services describing the possibility of a global pandemic impacting the firm's ability to perform services in a timely manner and then shifting the risk of such events to the client by excusing the engineering firm's delay in performing services under such circumstances. A skillful attorney can aid in negotiating a contract to effectively reduce or avoid liability.

By Ownership

Businesses and individuals can often limit liability by structuring ownership in a manner where future financial losses can be limited to the amount invested. For example, an independent surveyor would be wise to form a corporation or limited liability company for purposes of providing surveying services – in which case the surveyor would not risk the immediate loss of personal assets (i.e., home, savings, retirement funds, etc.) upon the occurrence of a claim by a client for breach of contract or malpractice. In addition, various ownership models – such as partnerships, corporations, trusts, and limited liability companies – allow businesses and individuals to organize their ownership and management affairs among multiple stakeholders. For example, an engineering firm that is employee-owned may have a structure of ownership that allows select employees to manage the day-to-day business of the engineering firm while protecting the ownership interests of all employees. Determining an entity benefits tremendously from the advice of counsel and/or tax professionals at the earliest stages of negotiations among owners.

Insuring over Liability

Businesses and individuals can limit the impact of liability by acquiring insurance associated with certain risks. Through a wide variety of insurance products available for many risks, such as general liability insurance, commercial property insurance, errors and omissions insurance, or professional liability insurance, businesses and individuals can ensure sufficient financial resources are available to satisfy legal obligations when they rise. It is important to note that insurance does not eliminate liability or reduce liability. Rather, insurance provides resources to defend against claims of liability or otherwise satisfy liability obligations when they arise. Obtaining the correct type of insurance with the appropriate terms and conditions and at affordable rates can often be facilitated through insurance professionals.

In Appendix A, Part 4 of the Millbrook Logistics Park, the client and engineering team transition from an intent to award to contract negotiation. This section includes a sample contract developed by EJCDC, with annotations representing an engineering firm's legal review. The contract outlines the scope of work, period of performance, and payment procedures as established by the information described in the narrative.

CHAPTER 5

Scope Definition and Quality Management

Before starting the work, make sure it's the right work.

CHAPTER OUTLINE

Management Essentials for Civil Engineers: A Practical Guide to Business, Communication, Ethics, and Risk, First Edition. Cody A. Pennetti, C. Kat Grimsley, and Brian M. Grindall.

Chapters 2 and 3 identify the project origin and the initial description of the project scope, which is formalized through agreements as noted in Chapter 4. While Chapters 2 and 3 have described the process of procurement, this chapter describes the importance of managing project scope, beginning with the process of collecting project requirements and a well-defined scope of work. This chapter includes descriptions of quality management, which is integral to scope management, and an overview of relevant adaptive methodologies.

5.1 Introduction

Project scope is the foundation for all project operations. How can you plan and manage resources, schedules, and budgets without scope? How can you check the quality of work if the scope is not well-defined? How can you determine the amount of work completed if there is no definition of when the project is done? The common challenges of scope management include ambiguous scope, scope creep from an eagerness to maintain a positive client relationship, and a lack of proper change management.

Ambiguity in project scope can come from a client unfamiliar with the project needs or from a consultant (architect, engineer, contractor, et al.) who has made broad assumptions about project requirements. In technically challenging fields, such as engineering, it is common that a client may understand the desired product but does not have (or need) the technical expertise to define the project requirements. Some of this ambiguity may come from early scope definition in the requests for proposals (RFP), which go uncorrected and unchecked before the contract is finalized. The teams, both the client and engineer, are often uninterested in harboring details of the scope or spending the time necessary to define the scope adequately during early procurement stages. While the project *can* begin without a well-defined scope of work, any conflicting requirements and ambiguity (unrealized or not) in the project and product scope will create issues as work progresses. Client and engineer relationships introduce challenges in managing scope as all parties want to maintain a civil working relationship to promote project execution.

It is common that a client may understand the desired product but does not have (or need) the technical expertise to define the project requirements.

5.2 Challenges in Scope Management

When the scope of work is not adequately defined or not clearly understood by all stakeholders, the project can see failures from multiple conditions. In the worst case, this could mean that the constructed product (i.e., site development, road, utility system, etc.) was not what the client envisioned or not what they believed they requested in the contract scope of work, which could lead to adverse client-engineer relationships (or litigation). Civil infrastructure and real estate projects involve intricate site development, zoning requirements, and environmental considerations and include many stakeholders, each with their own expectations and demands. These conditions establish challenges in defining and monitoring project scope, which often prompts the need for risk management (see Chapter 8 for more information on risk topics). In most cases, these challenges arise from client direction and engineer action, but both parties can facilitate scope management. The seven most common challenges associated with scope management include:

1. *Scope creep.* This is often considered the greatest threat to the project's success. Scope creep refers to a series of seemingly minor requests for extra work during the project that may be considered trivial initially and is often an undocumented change request (such as a verbal request for a design modification). Over the project duration, these changes (even if small) will add up and cause failures of schedule, cost, and resources. This can occur during informal plan reviews, where a client or stakeholder directs changes during a meeting or through email, often when an engineer is inclined to affirm that changes can be made without considering how it affects the scope, budget, and schedule.
2. *Undocumented change requests.* When scope creep occurs, it is likely because of changes requested through verbal, unauthorized, or otherwise untraceable changes made to the scope. This could be a client who notes a few superfluous features during a phone call. The client's unauthorized representative could also request the changes, which creates conflict with the project team's direction. Similarly, an engineer may suggest a change affirmed by an unauthorized representative. To avoid this, change requests should follow the prescribed change control plan.
3. *Gold plating.* When the engineering firm uses project resources and decides to exceed the project requirements, it is referred to as *gold plating*. While the intentions might be sincere, the client's requirements will

govern the appropriate use of the project budget and resources. Extra design elements or site features should not be developed just because there is "extra" time or budget. This challenge may also apply to interim deliverables, where an engineer could be inclined to provide more work and design information than required for a milestone. In many cases, milestones are established to provide time for client and stakeholder reviews that could inform future project work efforts.

4. *Unforeseen conditions*. Even with a reasonable effort of due diligence, it is likely that unforeseen conditions will arise that will force a change to the project requirements. For example, a construction project may not reasonably estimate the amount of subsurface rock, which could unexpectedly raise project costs or require design revisions to minimize excavation. Risk management is used to mitigate these conditions with contingencies and directives that establish parameters for unknowns.
5. *Regulatory interpretations*. Many regulatory codes, requirements, and specifications can be subject to interpretation and introduce ambiguity in the project scope. For example, with real estate development, it is common for each jurisdiction to have unique definitions (e.g., building height measurements) and processes, making it challenging to compare project requirements and predict the outcome of reviews.
6. *External requirement changes*. Many projects can take years to complete, which puts the project at risk of changes to external environmental factors (EEFs), such as changes to legal requirements, market conditions, resource and material availability, etc. These external conditions can force scope modifications. For example, the conceptual design may anticipate residential units in a mixed-use development. However, housing demand may become unfavorable while the project progresses through the final design, and the residential use must be removed.
7. *Unrealistic requirements*. When a client and the engineering team agree to unrealistic requirements, the project is at risk of failure. This could come from product specifications that can't meet the given budget, an overly aggressive project schedule, or a product scope that does not adhere to current material or resource availability.

Effectively mitigating these scope management challenges hinges on establishing clear communication among all stakeholders, meticulously documenting change requests, and conducting regular operational reviews.

This approach ensures consistent alignment between the project's progress and stakeholders' expectations of delivered value. A critical aspect of this process is the frequent and thorough review of the scope of work and the contract. Such reviews enable proactive adjustments, maintaining congruence between the project requirements and the work performed. It is important to recognize that early deviations in scope, if not promptly addressed, can set a precedent for poor scope management practices. These practices can escalate into more significant challenges later in the project, where scope changes tend to be exponentially more costly and disruptive. Therefore, early detection and rectification of misalignments are crucial for avoiding compounding difficulties and ensuring the project's success.

Early detection and rectification of misalignments are crucial for avoiding compounding difficulties and ensuring the project's success.

Scope management for predictive delivery methods will differ from adaptive methodologies. With most civil infrastructure and real estate development projects, the delivery method follows a predictive approach, meaning that the requirements are well-defined early in the project, the budget and time reflect these requirements, and any changes to the scope will be managed through formal change requests and will modify the budget or schedule. Changes in the scope can arise (as noted by the common challenges), but the scope is intended to be a constant. Adaptive project delivery methods will flip this condition – the budget and schedule are the constants, and the scope will adhere to those constraints.

While adaptive methods are more frequently seen in computer science, real estate projects may include a hybrid approach that mixes predictive and adaptive. For example, a highway project would be predictive, conforming to local and national design standards and construction specifications, which predefined the scope. A commercial project may use a hybrid approach, where the core site infrastructure adheres to a well-defined and constant set of requirements, but aesthetic features like building finishes and hardscape could follow an adaptive methodology. Based on the client's schedule and budget, the client would prioritize which architectural and infrastructure details to invest in, working through multiple concepts to arrive at a final design. Whether predictive, agile, or hybrid, the effective management of project scope begins with a well-defined set of requirements.

5.3 Defining the Scope

There are two categories of scope management defined by PMI: project scope and product scope. The project scope includes how the work is to be executed. For example, the project scope will include the objectives, deadlines, costs, the volume of product production, and other similar requirements. Within the project scope is the product scope. The product scope defines the features of the work produced by the project. For example, if a project includes the development of a roadway, the product scope will include the road materials, expected performance life, or other parameters. Figure 5.1 provides a summary of the product and project scope. Defining the project and product requirements begins with collecting the requirements.

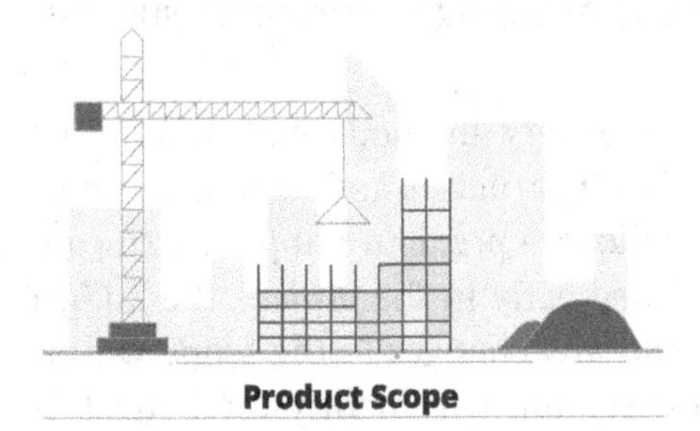

Product Scope

What are we building?

Product scope refers to the features, functions, and characteristics of the product or deliverable that the project is meant to produce.

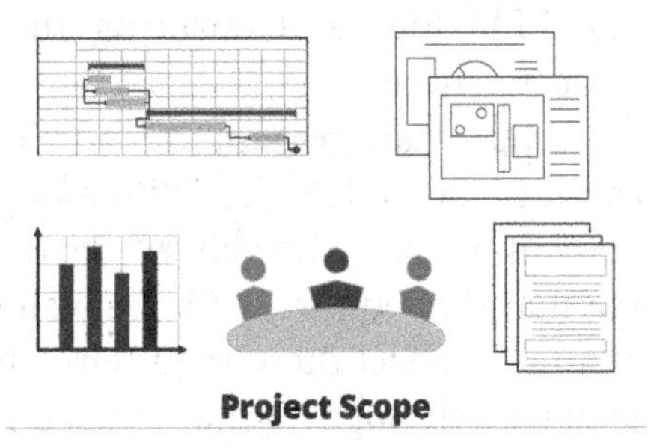

Project Scope

How are we going to build it?

The work that needs to be performed to deliver the product with the specified features and functions. It includes project objectives, tasks, deliverables, deadlines, and the processes needed to create the product.

FIGURE 5.1 Product and project scope.

5.3.1 Collecting Requirements

Collecting requirements is most effective when driven by actively engaged and thoroughly informed stakeholders. In predictive project delivery methods, requirements are established during project planning and initiation, then monitored throughout the project execution to ensure applicability. The International Council on Systems Engineering (INCOSE) classifies several sources of requirements, organized as those coming from external factors, the organization, and the project environment. INCOSE notes other external environmental factors to consider in the larger context of conditions that govern the project, such as laws, legal liabilities, existing technologies, labor conditions, and even public culture. This could include such requirements as jurisdictional permits, zoning regulations, available design software, and the technical capabilities of the workforce.

> Collecting requirements is most effective when driven by actively engaged and thoroughly informed stakeholders.

Beyond the external factors are organizational policies and procedures, guidelines, and the company culture, to name a few. These requirements consider how the client and engineering firms operate, including normal work hours, staff availability, accounting practices, formalized communication channels, and other documented corporate policies governing a project.

Finally, project-specific requirements are defined and negotiated with the project agreements and will adhere to the external and organizational requirements. Within those project requirements include such factors as requirements driven by the project team – how does the team operate together, how does the project manager best communicate with the team, and what expectations are there for team support and schedules? These project and team member requirements must be considered to understand the project's situational constraints and to inform the project requirements. Figure 5.2 depicts the hierarchy of project requirements for an example residential portfolio of projects.

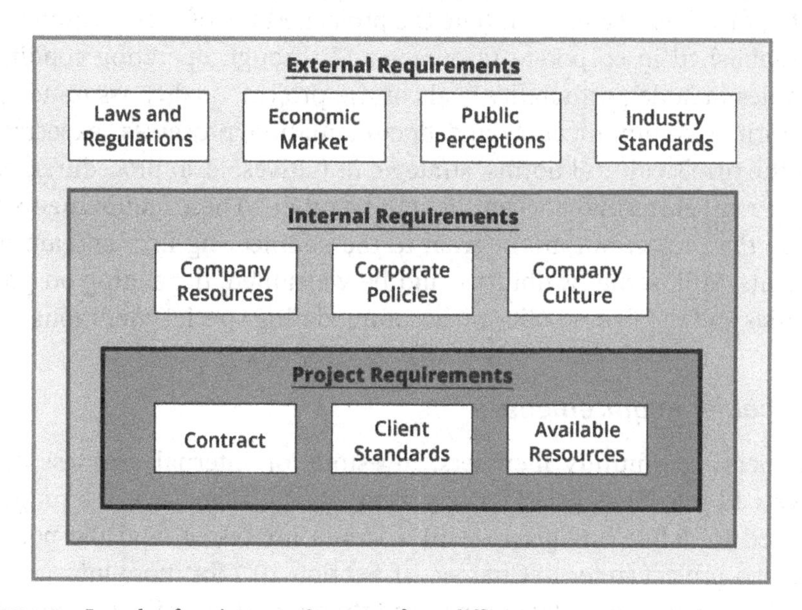

FIGURE 5.2 Sample of project requirements from different sources.

The requirements collected from different stakeholders and operating environments can be categorized as referenced in this section.

External Requirements

The project team has little to no control over the project's external requirements. These refer to larger elements such as federal and local laws and regulations, permitting processes, licensure requirements, utility purveyor specifications, industry standards, and others. For example, transportation design standards are established by research projects and documented in published works that define the standard policy (e.g., the American Associated Highway Transportation Officials book, *A Policy on Geometric Design of Highways and Streets*). The scope of external requirements is vast but generally less variable compared to other requirements, such as project-specific stakeholder requirements or even policies of an engineering firm. Many of these requirements are based on the project's geographic location and the associated zoning, engineering standards, and environmental laws.

Business Requirements

These internal requirements are associated with an engineering firm's policies and procedures and any project-specific requirements associated with *why* the firm has elected to initiate the project. Many of these requirements are established in corporate resources and through operating conditions. Examples include profitability goals of the project, quality assurance policies, formal communication and documentation processes, expectations from the firm's cultural norms, strategic initiatives, legal procedures, existing noncompete terms between clients, and others. These requirements exist outside the project but are internal to the engineering firm and are often constants. Still, some flexibility could be warranted, depending on project priorities and the firm's policy on accommodating special conditions.

Stakeholder Requirements

Customers, community members, investors, or internal business operators can all inform stakeholder requirements. *Stakeholder* is a term generally used to define any group with a vested interest in and the potential to affect a project (refer to Chapter 10, Section 10.7 for more information). The relative power of a stakeholder, and therefore their influence on project requirements, can vary throughout the project life cycle. For example, in the early phases of the project, the community members could potentially influence jurisdictional approvals of a real estate project through support (or opposition) of project permits (rezoning); however, as a project proceeds to construction, there is little room for community members to influence project requirements directly. If a stakeholder or an external factor controls a project's work or outcome, this would be considered an influence, not a requirement.

From a client's perspective, these requirements could be defined by corporate objectives or vision, potentially to build a portfolio of work with a given return on investment. A developer may establish a need to grow business revenue or geographic coverage and have established requirements based on strategic direction. These requirements might also be established only through a formal business case, and a project is approved to proceed under the condition such as a measurable benefit-to-cost ratio or if there are other nontangible values (brand recognition, social benefit, brand recognition, or others). These requirements establish some parameters of the project, which may include schedules and budgets developed through market research and projections of project revenue (such as land leases).

Solutions Requirements

Categorized as functional and nonfunctional, the solutions requirements often apply to a product. As defined by PMI, the functional requirements describe the behavior of the product. This might include site accessibility requirements, as dictated by roadway geometry or the grades of pedestrian walkways. Generally, the functional requirements are derived from the product's conditions, actions, and usage. PMI describes the nonfunctional requirements as those that focus on attributes such as reliability, sustainability, safety, resilience, and others. While the functional requirements determine what the product should do (via conditions and actions), the nonfunctional requirements apply to how effectively the product meets these conditions or actions.

Functional requirements determine what the product should do, the nonfunctional requirements apply to how effectively the product meets these conditions or actions.

Transition Requirements

When projects are expected to experience a transition (decommission, retirement, redevelopment, or others), the transition requirements will influence product development and operations. In the early phases of a project, the transition requirements may focus first on training or initiating project operations – such as understanding the operations of a building or energy system. Transition requirements may also consider how a project would transfer ownership and maintenance responsibilities. For example, transition requirements would vary significantly between a client that will own and operate a civil infrastructure project, as compared to a developer of a residential neighborhood who will sell housing lots. An early evaluation of transition requirements could significantly influence project parameters, such as complexity or accessibility.

Addendums

When requirements arise after the finalization of a contract, they are documented as addenda. These addenda can modify preexisting specifications or provide extensions, reductions, clarifications, or corrections to prior stipulations. Addenda are typically categorized sequentially (e.g., Addendum 01, 02, etc.). During preconstruction, the client may instruct the engineering firm to issue an addendum to clarify or correct design documents.

5.3.2 Data Gathering

Data gathering (from multiple sources) is used to establish project requirements and uncover and adequately define the scope of work. More data leads to a better definition of requirements, but gathering and organizing data requires time and resources. For this reason, it is necessary to recognize the value of data with respect to the appropriate level of scope development.

> More data leads to a better definition of requirements, but gathering and organizing data requires time and resources.

For example, a significant land development cost is earthwork – specifically, rock excavation. To define the project scope associated with rock excavation (which informs costs, schedule, and resources), geotechnical data can be gathered from the site with borings (i.e., an ASCE Standard Penetration Test). With more boring locations, the project team can better estimate the scope of work and price; however, each test has an incurred cost and time, and a high density of tests could be prohibitively expensive. For this reason, it is necessary to consider the cost-benefit of data gathering and the appropriate methods to define the project scope.

One source of requirements development is *expert judgment*, which comes from individuals experienced with the type of project envisioned. In the absence of individual expertise, organizational process assets can serve as a reference, such as requirements of prior similar projects. Lessons learned documents, including prior project issues, common sources of ambiguity, prior conflicts, and other similar data, can inform the requirements, assumptions, and exclusions of a project. With the rock excavation example, an experienced professional may have tacit knowledge regarding the local geotechnical conditions or lessons learned about the appropriate density of tests based on the proposed development. Figure 5.3 provides a few more examples of data-gathering methods.

For civil infrastructure and real estate development projects, data gathering can come from industry-standard sources that will be familiar to stakeholders. Much of this data informs engineering design and is included

Focus Groups
Structured meetings, often with moderators, to discuss objectives

Interviews
One-on-one conversations with pre-defined questions

Questionnaires
Distributed to a diverse and dispersed team to elicit input

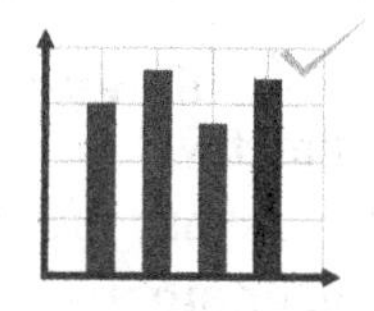

Benchmarking
Evaluating other existing project data and designs to inform current work

Brainstorming
Open group discussion to generate and analyze ideas

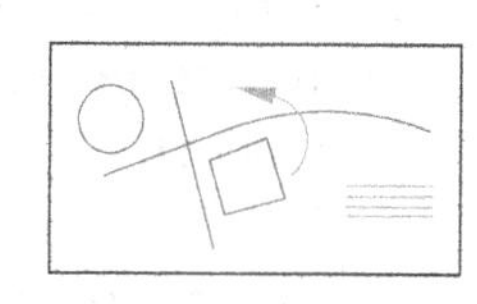

Design Charrette
Design-focused meeting to sketch and discuss the project program

FIGURE 5.3 Data-gathering methods.

in plans or in narratives of a report. For example, traffic studies or published traffic data (often from a transportation agency) can establish requirements for roadway design, pedestrian networks, intersection controls, and other site features. Some standard data sources are included in this section:

- *Market reports and analysis.* Real estate market reports and analyses provide important insights into local market conditions, including supply and demand, vacancy rates, rental rates, and sale prices. This data can help developers determine a project's feasibility and potential profitability.
- *Demographic and economic data.* Demographic and economic data, such as population trends, income levels, and employment statistics, can provide valuable information about the target market for a real estate project. This data can be obtained from government agencies like the Census Bureau or Bureau of Labor Statistics.
- *Zoning and land use regulations.* Zoning and land-use regulations can significantly impact the feasibility and design of a real estate project. Clients can obtain this data from local planning departments and other government agencies.
- *Site surveys.* Site surveys provide detailed information about the physical characteristics of a site, including topography, soil conditions, and environmental factors. This data can help clients determine the feasibility and cost of a project based on the existing conditions or land records.
- *Traffic studies.* Civil engineering projects may require traffic studies or a reference to previously published studies created by traffic engineers. These studies include traffic counts, expected trips, roadway speed conditions, forecasted growth, vehicle distribution, and other factors that influence engineering-related geometric and operational requirements.
- *Customer surveys and focus groups.* Surveys and focus groups can gather information directly from potential end-users or tenants about their needs and preferences. This data can help developers design projects that meet the needs of their target market. For real estate development, a client may perform customer surveys, or public meetings may provide data on desired uses, site features, landscape screening, or others.
- *Industry experts.* Real estate industry experts, such as brokers, appraisers, and architects, can provide valuable insights and advice on real estate development projects. Developers can seek these experts for advice and guidance on specific aspects of their projects.

5.3.3 Writing Requirements

Poorly written requirements will jeopardize a project. If requirements are not written comprehensively, they can inadvertently introduce scope creep, as incomplete specifications often necessitate continuous adjustments and additions. If requirements are written with ambiguity, it leads to multiple interpretations, creating inconsistencies and potential conflicts in project execution. Poorly constructed requirements, especially those that propose unrealistic or unfeasible goals, can set a project on a path to failure from the beginning. Therefore, it is crucial for the authors of these requirements to dedicate the necessary time to fully understand and accurately articulate all facets of the project and product requirements. This precision should be achieved through detailed data gathering and methodical analysis processes.

Poorly written requirements will jeopardize a project.

After the acquisition of the requisite data, the requirements must be written and communicated effectively. More information on written requirements of scope, as it pertains to contracts, is described in Chapter 4.

It is not always necessary to write elaborate project requirements. Through the data-gathering process, many requirements could be realized as existing operational or regulatory policies that predefine the project requirements. These existing requirements are commonly published and recognized through commercial institutions. For example, the International Organization for Standardization (ISO) has established numerous product and operational standards by collaborating with industry experts. These include quality, environment, energy, safety, and technology standards referenced by manufacturers, corporate organizations, sellers, and other groups that wish to refer to a well-defined and accepted standard. For example, instead of describing the details associated with a required quality management plan, ISO 9001 is the industry standard for quality management systems. Whether equipment or processes, these standards will serve as a grounded resource for establishing clear and attainable requirements through reference to the applicable ISO.

Like ISO, the AEC (architecture, engineering, construction) industry includes many resources for project performance and material requirements through specifications. Resources, such as the American Institute of Architects (AIA) MasterSpec, publish a template for establishing project requirements for all real estate construction and operations categories.

Established with a standard structure and reference sections (Divisions), MasterSpec provides a common language that can be tailored for each project without authoring new detailed technical content. These specifications include procurement, demolition, masonry, landscape, and many others. The specifications are often formatted to reference installation, inspection, quality control, and other parameters.

Still, other sources of pre-authored requirements may come from regulatory authorities, such as a state (or federal) Department of Transportation. Government agencies, such as the U.S. Army Corp of Engineers (USACE), also maintain an extensive library of construction specifications for their development and operations. A commercial organization may also manage specifications tailored to their development types based on corporate design and construction practices.

INCOSE notes that the following attributes should be considered for project requirements:

- *Unambiguous* – the stated requirement should have one clear meaning and not be subject to alternative interpretations.
- *Nonconflicting* – the stated requirement should not conflict with other requirements or other provisions found elsewhere in the agreement.
- *Uniquely assignable* – the stated requirement should be clearly directed to one or more particular parties.
- *Necessary* – the stated requirement should be an essential consideration for the subject work.
- *Achievable* – the stated requirement should be expressed in a manner that can facilitate successful achievement.
- *Traceable* – the stated requirement should consider a process or method facilitating the progress of work.
- *Verifiable* – the stated requirement should be expressed in an objectively ascertainable manner.

Ambiguity may stem from a lack of detail or the use of indefinite terms like "as needed," or "etc." or "as reviewed and approved." These statements leave room for variable interpretation. *Nonconflicting* refers to sets of requirements that would be impossible to achieve, likely written out of error (e.g., a site program requires 100 new residential units, but elsewhere says 60 townhomes and 60 single-family detached units). The unique assignability is essential for accountability and responsibility of project work – does the requirement dictate who is responsible for the work and that there is no overlap in the assigned responsibilities? Other parameters may seem intuitive, such as *necessary* and *achievable*. Still, the requirement author may not know how to determine what work is necessary or achievable for their desired objective. For example, a client may inadvertently include work for specific permits that are not applicable based on project geography, or the client may require an overly aggressive site program that is impossible given site constraints. Finally, the ability to *trace* and *verify* requirements is critical to document the progress and the work's origin and, therefore, establish verifiable acceptance criteria (e.g., how stakeholders can determine the work is complete).

Requirements often come from the client, as outlined in the project RFP; however, there are cases when the engineering team will be asked to author the scope of work and requirements. This is common for private-development projects where the client has described the project objectives but defers to the team's technical expertise to define the work required to achieve the objectives. In this case, engineers can follow the same guidelines for well-written requirements when authoring a proposal that includes the project approach. Thorough correspondence between the client and engineer will facilitate well-defined scope, especially as the engineer learns more about the context of the objective (why does the client need this?). A sample format for a requirement is shown in Figure 5.4.

```
Description: [Concise and Clear Description of the
Requirement. Use complete sentences.]

Acceptance Criteria:
[Measurable Criterion]
- [Detailed description of what constitutes fulfillment]
- [Any specific metrics or data points, if applicable]
- [List of expected deliverables such as plans and reports]

Dependencies:
- [List of any dependencies that must be met to begin]

Assumptions:
- [List of any assumptions made in defining this requirement]

Constraints:
- [Any known constraints that apply to this requirement]

Source:
- [Origin or purpose of the requirement]
- [Names of individuals or teams who authored the requirement]
```

FIGURE 5.4 Sample requirement format.

The content of Figure 5.4 would likely be organized by project topics (site plans, permits, meetings, etc.) and written in paragraph form (or sometimes as design specifications). Some elements, such as assumptions and constraints, may be included with a later section.

Open-Ended Requirements

While the importance of well-defined requirements cannot be overstated, there are specific contract types, such as Basic Ordering Agreements (BOA) and Indefinite Delivery Indefinite Quantity (IDIQ) contracts, where flexibility is beneficial. These contracts are designed with an intentional degree of scope ambiguity, allowing clients to shape a task or work order's scope (which would have detailed requirements).

A highly detailed contract, in some instances, can inadvertently limit a client's options. Especially for entities like government agencies, modifying a contract can be a bureaucratic and time-intensive process. In contrast, a more generalized scope within the initial contract can accommodate varied requests without time-consuming alterations. Consider the realm of environmental engineering: a client might request "general environmental engineering services" rather than specifying "wetland surveys, floodplain mapping, and streambank restoration." While the latter provides clarity, it restricts the client from later requesting services outside an explicit list. Despite the BOA or IDIQ contract's generalized definition of scope, the individual work orders within these contracts must possess clear and well-defined requirements to ensure mutual understanding and successful execution.

Considering Multiple Perspectives

Word choice is essential when authoring requirements. When creating or reviewing requirements, engineers can guard against scope creep, unexpected rework, and other costly misunderstandings by deliberately considering the possible underlying assumptions and interpretations of different stakeholder perspectives. For example, when pricing work for a site design that will be "modified as required," the engineer must anticipate under what circumstances work might be "required" to be redone and how many hours should be included in the budget for this rework to mitigate the indefinite term. The following are simplified examples but demonstrate four different (but reasonable) interpretations that might stem from the same language concerning necessary rework:

1. *The civil engineer's perspective.* A site plan is acceptable if it is technically accurate and allows all the project elements to fit on the site. Rework is only "required" if unexpected site conditions are subsequently discovered.
2. *The developer's perspective.* A site plan is acceptable only if the developer has the necessary community support for the project and/or interest from potential commercial tenants. Rework is "required" if negative feedback is received at community charrettes and/or from potential tenants that suggest other site layouts would better serve local preferences.
3. *The architect's perspective.* A site plan is only acceptable if it aligns all aesthetic goals, utility connections, accessibility conditions, and other necessary connections with the building. Rework is "required" if changes to the building program (as directed by the client, public sector, community, or otherwise) dictate a different orientation than the one originally anticipated by the site design team.
4. *Public-sector perspective.* A site plan is only acceptable – and will only be approved – if it is cohesive within the context of the local codes, the existing built environment, and public infrastructure and is consistent with area planning goals. Rework is "required" if any changes need to be made to obtain approvals.

Suppose the engineer assumes the designs will only require rework based on Perspective #1 (civil engineer). In that case, the number of hours needed will likely be underestimated, and the engineer may inadvertently underbid the project. Then, as the client demands additional changes that exceed the engineer's budget, the firm will suffer profit losses. However, simply overbudgeting hours does not guarantee that the engineer will accommodate the necessary modifications. Inflating the budget in this way also raises ethical questions, as the engineering firm is attempting to protect itself financially at the direct cost of the client. Additionally, overbudgeting hours and costs might cause the engineer's bid to appear unnecessarily high and incentivize the client to contract with a different firm that offers a more reasonable price.

Instead, one way to improve the clarity of communicating requirements is to introduce clear metrics. In this example, the engineer could specify that the pricing includes no more than 20 hours of rework to make plan modifications and that the client will incur additional costs for requests beyond this threshold. If metrics are not possible, the engineer can clarify what is and is not included as a reasonable request for rework in an RFP response or other bid. For example, the engineer may stipulate that client-directed rework based on preference can be accommodated if feedback is given before 30% of design deliverables.

As a second example, consider the challenges in defining the scope of work associated with communication and coordination during a project. Although communication is taken for granted as a supporting effort to the project work, it requires time and resources and is sometimes documented as a budgeted task in the scope of work. A measurable and unambiguous scope should include an estimated number of meetings, duration, and expected participants. These values would inform the cost estimation associated with the work. Further, an extended scope description could reference the anticipated forms of communication to determine whether meeting travel time is accounted for, if meeting preparation time is included, and if the scope includes time necessary to coordinate with other consultants. While many of these tasks are assumed (travel and meeting preparation), a well-written scope will remove ambiguity on how the work is compensated. For example, the client may assume that the design scope of work and the associated fee include all costs associated with coordination. In contrast, the engineering firm may intend for those costs to be tracked separately because they have little control over the client's aspirations for the frequency and duration of project meetings.

5.4 Ethical Considerations for Scope

While defining and monitoring scope are critical tasks, additional ethical considerations must be considered to ensure a robust and defensible final project scope. As with many ethical scenarios, engineers involved in developing scope may find themselves balancing the need to win new business with an obligation to raise concerns that may jeopardize the assignment by complicating the discussion or making their pricing seem less competitive. These conditions arise when the scope defined by the client may have conflicting requirements, ambiguity, or omissions that can lead to a potential ethical situation.

Scope defined by the client may have conflicting requirements, ambiguity, or omissions that can lead to a potential ethical situation.

For example, upon reviewing a statement of work (SOW) and accounting for the potential client's project goals, an engineer *suspects* that some work has not been fully accounted for or has otherwise been left out of the scope. This suggests that one or more change orders will be necessary, likely affecting the client's overall project cost and schedule; however, if the engineer includes cost increases to accommodate the necessary (but omitted) scope, their pricing may not be competitive. The engineer can approach the pricing exercise in several ways in this scenario:

1. Ask questions to clarify understanding.
   - If the engineer is not certain that the scope is incomplete, the best option is to seek clarification (if the process allows). Clarifying questions should be well-researched to avoid asking for information that has already been provided or clarified (e.g., asking why surveying services are not included when available information indicates the client has an existing on-call contract for survey work). For public clients, these inquiries are often shared with all bidders and would result in providing other competing firms with additional information.
   - Whether clarified through direct communication or through a process that requires sharing information publicly, clarifications will provide a balanced bidding process for all parties.
2. Revise the SOW and increase the bid price accordingly.
   - In this instance, the engineer decides that the best way to strengthen the professional relationship and build trust is by providing the client with an open analysis of potential shortcomings in the scope. The engineer is willing to accept the risk of losing the business if the client disagrees. Additionally, if the price is raised to accommodate the omitted work, it may not be competitive.

- This response may be easier with private-sector clients because the client and engineer can have informal discussions to clarify omissions or misinterpretations. This is more challenging with public-sector clients that are upheld to policies on communicating scope changes, but most agencies provide some scheduled time to accommodate questions and clarifications.

3. Do nothing and price the SOW as-is.
    - The engineer may prioritize the obligation to win the business and proceed without further action, assuming that the client's scope is accurate. In this case, the omission may be documented (perhaps internally), acknowledging that the client has intentionally omitted the work (perhaps included with a different work order). The primary motivation behind this approach is to shift the responsibility of catching any client-related mistakes or omissions back to the client. The engineer's rationale for this decision is to avoid potential risks associated with making changes to the scope that may later prove incorrect or unnecessarily increase the project cost for the client.
    - In essence, the engineer's decision to prioritize winning the business and proceed without further action stems from a desire to increase the probability of winning the project and to protect their firm's reputation, as they aim to avoid any unintended negative consequences that may arise from altering the client's original project scope.

As with most ethical situations, it can be challenging to identify a single "correct" response because interpretations can often depend on specific context. For example, these responses could be different if it is obvious that critical tasks are missing rather than something the engineer suspects. Similarly, the responses could be different if the magnitude of the suspected problem is significant versus a single minor potential change order.

Examples of other ethical scenarios related to scope that might follow a similar decision pattern include:

- *Representing capacity and capabilities.* Engineers must balance being overly optimistic about their ability to deliver work for a particular project with being realistic about the team's capacity and capabilities. While presenting the team with some marketing flair is acceptable, flagrantly misrepresenting capacity and capabilities to win business is unethical.

- *Conflicts of interest.* Engineers must recognize what constitutes a conflict of interest and when disclosure is required, such as when design choices benefit another client, like an adjacent property owner. These potential conflicts should be communicated, and measures should be taken to mitigate potential effects. For example, different project managers (even in the same company) may represent competing clients.
- *Questionable scope.* Engineers must balance their obligations to clients with larger social and regulatory responsibilities. For example, the engineer will need to determine the appropriate action if a client presents a scope (or instructions for a scope) that appears to fall in an ambiguous area of code requirements concerning safety, accessibility, and sustainability. In this case, it can be helpful to ascertain client motives if possible. The engineer may decide on a different course of action if the client is unaware of specific requirements versus if there is reason to believe the client is deliberately trying to circumvent regulations. In the latter's case, the engineer should consider whether working with this client is in the best interest of the firm and their own career. Some engineering practices are nonprescriptive and rely on engineering judgment, which may need to be defended if a client challenges the engineer's choices.
- *Representations to the public.* When representing project impacts, scale, or scope to the public and public-sector officials, engineers must balance an obligation to follow client instructions to support the project with the imperative not to violate public welfare. This is an especially important ethical consideration if the engineer feels that potential misrepresentations are being directed or encouraged by a client. The engineer may decide on a different course of action if the client is simply overenthusiastic versus if there is reason to believe the client is engaged in a deliberate effort to mislead. Flagrant misrepresentations to any stakeholder group pursuing approvals are not considered ethical practices.

5.5 Work Breakdown Structures

A *work breakdown structure* (WBS) is a project management tool that disaggregates a large project into smaller, manageable components, often visually representing the project's scope and tasks. The WBS is typically organized hierarchically, with the highest level representing the overall project and each subsequent level breaking down the project into smaller components until each component represents a specific task or deliverable. Tabular formats of WBS are also common, established through a hierarchical list.

Civil engineering projects often span several years and transition through project milestones as the project development moves through site selection, planning, design, construction, site marketing, site occupancy, and property management. In the context of civil infrastructure and real estate projects, a WBS is an effective way to manage the various stages and tasks involved in a project, from site selection and design to construction and occupancy. An overarching part of this work is project management, which is often a component of the WBS. Without a WBS, the project scope may be challenging to comprehend and communicate across various stakeholders. The WBS is often used for other project management processes, such as schedule and cost management. For example, a WBS can be used to list different tasks and their associated durations with the schedule. Similarly, costs can be assigned to different components of the WBS. A high-level example of a WBS for a real estate development project is shown here:

1. Project Management
  1.1. Establish the project team
  1.2. Develop a project plan
  1.3. Define project scope and objectives
  1.4. Manage project budget
  1.5. Manage project timeline
2. Site Selection
  2.1. Identify potential sites
  2.2. Evaluate sites for feasibility and suitability
  2.3. Select the site for development
3. Design and Planning
  3.1. Develop architectural and engineering plans
  3.2. Obtain necessary permits and approvals
  3.3. Finalize site plan and building design

4. Construction
   - 4.1. Prepare the site for construction
   - 4.2. Procure building materials and equipment
   - 4.3. Manage the construction process
   - 4.4. Conduct quality control and safety inspections

5. Marketing and Sales
   - 5.1. Develop marketing materials and strategies
   - 5.2. Conduct market research and analysis
   - 5.3. Manage sales process and contracts

6. Occupancy and Property Management
   - 6.1. Prepare property for occupancy
   - 6.2. Manage tenant relations
   - 6.3. Conduct property maintenance and repairs

The team can better track progress, allocate resources, and identify potential issues or roadblocks by breaking the project into these components. The WBS will also help ensure that all tasks are completed in a timely and efficient manner, ultimately leading to a successful project. For example, the WBS for the real estate example could continue with additional levels for the scope that include architectural and engineering plans, perhaps including the plans associated with different design disciplines (civil, mechanical, electrical), the different phases of the project (50%, 75%, 100%), the different sections of disciplines (civil roadways, civil storm sewer, environmental, etc.).

> The team can better track progress, allocate resources, and identify potential issues or roadblocks by breaking the project into components.

The level of work breakdown will vary, as determined by the needs and resources of the project team. A detailed WBS would require significant time and effort and may overcomplicate documents for project schedules, costs, and resources. While complex projects might necessitate a lengthy WBS, for many projects, there is a diminishing value with additional levels of detail (e.g., the level of detail that would identify tasks that take less than a day to complete).

The first WBS sample applies to the project work; however, a WBS may also be applied to a product. For example, instead of breaking down the phases of design and construction (prepare, procure, manage), the WBS might be better served by considering the components of the deliverable. An example of this for a land development project is shown by categories of site plan deliverables, permit documents, and project management:

1. Site Plan
- 1.1. Cover Sheet and Key Map
- 1.2. Legend and Notes
- 1.3. Existing Conditions
- 1.4. Demolition Plan
- 1.5. Site Plan
- 1.6. Environmental Plan
- 1.7. Utilities Plan
    - 1.7.1. Stormwater
    - 1.7.2. Sanitary Sewer
    - 1.7.3. Water Systems
- 1.8. Landscape Plan

2. Permits
- 2.1. Roadway Permits
- 2.2. Environmental Permits

3. Project Management
- 3.1. Client-Engineer Coordination
- 3.2. Monthly Meetings
- 3.3. Project Documentation
- 3.4. Subconsultant Coordination

By understanding the components of the product, the project team can determine resource requirements. For example, the site plan (as the project's product) can include a WBS to determine the utility design necessary for the project, which informs which designers, reviewers, utility providers, and other stakeholders will be necessary for the project design and construction. This WBS format can support the project manager in assigning the budget and schedules to the various deliverables and tasks.

5.6 Monitoring and Controlling Scope

A well-defined scope of work will facilitate project success, but the project team must monitor and control the scope throughout the entire project duration. As project constraints and opportunities change, it is necessary to adapt the project. While adaptation can benefit the project stakeholders, it is challenging to identify when a formal contract change is required compared to minor refinements that fit within the original scope.

> The project team must monitor and control the scope throughout the entire project duration.

While it may seem obvious, all team members must understand the scope and have access to the written project requirements. In many cases, the scope of work is established with the contract and then archived before the team begins project execution, often relying on prior project experiences when completing tasks. Even if the project team understands how to perform the technical work, revisiting the requirements will remind the team of the project goals, objectives, success criteria, and measure of value. Each project is unique, and while many projects will share similar workflows and tasks, the team must refer to the scope when creating the deliverables. The project team benefits from reminders on the purpose of tasks, often established by project requirements – similarly, the team should understand when and why scope changes are made.

5.6.1 Change Management

Engineers are often eager to support client requests to maintain a good relationship and support the client's objectives. Establishing a change management process effectively mitigates common challenges, such as scope creep, and improves the team's efficiency. The change management process would be implemented when a request for services or deliverables is outside the current scope or if work is being removed (descoped). Implementation of this process should be disciplined, such that the client and engineer have consistent expectations when a change management process is required. While unforeseen conditions or external environmental factors can influence the project scope, the client and engineer must be consistent about adapting to change. Documenting a requested change helps engineers evaluate impacts on schedule and budget. It provides a record of why the work product may differ from the original contract scope. Any changes in scope (or schedule or cost) should be documented as addendums to the contract.

The critical steps in successful scope change management are (1) recognizing the need for a change, (2) agreeing that the requested work is outside the original project scope, and (3) agreeing that the work is necessary for the project. Next, the client and engineer will better define the proposed change concept, which may include a narrative or sketch identifying what is expected. With an agreement on the concept, the engineer can better define the scope and impacts to schedule and the cost to incorporate the change. Although this change request process is intuitive, proper change management often competes with the urgency for implementing the requested changes. However, succumbing to the immediate pressure for modifications without engaging in a formal change management process can lead to significant issues. This oversight may result in discrepancies between the actual work performed and the initially agreed-upon project scope, leading to efforts that are not compensated with potentially contentious discussions at the project's conclusion regarding deviations from the original plan.

In some cases, new work may mean that other scope work is no longer needed, so the engineer would credit some efforts while also identifying the cost of the new work to establish a balance of the cost. Suppose the client decides they prefer one stormwater pond on the site instead of multiple rain gardens. The engineer would establish the cost of the new pond design while also determining the credit for not designing the other rain gardens. Work in progress, even if no longer required, is not credited back to the client with a change. For example, if the engineer was 50% complete with the rain garden designs, the client would only receive credit for the remaining 50% of the work. Timing of the change request is crucial as well – a late change will require additional rework to coordinate the cascading effects, such as revised details, narratives, computations, annotations, and the associated quality control. Finally, the client approves the new scope and any adjustments to the schedule or budget, and the engineer is authorized to complete the new scope of work.

In construction, modifications to scope or schedule are issued through a *change order* (CO), which is an amendment to the contract, often associated with additional labor or materials necessary to complete a directed change. Suppose a building tenant decides to upgrade hardwood floors after seeing a sample of the material. In that case, a change order can be issued to cover the additional costs and possible changes in time and materials. These change orders may begin with a client-authored RFP (specific to the requested change), and the team responds with a potential change order (PCO) or change request. The conversation around the PCO might identify the anticipated costs associated with an inquiry on new material pricing, unforeseen conditions, or additional work caused by design errors and omissions.

In design, the effort of changes can be more abstract (when compared to construction change orders). The cost to realign a road on a computer is much faster and cheaper than moving one in the field – and there are often no new materials required for design changes (compared to buying more asphalt). Minor edits are often easily accommodated, especially in early design phases, and refinements are considered part of an iterative design process. However, late changes in the design (even minor edits) can disrupt the project schedule and substantially increase the project effort. A formal change management process can benefit the designers and project owners by establishing clear boundaries around what constitutes a CO and what might be considered part of routine iterative design.

Instinctively, many engineers will quickly accommodate change requests and may do so without spending adequate time to assess the costs or schedule impact. While this may initially feel like customer service, it could inadvertently place undue pressure on the design team, as well as lead to poor-quality deliverables, cost overruns, or schedule delays that the client did not anticipate. For example, as the client reviews the 50% design documents, they may decide that they want a slight rotation in the building to establish better road frontage. While the initial adjustment (rotating a building pad) is quick, the cascading edits for annotations, plan dimensions, environmental calculations, zoning exhibits, and other factors will require additional edits.

> Many engineers will quickly accommodate change requests. While this may initially feel like customer service, it could inadvertently place undue pressure on the design team, as well as lead to poor-quality deliverables, cost overruns, or schedule delays that the client did not anticipate.

As a change is requested, the engineer should follow the formal change management process to estimate the magnitude of the change and communicate it in a concept (with cost and schedule estimates) before the change order is formally processed. A hasty response by the engineer to accommodate a change could mean the client does not have the requisite information to determine if it is the right decision. For example, if the client's request to change the building orientation is later determined to result in a three-week project delivery delay, the client may not have elected to proceed. Considering the technical complexity of the work, it's important for the engineering team to engage in early discussions with the client. These conversations should focus on clarifying which aspects of the design are easily adaptable and which ones become significantly challenging to modify as the project progresses.

This change management process of initiating, evaluating, reviewing, and implementing the change order requires project resources. Writing the CO documents and estimating costs requires time. For this reason, many projects will include a budget or task associated with managing change requests.

5.6.2 Acceptance Criteria

Acceptance criteria is another term commonly used with construction projects and applicable to various project deliverables during the design phase. *Acceptance criteria* define when a project deliverable has met the established requirements. If the criteria are written too generally, there is potential for ambiguity between the contractor and the client. Acceptance criteria should adhere to guidelines that establish measurable, testable, and relevant project conditions (refer also to the INVEST mnemonic with User Stories in Section 5.8.1). In construction, project specifications often dictate tests establishing acceptance criteria, such as laboratory tests for structural fill or sieve tests for aggregate. The acceptance criteria can be more challenging for design work, such as developing plans and permit documents. Some projects may include criteria for testing the design with modeling and simulation software (such as running a clash detection of site and building features).

Acceptance criteria can be associated with project payment processes. Most professional engineering projects are paid with monthly invoices based on completed work. The client would issue payment if there was evidence that the work is complete (or progress is measurable), or payment may be withheld if only some of the work meets the acceptance criteria. The acceptance criteria might include categories based on the work's geographic region, project phase, or discipline. For example, a road project may establish separate acceptance criteria for pavement, sidewalk, drainage, and landscape – and each portion of work would have a unique acceptance criterion.

Definition of Done

Acceptance criteria will establish stakeholder concurrence for the *definition of done* – meaning, what conditions must be met for the engineering design work to be considered complete? The definition of done is a term from adaptive methodologies that ensures all work is categorically complete. A well-written project requirement will include acceptance criteria to establish the definition of done, which would clarify ambiguity as to what conditions are required to determine when work is complete. For example, is the engineering design done when (a) the team has completed the work, (b) it's been delivered to the jurisdictional review, (c) it's been approved by a review agency, (d) it's been delivered to the client, or (e) when the client has validated that the design meets the project objectives?

The criteria may also extend to concepts of documentation delivery, meaning that the design is complete and the appropriate documentation has been provided to the client. As with other scope elements, the client and project manager must establish the definition of done with the requirements for a mutual understanding of when the work is complete. The longer a project remains unfinished, the more opportunity there is for scope creep, changes in external factors, stakeholder expectations, and other project risks.

Verification and Validation

According to definitions by PMI, the *verification* process considers whether a product, service, or result complies with a requirement, regulation, specification, or imposed condition. This is often associated with an internal process as part of quality management and relies on well-written requirements and specifications. Verification is often performed through checklists or internal reviews. For example, many jurisdictional authorities require a checklist to be submitted with a plan to verify that the design complies with local codes. This process seeks to answer the question: "Is the plan designed per requirements and standards?"

Alternatively, PMI defines *validation* as an assurance that the product, service, and result meet the needs of the customer (and stakeholders).

Verification
Process-oriented, compliance with specs

Are we building it right?

Ensures that the product, system, or component is being developed or produced correctly.

Validation
Product-oriented, stakeholder satisfaction

Are we building the right product?

Ensures that the product, system, or component meets the needs of the client and fulfills its intended purpose.

FIGURE 5.5 Validation vs. verification.

This is notably different from verification. A design or construction product could be verified internally by the engineering team as having met the technical requirements, but validation is externally facing and must come from the client. For example, the requirements for a new development project could include a layout for a retail center and the associated parking area and open space. Each feature could be well-defined with building dimensions, the number and size of parking spaces, and landscape features of the open space area. The design team can review the plans and verify that the requirements have been met, but the client will validate that the design meets expectations. Validation seeks to answer the question: "Does the design satisfy the customer and stakeholders?" A summary of these two terms is shown in Figure 5.5.

5.6.3 Visual Task Boards

The practical methods for monitoring and communicating the project scope will vary by team size and organizational structures. Direct communication (email, chat, or voice) might be a reasonable method of communicating scope and tasks for a small team and project, but larger teams and complex projects benefit from improved visibility of workflows retrievable to the team members. A visual task board openly communicates work to be complete, work in progress, and completed work. Instead of a list of tasks stored on personal notes (or in memory), these visual

Larger teams and complex projects benefit from improved visibility of workflows retrievable to the team members.

task boards should be openly displayed (physically or digitally) for the entire team to see.

A *Kanban board* originates from Japanese manufacturing and is a common visual task board to depict project workflow. Building from the concepts of user stories (Section 5.8.1), the Kanban board displays concise descriptions of project tasks. The board is continuously updated throughout the project life cycle, often daily, as tasks are completed and new tasks are identified. The format can vary by organization, and not all tasks may need to be tracked this way, but this tool can help the team better manage their tasks. Figure 5.6 shows an example of the structure of a Kanban board with notes identifying task descriptions and the status.

The project manager will establish a backlog of tasks (the *to do* column in Figure 5.6). Each team member will identify which task is currently in progress and when it is done. As noted in Section 5.6.2, the team must have a clear understanding of the work's acceptance criteria before moving a task into the *done* column. A task considered "done" with qualifiers ("it's done, except...") is still incomplete and should not be in the *done* column. A common practice is that these boards are intentionally developed with physical media, such as paper notes on an office wall, but digital versions are often necessary for large multidisciplinary teams and remote staff. The intent is to provide an at-a-glance view of current project activities. In this format, the task board can also be used to engage with other project stakeholders and managers to communicate the progress, status, and workload of the team.

To Do	In Progress	Done
Add high and low point spot shots Label contour lines Create detail sheets for curb ramps Detailed grading for storm headwalls	Check road profiles and grading	Develop rough grading plan
Create storm computation table Check outfall protection requirements	Optimize storm sewer pipe sizes	Establish land cover and runoff conditions
Final check of intersection sight distance Develop road profile sheets Create road typical section sheets	Create storm computation table	Draft road alignments Determine road classifications and speed Initial check of intersection sight distance
Create landscape planting table	Label landscape plantings	Create a list of possible plantings for client Design draft planting layout
Update project plan Prepare monthly invoice	Send change order for site layout revision	

FIGURE 5.6 Structure of a Kanban board.

The Kanban board, or similar workflow chart, is most effective when coordinated with information about which team members are accountable for the tasks (such as with a RACI chart described in Chapter 6, Section 6.5). All team members must understand task responsibility with new assignments, especially with large and multi-tiered teams. In some cases, these visual task boards include *swim lanes* that establish rows cutting across the columns to differentiate tasks assigned to each team member. Rows may also differentiate categories of tasks (e.g., design, management, coordination, etc.). Other options include color-coded tasks, which may establish urgency, work category, responsibility, etc.

Many tasks in civil engineering require coordination across multidisciplinary teams. The transparency of task workflows with tools like a Kanban board will provide peripheral awareness of task interdependencies for each team. For example, the stormwater management team will have a set of tasks dedicated to the hydrologic analysis of land cover. As the transportation design team works through their tasks, the design choices made for roads and sidewalks would affect the land cover conditions of the site and the calculations of the stormwater team's work. The Kanban board can work well for design tasks where the product evolves, and work interdependencies are revealed over time.

5.7 Quality Management

Quality management is a crucial internal process in engineering projects, fundamental to ensuring that the deliverables consistently meet the required standards of care. This process starts with establishing well-defined requirements. Without these, there is no clear baseline for verification, nor is there a set criterion to evaluate whether the work aligns with the intended scope. It is essential to understand that the onus of quality management lies within the project team; it is not a responsibility that should be passed onto clients or external review bodies. For instance, a jurisdictional agency tasked with reviewing design and construction products does not bear the responsibility for quality control. Highlighting quality management as an internal function underscores the imperative for the project team to proactively uphold quality standards throughout every phase of the project's life cycle.

Quality management has two facets: quality assurance (QA) and quality control (QC). *Quality assurance (QA)* pertains to managing the project to ensure it meets predefined requirements, focusing on the methods, procedures, and processes necessary for producing quality work. It's a proactive practice, continually evaluating and improving processes to produce high-quality results. In comparison, *quality control (QC)* is reactionary and involves the review of completed work to verify its compliance with requirements. For instance, a project plan might outline QA processes like independent reviews and root cause analysis for issues, while QC would involve specific activities like reviewing design documents with checklists.

Quality management standards, typically set forth by the engineering firm or client, provide a framework for QA processes. At the project level, it's essential for the team to specify how these standards will be put into practice. This includes deciding who will conduct quality control reviews, establishing a process for documentation, selecting appropriate checklists, and other related procedures. Additionally, the project's scope of work significantly shapes its quality management approach.

For effective quality monitoring, project requirements should be defined in measurable terms. This enables consistent tracking and improved quality of the deliverables. The project team can use clear quality metrics to assess performance and identify operational improvement areas. These metrics should encompass more than just typographical or mathematical errors; they need to be customized to align with the specific goals of the program and project. Tailoring these measures ensures that quality control focuses on adherence to standards and contributing to the overarching objectives of the project, often referred to as the dimensions of quality. Figure 5.7 shows a summary of several quality dimensions.

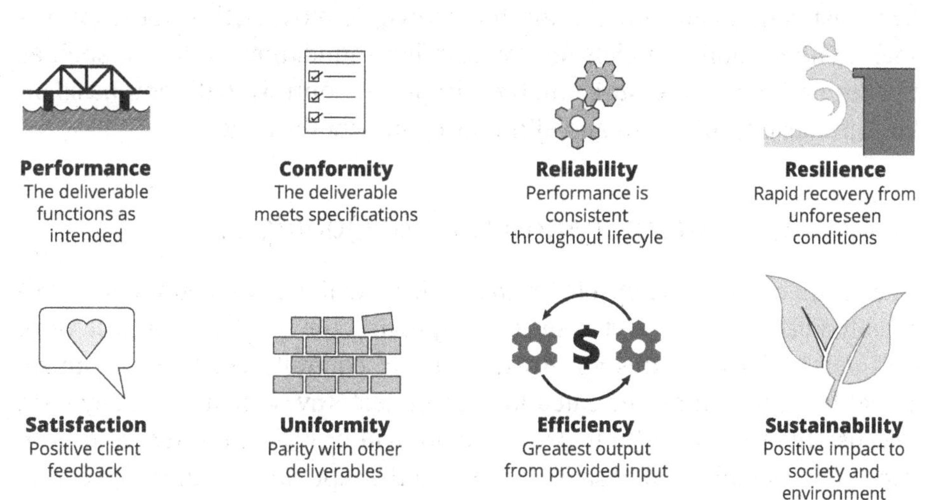

Performance
The deliverable functions as intended
Conformity
The deliverable meets specifications
Reliability
Performance is consistent throughout lifecyle
Resilience
Rapid recovery from unforeseen conditions
Satisfaction
Positive client feedback
Uniformity
Parity with other deliverables
$
Efficiency
Greatest output from provided input
Sustainability
Positive impact to society and environment

FIGURE 5.7 Example list of a project's many dimensions of quality.

Performance, uniformity, and conformity are the most common metrics determining if project designs will provide the intended product and meet specifications. Project features like reliability, resilience, and sustainability may also measure project quality, dependent on the project objectives. For example, the project may meet local jurisdictional requirements, but does it achieve the client's needs for sustained operations during disasters? Does the client have sustainability goals that must be met? Is the design aesthetically pleasing and comfortable to use? These objectives connect to other quality metrics, such as client satisfaction and project validation.

Continuous validation of work is essential for quality management. Client reviews and feedback can inform the project team's service and deliverables, not just after a project is complete but throughout the life cycle. Additionally, quality measured by efficiency is an integral part of how the team may define success. For instance, was the project completed with the best processes to meet the scope requirements while simultaneously adhering to the schedule and budget?

Quality management is a cyclic process, often referencing the Plan-Do-Check-Act (PDCA) process credited to W. Edwards Deming. The first step, *plan,* begins with the project team understanding the project's requirements and success criteria and defining the process to complete the work. Next, the work is executed based on the project plan with a commitment that designers self-certify their work before independent reviewers perform a formal quality control check. The final step, the *act,* addresses nonconformance issues but, most importantly, seeks to address the source of the issue. This last component, identifying the source, is most critical for improvement. Other quality models follow a similar foundation to PDCA, such as Six Sigma (define, measure, analyze, improve, control), with emphasis on continuous improvement and eliminating nonconformance.

5.7.1 Accountability and Cost of Quality

The entire team is responsible for the project quality, but accountability ultimately falls on the project manager. Improving the quality of project work and deliverables requires an investment and commitment at an organizational level, with time dedicated to each project. Investment is the key term because the time spent to improve quality will provide a positive return on project deliverables, client satisfaction, and the reputation of the engineering firm and project team members. Conversely, poor quality work often results in project setbacks, necessitating costly rework and leading to waste. Such quality failures not only incur liabilities but also risk future business, as clients may begin to associate the engineering firm and its contractors with substandard quality.

> The entire team is responsible for the project quality, but accountability ultimately falls on the project manager.

Investments in preventative measures and inspections will mitigate quality issues. Preventative measures focus on the root cause and require investments in project planning, training, and improved processes. Appraisal costs include the inspection of work in progress, deliverables, and the time and resources necessary to perform these tests.

Preventative investments focus most on organizational standards, tools, and policies. The project work is completed by people and their resources, and any investment in those people and resources will improve the deliverables. The challenge is allocating the necessary time to provide these investments in prevention measures. Some common challenges include:

- *Limited resources.* An organization may not have in-house resources or knowledge to develop quality improvement plans or skills to train staff in proper management and quality assurance practices. There may also be challenges to deciding on the appropriate policies and then implementing those processes.
- *Resistance.* Improving quality often requires changing some processes, which can be challenging for project teams that have become familiar and comfortable with traditional practices. Other resistance may come from teams feeling they can't adapt to a universal practice or process.
- *Metrics.* Without effective metrics, it is difficult to monitor the progress of improvement. This makes it challenging to realize the value of implementing new quality-improvement programs.

The first two listed challenges (limited resources and resistance to change) must be tackled at an organizational level, acknowledging the upfront costs necessary for enhancing quality. Resistance to implementing new quality management practices can be justified, particularly if these practices lack adaptability across various project types. For instance, a standard procedure might mandate a comprehensive quality management plan, which may not be suitable for every project. While such a detailed plan is beneficial for extensive projects like a new multi-mile rail corridor, it might not be as effective for smaller scale projects. Take, for example, a modest project like a written feasibility study; in this case, an elaborate quality management process with a multi-page checklist might add little value, especially if it cannot be scaled down or tailored to the project's size and scope.

Metrics associated with project deliverables enable the teams and organization to monitor progress and the effectiveness of quality improvement processes. For example, a checklist that tracks error rates, documenting the number of design-related change orders, the feedback from project team members, scores from client satisfaction surveys, the number of safety incidents, and other metrics can be practical. These metrics need to be balanced, meaning that nonconformance errors should be helpful indicators of nontrivial issues that can be traced back to a root cause and addressed accordingly. For a manufacturing project, control limits could define metrics – evaluating product uniformity, for example. Control charts are harder to adapt to design work but could still communicate performance, such as the number and severity of plan errors. For instance, if the project experienced several utility clashes during construction, the team may choose to implement new software that automates clash detection. After training and implementation, the project team could monitor the decrease in clashes during construction to inform the efficacy of the quality improvement process.

5.7.2 Planning for Quality Management

The project team must plan for quality management during the initiation of work or the team may not have enough time or resources to accommodate the required practices. Effective quality management begins at the project's initiation. Integrating quality processes early is critical, as it ensures sufficient resources and time for quality management processes, preventing last-minute scrambles and noncompliant work.

It's critical to distinguish between internal quality control deadlines and external submission dates.

It's critical to distinguish between internal quality control deadlines and external submission dates. The project manager should allocate time for quality control review and the project team's corrective actions prior to a submission deadline (such as delivery for jurisdictional review and permit). The QC schedule should be communicated to the project team to avoid misalignment in task management. Internal deadlines (sometimes called the *pencils down* deadline) should be set well before the final submission, allowing ample time for thorough QC reviews. In most cases, the project plan or the engineering firm will dictate the duration and process of the quality control review. Still, it generally requires a process such as depicted in Figure 5.8 and defined in this section.

1. Self-Certification
   a. Before the QC review, the project team should self-certify their work using checklists to verify adherence to project and engineering plan requirements. This self-certification should include a narrative for the current project status, indicating progress on interim deliverables (e.g., what should be included with 50%) and noting any unresolved items.
2. QC Submission
   a. This QC submission and initialization should occur days or weeks before the formal submission and should include all relevant plans, specifications, and other deliverable content required for review.
   b. This submission process may benefit from initial discussions between the designers, project team, and QC reviewers to provide context – but in some cases, this process is intended to operate completely independently of information received from the project team, which could better facilitate review and comment that could occur from external stakeholders (clients and review agencies).

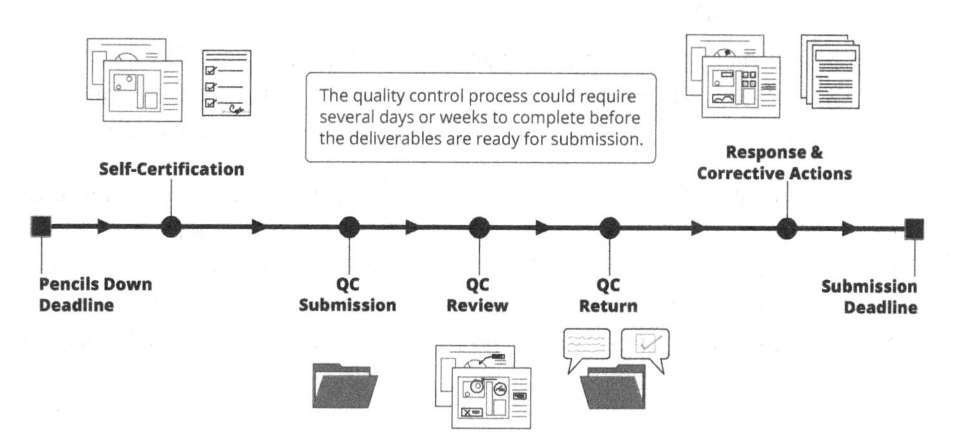

FIGURE 5.8 Quality control process.

3. QC Review
   a. Based on an agreed duration, the quality control team will review the submitted content, often through an independent review session, and provide comments on how the deliverable conforms with the project requirements.
   b. The review and comment process benefits from a standard practice and format. For example, reviewers may be required to indicate the authorship of comments and provide reference to project requirements, or there could also be a need to indicate when comments are preferential (suggestions).
   c. Even with independent review processes, critical flaws should be discussed early with the project team. This gives the team more time to respond to the issue or explain why the content appears noncompliant, which could prevent misunderstandings, costly rework, and delays in project timelines.
4. QC Return
   a. As deliverables are returned to the project team, the parties should clarify any required actions or review comments. The QC team must adhere to the established schedule for review and then return to provide the project team with enough time to address the comments.
5. QC Response and Corrective Actions
   a. Any comments made through the QC review should be accepted (and corrected), clarified (indicating how the comment is already addressed), disputed (with a reason), or deferred (until the next submission).
   b. The project team benefits from a standard process to indicate how a comment is resolved (using a color-coded system) and when the comment has been addressed (action taken and re-checked).

This section outlines an engineer-centric approach to quality management for scenarios where an engineering firm prepares deliverables for a client, which are subsequently submitted to a government agency for review and permitting. However, it's important to note that clients often play a key role in the quality management process, particularly in design validation.

During the QC review stage, the client can also review the plans. This step ensures that the plans align with both the client's and end users' needs before submission to a plan review agency for permitting. While major scope changes are not expected at this stage, client involvement allows for early detection of any misalignments or necessary adjustments. Sometimes, the client's review might occur simultaneously with the engineering firm's internal QC review. However, starting with a client review can benefit the review process because the client's validation and any feedback provided can be seen and incorporated into the subsequent internal QC process. This method helps avoid conflicting comments arising from the two separate reviews.

Integrating the client's feedback with the independent QC review allows the project team to address a comprehensive set of comments, ensuring a more thorough and cohesive review process. This collaborative approach enhances the overall quality and alignment of the final deliverables with both client expectations and regulatory requirements.

After all comments have been resolved, the project manager (or other reviewer determined by the quality management system) should validate that the corrective actions have been taken. This often includes reviewing the final deliverables with the QC review set to check the original comment, the response, and the final resolution.

While this process provides a general framework, each project or engineering firm will have its unique approach. Project managers must allocate sufficient time and resources for these quality control steps, ensuring the team is prepared well before submission deadlines. Some of these steps may require multiple iterations, such as a self-certification process that identifies issues that must be corrected by the project team before the deliverables can be sent to the QC review team. This process also depends on the availability of independent reviewers, who may not be intimately familiar with the project's timelines, highlighting the importance of incorporating their schedules into the project's planning phase.

5.7.3 Root Cause Analysis

As errors and omissions are identified in the project, directly addressing nonconformance is only one small step for project quality management.

Project teams and organizations should seek continuous improvement to prevent future errors.

Project teams and organizations should seek continuous improvement to prevent future errors. *Root cause analysis* digs into the origin of the quality control issue. There are two standard practices: *Five Whys* and *diagramming*. The Five Whys simply ask "why" the error occurred enough times until the root cause can be identified (recognizing the number five is not to be interpreted literally). For example, if the contractor encounters a clash when building new utilities, the question begins with:

- Why? The plans depict the utility designs for two utilities but omit the clash location.
- Why? The quality control reviewer missed the omission.
- Why? The reviewer assumed that clash detection was automated with design software.
- Why? The designer used software capable of clash detection but did not run the process.
- Why? The designer was not aware of the process.
- Why? The designer was never trained in the software (root cause).

In this example, a construction-related issue was traced back to a lack of training for a design team member. This insight would inform the project team and organization that there is a need for training on the software. Quality management is about supporting the project team to improve the deliverables created by team members.

Root cause analysis is also performed through diagrammatic methods, such as Ishikawa (or fishbone) diagrams and cause-and-effect diagrams. The quality-control issue is identified at the head of the diagram, and different categories are defined as branches. Categories vary by the analysis but often include processes, equipment, staff, training, external, and management. Issues are identified along each branch to visualize relationships and help identify opportunities for improvement. Figure 5.9 shows an example of a fishbone diagram tracing the causes that led to the issue/effect of a construction clash. The causes share similar themes, but the branches reveal how different factors led to the issue. These tools are best utilized by stakeholders with different perspectives to review the cause and effect thoroughly.

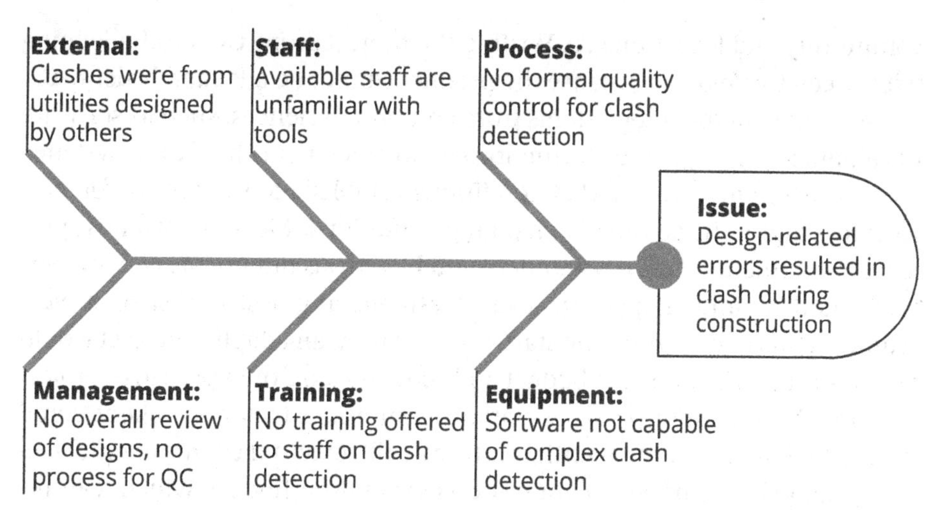
External:
Clashes were from utilities designed by others
Staff:
Available staff are unfamiliar with tools
Process:
No formal quality control for clash detection
Issue:
Design-related errors resulted in clash during construction
Management:
No overall review of designs, no process for QC
Training:
No training offered to staff on clash detection
Equipment:
Software not capable of complex clash detection

FIGURE 5.9 Fishbone diagram to evaluate the root cause of an issue.

5.8 Adaptive Methodologies

Most civil infrastructure and real estate projects will be structured with predictive requirements – a well-defined scope based on local requirements and standard practices for development. This requirement format has worked well for the traditional design-bid-build process. Specifically, during the construction phase of a project, a design-bid-build project will utilize permitted design documents with specifications that outline the details of the project work. However, adaptive requirements may benefit certain project delivery types (design-build) or projects with complex or uncertain design conditions. Adaptive (and agile) practices recognize that the scope can be fluid, especially early in the project life cycle.

> Adaptive requirements may benefit certain project delivery types or projects with complex or uncertain design conditions.

For example, consider the scope definition for a new mixed-use development project requiring a rezoning application (or similar entitlement processes). The site infrastructure and building design will adhere to published codes, but the site planning begins with some ambiguity. The number of buildings, types of buildings, parking areas, green space, and other site elements are yet to be defined and may require negotiation with the community and local officials. Writing the project scope can be challenging with predictive models because the product cannot be adequately defined.

Adaptive methodology stems from computer sciences, such as software development. The pace of technological advancement, hardware systems, user expectations, and market conditions established a necessity for projects to adapt to these rapidly changing conditions. Most importantly, project development teams must understand how to accommodate client needs that could rapidly change. Instead of establishing a strict list of project requirements and scope at the start of the project, an adaptive project would work within an established budget and schedule and then prioritize project features. Prioritization is an essential requirement factor – a client could have an extensive list of potential features. Still, external environmental factors (coupled with project team resources) would influence which features should be developed and implemented first.

Civil infrastructure and real estate development projects benefits from implementing these adaptive practices, especially in the early phases of project planning or on some project delivery methods, such as design-build, which benefit from responsive project development. Without a clear definition of the proposed project elements (buildings, uses, infrastructure), it is unreasonable to establish a detailed project scope for the final engineering, architecture, and construction work. Instead, an adaptive approach can be applied to the early planning phases, with predictive models used in later development, thereby establishing a hybrid project delivery method.

Considering the initial example of this section (a mixed-use development with a rezoning requirement), the project's adaptive scope could describe several potential iterations based on the schedule. Then, the developer and design team would prioritize which design options should be created and presented to the stakeholders first, refining the site layout and program with feedback received. The site layout will adapt to the feedback, and the design team must remain responsive while managing the project schedule and budget constraints.

5.8.1 User Stories

The term *user story* is borrowed from computer science and agile project delivery methods. The purpose of a user story is to communicate how the project scope will deliver value to the client. Recognizing that many engineering projects can include technical and nontechnical team members, it is expected that the value of project work could be misunderstood. With a computer science example, the user story would define the desired value and features of a product instead of creating a technical list of code requirements. The user story promotes communication between clients and engineers. It prompts engineers to ask the right questions about *why* a requirement has been established to understand best how to achieve the desired *value* for the client. Without the value condition, a technically correct solution may not be the client's preferred option.

Figure 5.10 shows one of the most common formats for a user story, intentionally structured to be concise, as if it could fit on a small card.

This format is deliberately simple. The *role* defines which stakeholder perspective is being considered. The *action* speaks to a product's performance requirement. Finally, *value* is a critical element in defining the purpose of the feature. The user story is incomplete if the role, action, and value are not defined, and the user story becomes less tangible if the format is altered significantly from the example outline. Generally, these user stories are authored by the client and the project design team to develop a consensus for defining components such as the intended action and the desired value. The *value* intention can significantly influence the team's design objectives.

In civil infrastructure and real estate development projects, as with other technical projects, a user story can better communicate the intended value and purpose of the site and building features, which might otherwise be miscommunicated through technical conditions. As an example, a development project might require some form of environmental stormwater management. While the developer recognizes that this is a requirement, the design team is left with many possible solutions to satisfy the design requirement for stormwater management. The design team could prioritize cost and design a simplistic detention pond in a location that minimizes earthwork (perhaps at the expense of locating it at a highly visible area of the site). Alternatively, the stormwater system could be designed with enhanced landscaping and located on the site to serve as a public amenity. This is where the importance of *value*

As a <role>, I want to <action> so that <value>.

FIGURE 5.10 Structure of a user story.

As a developer, I want to build a stormwater system, so that the site has a natural amenity.

FIGURE 5.11 Example user story.

is established and can best define the project requirements beyond the technical solution. Figure 5.11 provides an example of a completed user story.

In this example, the *role* is of the *developer*, and the *action* is to *build a stormwater system*. Initially, the user story informs a requirement for an environmental protection system – the value statement better communicates the team's design requirement. Recognizing that this developer seeks to have a natural amenity will influence which technical solutions the design team will implement.

With agile practices, user stories are just one element of a larger hierarchy. While the lexicon can vary, multiple stories can be associated with a single product *feature*, and multiple features can establish an *epic*. For example, several user stories could establish tasks to design a building with sustainable building features, and multiple features could lead to a net-zero building condition (or achieve a similar environmental sustainability accreditation).

In addition to formatting a user story with the role, action, and value, there is an important mnemonic with user stories. Bill Wake, a consultant on Extreme Programming (XP) and Agile software, created INVEST as a standard checklist for a well-authored user story. INVEST stands for Independent, Negotiable, Valuable, Estimable, Small, and Testable. A user story must meet these conditions to successfully implement and realize the intended features while supporting the design team with the appropriate tasks. While these are established for user stories, these conditions are also best practices for any scope item. These elements of a well-defined requirement help establish acceptance criteria (defined in Section 5.6.2).

5.8.2 Scrum Framework

Scrum is an agile framework that embodies adaptive project delivery. While Scrum has well-defined team roles and events, its design inherently facilitates ease and flexibility of implementation. As an adaptive framework, Scrum can work well for projects requiring agility in scope and execution. Some projects may benefit from a hybrid approach, using a few components of the Scrum framework throughout the project. Alternatively, some projects might find value in applying Scrum's more structured approach to specific phases, such as early conceptual design. This section provides an overview of Scrum.

Scrum organizes the execution of project work into sprints – short, consistent work cycles – allowing teams to break down complex projects into manageable tasks completed in a fixed duration. The duration can vary, but each sprint is usually two to four weeks long. Each sprint is focused on delivering specific portions of the project, facilitating client (and stakeholder) feedback, and allowing for rapid adjustments based on evolving project conditions. Once the duration of a sprint is established for a project, it's typically a fixed length to provide a consistent cadence of planning, execution, and review of work completed during each sprint.

For example, if the civil engineering design for a new road project has a nine-month design schedule, the engineering team may determine that each sprint will be four weeks (about 10 sprint cycles). Each sprint would focus on deliverables that have a recognized value to advancing the project, achieved by completing tasks that progress the design through milestones or portions of the project work. At the end of each sprint, the team should produce components of the product, such as engineering plan sheets, to demonstrate progress defined with each sprint. The deliverables associated with each sprint are intended to be reviewed with the client, encouraging transparency in progress and establishing a consistent communication process.

This list provides an example of possible key themes and the possible deliverables of 10 sprints for a civil engineering road design plan.

1. Preliminary horizontal design and environmental risks: Initial assessments on the preliminary road design, such as the alignment, profiles of existing grade, and typical sections. Early identification of environmental constraints or risks that may influence the design.
   - Deliverables: Plan sheets depicting the preliminary alignment and locations of possible environmental risks.
2. Preliminary vertical design: Initial assessment of topographic conditions influencing the roadway's vertical geometry, sight distance, and early earthwork computations.
   - Deliverables: Profile sheets with initial vertical geometry and rough earthwork calculations based on early design models.

3. Conceptual design: advancing the road design to include early concepts of intersections, drainage components, and vertical road geometry. Early considerations for demolition, maintenance of traffic during construction, landscape plantings, and others.
   - Deliverables: The refined plan sheets with preliminary alignments, which include additional detail at intersections and approximate areas of stormwater features. New draft plan sheets for demolition, erosion control, and conceptual landscape plantings.
4. Geotechnical and structural preliminary assessment: Review of geotechnical reports for pavement subbase conditions and structural foundations (bridges, retaining walls, etc.)
   - Deliverables: Geotechnical and structural memorandums and recommendations based on a review of preliminary and conceptual designs, as well as road geometry refinements based on recommendations.
5. Conceptual hydrologic and hydraulic design: Determination of land cover conditions and drainage patterns, along with initial sizing of drainage catch basins and conveyance systems to stormwater management facilities.
   - Deliverables: Conceptual design of drainage network and early hydrologic model computations.
6. Final environmental mitigation design: Updates to environmental factors (from Sprint 1) that influence final road design details, such as sound walls, vegetation protection, wetland mitigation, and others.
   - Deliverables: Final plan sheets depicting environmental design elements, reports, and permit applications for environmental impacts (coincident with 50% design submission).
7. Final design (Part 1): Finalization of road geometry for both horizontal and vertical geometry, storm drainage design, and utility services.
   - Deliverables: Updated road design plans with final geometry and plan view of utility system designs.
8. Final design (Part 2): Additional annotations and reports for road plans, profiles, utility networks, hydraulic calculations, and structural designs.
   - Deliverables: Finalization of utility system profiles, stormwater management calculations, and structural details.

9. Final Design (Part 3): Ancillary road design elements, such as signage and pavement markings, landscape design, planting lists, hardscape details, and maintenance of traffic plans.
   - Deliverables: New and refined plan sheets for traffic engineering and landscape design.
10. Final design review and documentation: Comprehensive review of all designed content to confirm compliance with requirements, project objectives, and quality control standards.
    - Deliverables: Construction packages, including design drawings, specifications, narratives, calculations, quality control checklists, and other bid documentation.

The example list of 10 sprint topics and deliverables demonstrates the iterative nature of the Scrum framework; however, in practice, the sprint planning process is adaptive as the project progresses. The sprints are not predetermined but are informed by an iterative and incremental development process.

Sprint planning is a core principle of the framework and operates differently than traditional project execution operations. At the start of each sprint is a planning session, where the team assesses their objectives and resources to determine what work can be completed during the sprint. Estimating the amount of work that can be completed is complex. It relies on expert judgment while understanding how the team can measure the progress of completed work (refer to Chapter 6, Section 6.4.4). Still, the team intends to improve their estimating capabilities with each sprint to establish an accurate velocity of work that can be completed with each cycle.

During the sprint, the team has brief (15-minute) daily meetings to review the work completed, the work in progress, and the backlog of work for the sprint. These short-duration meetings are often strictly enforced, requiring team members to come prepared with a concise update to communicate their progress. This emphasizes the team's autonomy and individual accountability. Instead of a project manager periodically asking for information from each team member, each team member owns their progress. Similarly, the team has the autonomy to decide how best to complete their tasks rather than following explicit instructions from a manager. Still, the team leader (i.e., project manager) can serve the team by validating the reasonable amount of work that can be completed in each sprint and monitoring any external factors that could influence the scope or execution.

User stories (Section 5.8.1) in Scrum facilitate a client-centric approach to project planning, ensuring that each task directly contributes to the project's overall goals and provides a value proposition. By breaking down complex projects into smaller, manageable tasks, Scrum promotes more accurate estimations, better resource allocation, and more effective risk management. Scrum's emphasis on iterative development, continuous stakeholder engagement, and team collaboration makes it a valuable framework for managing the dynamic conditions of civil infrastructure and real estate development projects.

At the end of each sprint, the team conducts a sprint review, or *retrospective,* in which the team evaluates performance, such as client satisfaction with progress, an alignment of goals, the velocity of work completed (to inform future sprint planning), and a discussion on what could be improved with the next sprint. The sprint retrospective, often invites the client and other stakeholders to review work completed during each sprint. This way. each sprint should critically evaluate the team's project execution, seeking to incrementally improve their efficiency, estimation of work velocity, and correct any variances in client validation.

Scrum also benefits from visual task boards (Section 5.6.3), called Scrum boards. These boards share similarities with Kanban boards but are more structured because the listed tasks are linked to each sprint. This means that the backlog of activities on the Scrum board only includes the work planned for the sprint, and then the board is reset at the end of each sprint.

Scrum works best for small project teams, typically with fewer than 12 participants. In the context of civil engineering, where projects can be large and multidisciplinary, it's common for several smaller teams to adopt Scrum independently. With larger teams, a mechanism known as the *Scrum of Scrums* is implemented to ensure coherence across these diverse groups. This approach involves representatives from each specialized team – such as architects, transportation engineers, geotechnical engineers, landscape architects, and environmental engineers – coming together. These representatives, typically one or two from each discipline, convene in a separate, cross-disciplinary meeting to synchronize their efforts and share updates. This multidisciplinary discussion, or Scrum of Scrums, facilitates effective collaboration and coordination among the various teams working on a complex project, ensuring that all facets of the project align seamlessly.

By fostering a structured yet flexible work environment, Scrum enables teams to focus on high-value tasks, improving performance and efficiency.

This framework encourages open communication between clients, early stakeholder feedback, and continuous improvement, prompting a proactive and responsive project management approach. While less common in civil engineering, the benefits of enhancing project outcomes and team dynamics are well-documented in other industries, making it beneficial to consider hybrid or complete implementation.

In Appendix A, Part 5 of the Millbrook Logistics Park, the storyline advances with the engineering team handling the early project work. A focus is placed on the engineering team's increased responsibilities and involvement during a phase where the engineers have only limited responsibility. This challenge prompts discussions on project scope ambiguity and the fact that work is shared across different team members. The narrative delves into scope negotiation and the alignment of client expectations with contractual terms, underlining the practical aspects of civil engineering projects in maintaining project coordination, defining a clear scope of work, and ensuring effective team collaboration.

CHAPTER 6

Project Planning and Scheduling

Two ovens can't bake the cake twice as fast.

CHAPTER OUTLINE

Management Essentials for Civil Engineers: A Practical Guide to Business, Communication, Ethics, and Risk, First Edition. Cody A. Pennetti, C. Kat Grimsley, and Brian M. Grindall.

This chapter describes project planning concepts, beginning with how project teams can be structured. Next, this chapter describes the primary project life cycles, predictive and adaptive, that establish the core processes of project workflows. Based on the team structure and life cycle, project schedules can be created and monitored throughout the project. This chapter also includes a sample project plan template with summaries of core topics.

6.1 Introduction

Project plans are a critical resource for project execution. The well-prepared plan will communicate the scope, schedule, budget, strategy, resources, risk, and other project elements (a sample project plan outline is included at the end of this chapter). The initial development of a project plan occurs at the start of the project; however, a project plan is dynamic and requires updates throughout project execution and when external factors influence the project. Project plans must be tailored to the project scope to provide the requisite information while considering the time and resources required to develop the document. An engineering firm may prescribe a project plan outline, which may exist in different formats based on the project size. For example, a project budget of less than a certain amount may only need a summary document. Larger projects will require a formal project plan, and the document serves as a primary communication and management tool for the project team. In this way, the team can refer to the project plan for schedule updates, clarification on requirements, or check responsibilities based on the organizational chart. This chapter includes information on project team structures, schedules, resource management, and an outline of a project plan.

A project plan is dynamic and requires updates throughout project execution and as external factors influence the project.

6.2 Project Team Structures

Many projects depict the project team members in an organizational chart, identifying a project manager supported by a team of professionals in a hierarchical format. While the project resources are organized this way, an engineering firm's staff is generally not structured by projects. For example, engineers and architects from multiple business units (traffic, hydrology, landscape, geotechnical, etc.) have a corporate hierarchal structure with individuals supervising the personnel working on multiple projects. For this reason, project-specific team structures can vary by project size, complexity, and a firm's standard practices.

The choice of team structure is influenced by both the project requirements and organizational structure and policy. The authority, role, and time commitment of the project manager will vary based on the team structure. This ranges from a full-time job designation with almost total control (e.g., projectized) to part-time and only moderate authority (e.g., functional or matrix). Descriptions of the teaming structures are included in this section, beginning first with an overview of the parameters that influence project team operations. Note that these structures and the operations must consider the broader teaming agreements, such as the use of subconsultants (as described in Chapter 3 and Chapter 4).

Each team structure has strengths and weaknesses that include trade-offs. These trade-offs often affect operations beyond a single project – a team structure may benefit one project's efficiency but restrict and pressure the resources required for other projects. For example, a projectized team requires dedicated personnel for the project, providing consistency in resource availability for that one project but limiting team members' flexibility to support other projects when needed. Project managers will need to consider a team structure that is appropriate for the project while considering the engineering firm's overall operations.

6.2.1 Parameters for Teaming Structure Operations

There are several considerations for selecting the right project team structure. Project operations can be impeded by various team structure conditions, which must be considered by the project manager. Almost all production resources on civil engineering projects are personnel, which means the project manager must consider several operating conditions relevant to human factors. These operating conditions include team member familiarity with the project, dynamics of a project team, and communication complexities. Many of these are inherent to any project work, but evaluating these conditions can help managers mitigate and plan for specific production conditions.

A few parameters for teaming structure operations and sizing are listed here, including the inherent ramp-up penalties, team performance, complexity of communication channels, multidisciplinary collaboration, and resource availability and allocation.

Ramp-Up

Ramp-up is the time it takes for personnel to become familiar with project objectives, processes, policies, and context. This ramp-up period could be a few days or several weeks, depending on the complexity and the current progress of the project. Ramping up during the later phase of the project will take more time, which creates challenges when assigning new team members to a project that is nearly complete.

Even frequent operations require team members to ramp-up as they establish daily priorities and schedules. Task switching, such as alternating focus across multiple projects throughout the day or week, can also impede the productivity of team members. Other impediments contribute to ramp-up penalties, such as interruptions of the thought process and the time required to set up project equipment and materials (e.g., opening the relevant design files and reports).

Performing as a Team

It takes time for a new team to reach the stage of efficient performance as relationships and dynamics mature. Bruce Tuckman's stages of group development refers to the stages of teamwork as *forming, storming, norming,* and then *performing* (the a fifth stage, *adjourning*).

In the early stages of team development, there is uncertainty about the capabilities and roles of each team member but, ideally, team members will eventually transition into a well-performing team. Trust, leadership, and respect must be earned between team members. While the steps in Tuckman's Stages of Group Development are primarily linear, setbacks do occur. Additionally, introducing new personnel to a well-performing team can require teams to retrace steps.

Communication Channels

Larger teams require more communication channels, influencing the efficacy and efficiency of guiding the team and sharing information. Larger teams also risk communication omissions or suffering from extraneous

Communication Channels:
[N x (N-1)] / 2

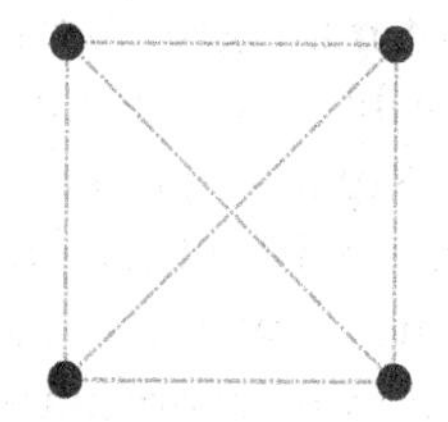

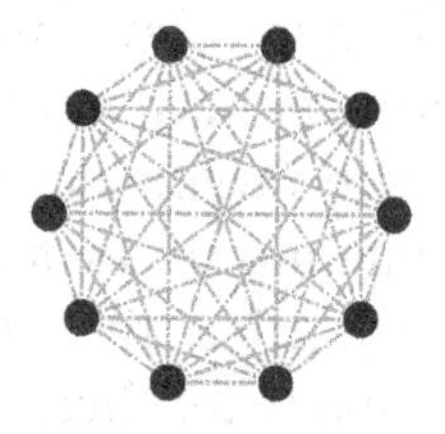

Team of Ten, N = 10
Communication Channels = 45

FIGURE 6.1 Expanding communication channels.

communication interference (i.e., noise) as it becomes difficult to verify that all team members understand their current responsibilities and are aware of the changed conditions and the responsibilities of other team members.

The number of communication channels is based on the number of team members and their need for bi-directional communication. Mathematically, this is measured as $N \times (N - 1) / 2$, where N is the number of team members. As depicted in Figure 6.1, a team of 4 will have 6 communication channels, whereas a team of 10 will have 45. Six new team members add 39 new channels of communication, significantly increasing communication complexity and risk of error. Additional considerations for team communication are covered in Chapter 9 (Section 9.4).

Multidisciplinary Collaboration

Civil engineering projects require collaboration between professionals from various disciplines (e.g., site-civil, geotechnical, structural, architectural, electrical). Coordinating and integrating these specialties is complex and requires careful planning and communication. Most team members, including the project manager, have only a limited understanding of the process, dependencies, and requirements for other disciplines. This can be more challenging when the other disciplines are subconsultants from a different department design firm or operate in a different department, where the project manager has limited oversight.

Resource Availability and Allocation

Competing demands for resources (e.g., team members) across multiple projects can lead to delays and conflicts if not managed effectively. Many civil engineering project managers are responsible for multiple simultaneous projects with varying demands based on disparate project phases. The demands of one project manager may compete with other projects when resources are shared. In this case, a project manager would need to evaluate resource availability for their own projects and must coordinate with other supervisors and functional managers. In this context, safeguarding team members from burnout and uncertainty about priorities becomes critical, especially within a matrix team structure where resources are shared across projects (and between project managers). A project manager must evaluate resource availability, coordinate with other supervisors, and actively work to protect their team members from being overwhelmed. This involves clear communication about project priorities, establishing boundaries, and ensuring that workloads are manageable.

6.2.2 Functional Managers in Team Structures

While there is only one project manager, in some team structures, there can be both a project manager and a functional manager (or multiple functional managers). A functional manager is an individual who supervises resources that the project manager does not directly control. This is common for large projects in an engineering firm when team members from multiple departments are involved. For example, for a new highway project, the project manager is likely from the transportation department and will directly supervise transportation engineers working on the project; however, the project will require stormwater management skills and personnel from outside the transportation department. In this case, the supervisor of the stormwater personnel serves as a functional manager who allocates resources and coordinates with the project team.

A functional manager may have different roles and authority based on the project team structure and an organization's policies. The functional manager does not always report to the project manager and may not appear on the project plan's organizational chart, but they greatly influence project operations through resource allocation and oversight.

6.2.3 Projectized Team

A *projectized* team structure is often reserved for large, complex projects that require constant effort from the project personnel for an extended period. This is a simple structure for project operations, with a team assembled and exclusively focused on the project through the life cycle. There is one project manager to whom all personnel report, often through a few levels of communication. The projectized team structure is reserved for large projects because there must be enough work to maintain the personnel. For example, projectized teams would be suitable for a large highway project that requires full-time work from a large team for several years.

The projectized team structure is reserved for large projects because there must be enough work to maintain the personnel.

In most cases, the team is tailored to the project, and personnel will come from multiple departments to provide the resources necessary for that project. These resource requirements must be well-communicated for each department involved, and there should be a shared understanding that personnel are dedicated to a specific project. Still, there is likely a transitional period at the start and end of the projects where personnel are not exclusive and must support other work. Once established, the project manager maintains full authority over the project resources (but coordinates with functional managers as needed).

The benefit of a projectized team is the focus and commitment of the team members. The project can be more efficient because there is predictability in who is available to work on the project, and the team members continuously work on project content without penalties caused by switching between projects and project teams. This structure is effective at overcoming ramp-up penalties, and uncertainties in resource availability and likely strengthens team collaboration.

The challenge for project managers of projectized teams is controlling the project resource supply and demand to establish a constant work effort, thereby proactively smoothing any demand peaks or lulls that would otherwise overwork or underwork the team members. For example, as resource demands diminish during plan review periods, the team must have enough

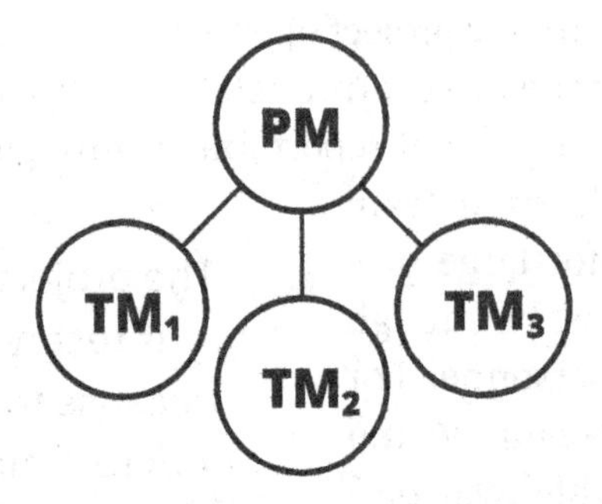

FIGURE 6.2 Projectized team structure.

work to maintain full-time involvement in the project (or else personnel utilization suffers). Figure 6.2 depicts how a project manager would have direct access to the team members under their supervision, and those team members would be assigned only to that project.

6.2.4 Functional Team

A *functional* team structure shares similarities with projectized teams, except that specialized personnel will report to their functional manager instead of the project manager. Functional managers are responsible for their specialized teams, often by discipline (traffic, landscape, geotechnical, etc.). In this way, the project manager communicates through the functional managers who assign and manage the work to their personnel.

Similar to projectized teams, a functional team structure is used for large, complex projects; however, functional structures are used when the project requires significant support from multidisciplinary personnel. In such cases, the project manager must rely heavily on resources controlled by other supervisors. For example, a large site design project that requires expertise from the civil, mechanical, electrical, landscape, communication, and other disciplines would benefit from a functional team structure, with each discipline led by a functional manager who coordinates with the project manager.

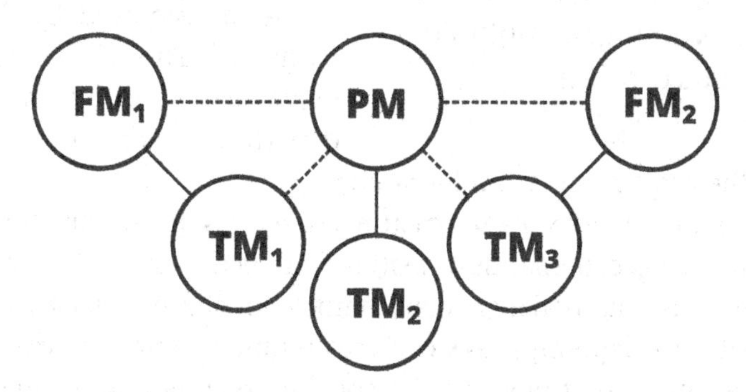

FIGURE 6.3 Functional team structure.

This team structure is effectively necessary for large, multidisciplinary projects because the project requires a team of experts. While this format will benefit each discipline's internal process and efficiency (because they know their specialties well), communicating across functional teams is often challenging.

Each functional manager coordinates with the project manager, meaning the project manager must coordinate project operations and schedules across many teams. For example, the landscape architecture team may want to rapidly complete their design to finish the work and be as efficient as possible; however, as the utility teams iterate through their designs, they will need to negotiate the utility and landscape conflicts, causing at least one team to rework their designs. The project manager and functional managers need to invest additional effort to communicate project goals and constraints across disciplines to streamline and coordinate the work effectively. Figure 6.3 depicts how a project manager would work with team members and their functional managers as compared to a team member under their direct supervision.

6.2.5 Matrix Teams

Matrix team structures are typical for engineering projects requiring subject matter experts from multiple engineering disciplines and teams working on multiple simultaneous projects. In a matrix structure, many personnel

Matrix team structures are typical for engineering projects requiring subject matter experts from multiple engineering disciplines and teams working on multiple simultaneous projects.

(resources) are shared across multiple projects as guided by a project manager. The project team can be fluid, adding and removing personnel as required by the project demands. This can occur in one department or across multiple departments of an organization based on the project's size and hierarchy.

The benefit of matrix team structures is that resources can more easily be distributed as project demands transition from peaks and lulls of work effort. The challenge is that managers may compete for shared resources. Different projects with overlapping peaks can create unfavorable resource demands, forcing overtime conditions and stressors, or creating risks of missed deadlines because there are not enough people available to work on the projects.

According to PMI, a range of weak to strong matrix structures refers to the strength of the project manager's authority. In a weak matrix structure, the project manager's ability to acquire resources is contingent upon the support and authority of the functional manager. Conversely, in a strong matrix, the project manager holds greater authority over personnel and resources.

A matrixed structure is common with real estate development projects because of the variations in project sizes, schedules, and scope. For example, resource demands are high for a project before a site plan submission, but once the plans are submitted for jurisdictional review, there is a low demand for resources on that project. When one project's demands are low, the team members will shift to a different project. This frequently happens during production periods when the team waits for information from other sources. The iterative process of a land development project means that production work is rarely linear and predefined. For example, landscape architects wait for the road design, and then the utility design must be coordinated with the landscape planting plans.

Matrix teams are also more casual in their structure, meaning they can appear deceptively easy to manage; however, matrix operations require planning for future personnel needs and rely on strong communication channels across managers. With shared resources, there are likely situations where a critical phase of one project will demand personnel that were otherwise assigned to another project. This condition also means that team members constantly operate with multiple project managers and could struggle to balance demands from multiple sources. Project managers in a matrix structure must collaborate effectively to avoid overloading the production staff. Without proper coordination, staff members may face undue stress and confusion, struggling to determine which project should take precedence in their priorities. The production staff must then work with managers to prioritize projects, establish a plan for overtime work, or acquire additional resources. Figure 6.4 depicts how three project managers would share a fraction of time from each of the pooled resources (team members) to accommodate project demands.

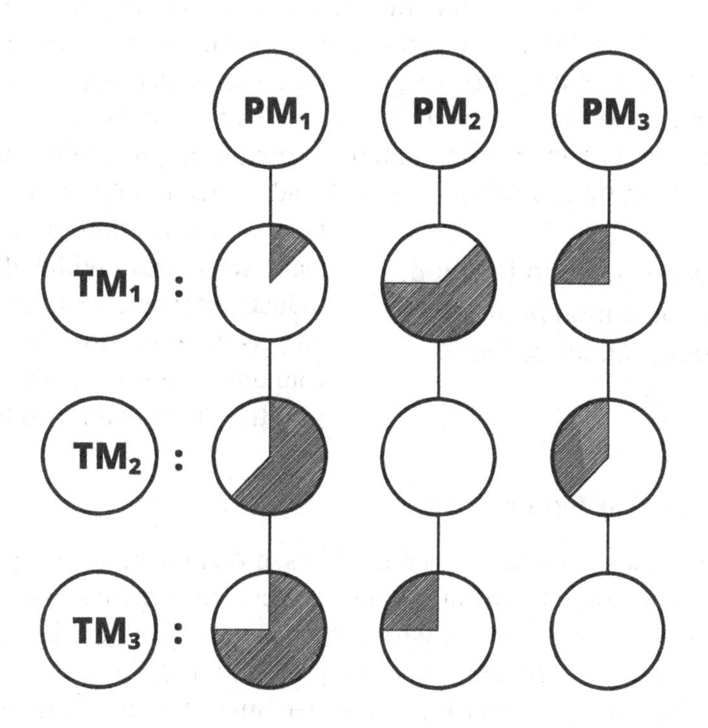

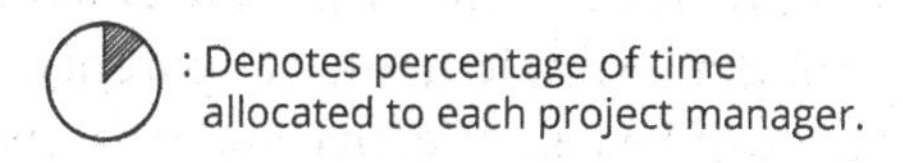

FIGURE 6.4 Matrix team structure.

6.2.6 Hybrids and Other Structures

An organization will likely use multiple team structures based on the project size and the personnel assigned. As a best practice, team members' personalities and the project manager must be considered when establishing the team structure. Some team members may prefer the decentralized approach of functional teams or the flexibility of matrix teams, or they may expect a rigid chain of command as provided by projectized teams. If the project can accommodate different options for team structures, the decision should consider the personnel preferences. Similarly, teams with subconsultants will likely operate through varying structures, and the parties of the teaming agreements will benefit from a mutual understanding of the team structure.

Many projects may also transition through team structures based on schedule and resource availability. Even with a large project, the resource demands may have peaks and valleys. For example, engineering design work would be minimal in early phases as survey information is collected, then peak during the design phases, and reduce again after a plan submission while the team waits for the review and comment. Early and late phases could benefit from a matrix structure to provide flexibility for personnel assignments. When the workload is greater and consistent, the team may transition to projectized team structures. While there are industry-standard team structures, all personnel must understand the communication plans and responsibilities of the team members.

> All personnel must understand the communication plans and responsibilities of the team members.

Self-Organizing Teams

The project manager role is less formal in self-organizing teams, providing personnel autonomy to operate as they see fit. This organic-type structure stems from computer science, where the team members work together to self-assign roles for different tasks. The project manager is a servant leader (refer to Chapter 10, Section 10.3 for leadership styles), working to remove constraints and provide resources and information as required by the team. This structure might be implemented as part of a hybrid structure where the project manager can be less directive and trust the team members to coordinate tasks as needed. This structure should not be used for less-experienced teams that require and expect direction and guidance with project work.

Remote Teams

A remote teaming structure can follow the primary structures (projectized, functional, matrix, etc.) but includes a few operating distinctions because of the geospatially diverse environment. Many project structures may have at least one remote functional team (or remote individuals) but otherwise operate best when the team members are nearby.

There are inherent benefits from teams working in the same geographic locations, specifically working close enough to overhear conversations and benefit from open discussion with peers and managers. In a remote setting, communication requires formal actions (calling someone or scheduling a meeting) that can impede the organic conversations seen in on-site working environments. These teams often need more tools to communicate tasks and schedules and benefit from formally scheduled meeting times and open discussion forums. Additionally, geographic dispersion could mean that operating hours, holidays, disruption, and other conditions vary and must be considered when planning meetings and deadlines.

6.3 Project Life Cycles

The project life cycle defines the mechanics of progressing from project initiation until closure. Traditionally, this process is considered linear as the team works through consecutive tasks until all requirements are met and the work has been verified and validated. This process would be considered predictive, which is appropriate for projects with a well-defined and constant scope of work (SOW). Alternatively, projects may follow an adaptive life cycle tailored to be agile in the face of changing conditions (from internal or external sources). The International Council on Systems Engineering (INCOSE) provides a modified categorization of three models: prespecified and sequent, evolutionary and concurrent, and interpersonal and emergent. The first category follows predictive methods, the second is hybrid, and the third is adaptive.

Traditional life-cycle phases for manufacturing and software development include a testing and validation step, but it is infeasible to construct and test a civil infrastructure or real estate development project. Instead, design models for hydrology, hydraulics, structures, traffic simulations, and virtual design and construction provide some capabilities for testing. Still, civil infrastructure projects rely more on design-phase feedback and expert judgment before the project transitions into the construction phases.

Many projects may use a hybrid of life cycles, alternating between predictive and adaptive based on the project phase or scope of work. This section describes the primary life cycles: predictive and adaptive.

6.3.1 Predictive Life Cycles

Predictive life cycles (also known as traditional or *waterfall* project management models) are characterized by their sequential and fixed nature and, again, appropriate for projects with a well-defined and constant scope of work. In these models, the full scope of work, along with the associated schedule and costs, are defined at the beginning of the project, with the expectation that changes will be minimal as the project progresses.

Defining the scope of work clearly at the project's initiation allows the project manager to develop a realistic schedule and budget that align with the project's requirements. This approach establishes accountability for the team to meet the set deadlines and budget constraints, provided the scope remains unchanged. Predictive life-cycle models are particularly effective for projects where risks are mitigated through comprehensive due diligence and for projects following a standard design-bid-build process. For instance, in a transportation agency's new highway project, if preliminary engineering has been completed, a predictive life cycle approach is appropriate for the design team tasked with final engineering. A thorough definition of scope ensures that the work is well-defined, allowing the team to confidently plan the schedule, costs, and resources to complete the project.

Predictive life-cycle models are particularly effective for projects where risks are mitigated through comprehensive due diligence.

However, predictive life cycles have a limited tolerance for changes. Once a project commits to this model, resources are allocated, and design conditions are set with the expectation of a constant scope. This approach, while efficient, can create challenges if changes do occur. Any alterations in scope or scheduled deliverables require a reassessment of the project's budget and resources. Even minor changes can lead to significant ripple effects, impacting extensive portions of the design and management processes.

In predictive project environments, where the scope is expected to remain constant, the project team often operates under the assumption that once a task is complete, they can move on without revisiting it. This leads to many design tasks being treated as discrete and independent, with the team finalizing various design elements when project conditions are accommodating (i.e., personnel are available to perform the work). However, this approach can become problematic when changes occur later in the project.

For instance, consider the progression of complex design calculations, such as stormwater modeling, in a site development project. A common best practice is to defer these complex efforts until all other site design elements are set. This sequencing minimizes the need for reworking and recalculating stormwater models in response to changes to the site layout. However, in a predictive environment where the emphasis is on meeting preestablished schedules and utilizing resources efficiently, the project team might opt to advance the stormwater design earlier in the schedule.

If the stormwater design is completed early based on resource availability or scheduling priorities, any subsequent modifications to the site layout, even minor ones, can lead to considerable rework. A small change in site configuration, such as the relocation of a parking area or adjustment of building footprints, can significantly alter drainage patterns, necessitating a complete overhaul of the previously "finalized" stormwater design.

Without iterative reviews and continuous stakeholder input, the project team is not structured to refine the design incrementally, leading to a scenario where late-phase changes have a disproportionately large impact. From a client's perspective, proposed changes could be seen as a minor adjustment, but the amount of rework is dependent on how the project team has ordered various tasks. This underscores the importance of carefully managing scope and stakeholder expectations in predictive projects and highlights the potential benefits of more flexible, adaptive approaches.

Because the predictive life-cycle model is based on a reliance on initial requirements, challenges are introduced if there is an incomplete understanding of the project's constraints and opportunities. For instance, detailed surveys and geotechnical reports might not be available during the early conceptual design phase. As new information emerges, it can substantially impact the design. Furthermore, this model presupposes that clients and other stakeholders have a clear vision of the project's progression, which is unlikely. Most stakeholders prefer to review and refine intermediate deliverables, finding it difficult to visualize the outcome from early concepts.

While requirement changes are possible within a predictive model, they are typically handled through formal change management processes. In most cases, projects will benefit from some form of adaptive (or hybrid) life cycles that promote iterative deliverables and client feedback and accommodate design refinements.

6.3.2 Adaptive Life Cycles

Adaptive life cycles represent a dynamic and flexible approach to project management and are particularly suited to environments where change is expected and embraced. Unlike traditional predictive models that rely on stable requirements and sequential tasks, adaptive life cycles thrive on their ability to respond to evolving conditions and stakeholder feedback. Many of the adaptive methodologies stem from computer science, where software development is required to quickly adapt to changes in technology, hardware, customer expectations, and other factors.

Adaptive life cycles are particularly suited to environments where change is expected and embraced.

Adaptive practices are relevant to civil infrastructure and real estate development work, as shifting market conditions, material costs, labor availability, and community perception will influence the project strategies. For example, in the early phases of a rezoning project, the team must be prepared to evaluate and accommodate feedback from a multitude of stakeholders that will influence the project scope. Instead of expecting predefined and stable requirements (as with predictive), the adaptive life cycle anticipates changes, and the project manager will allocate resources and modify schedules as needed.

One of the core concepts of adaptive life cycles is a focus on *iterative* planning with incremental development. Iterative planning begins with high-level concepts that become more detailed over time. Instead of investing time in complex calculations in early phases, the project team maintains a comprehensive but high-level approach to the design to elicit feedback from stakeholders. The project manager must determine the appropriate level of detail and focus on high-value tasks that will be most informative to stakeholders to guide future design iterations.

For example, a client is likely more interested in the location of site buildings and an approximate number of parking spaces and less interested in the exact count of fire hydrants or the linear feet of storm sewer (especially in early project phases). Iterative development is familiar to civil engineering practices, often including milestone deliverables based on the percentage of completed work (e.g., 30%, 60%, 90%, final).

The *incremental* approach, while aligned with iterative development, concentrates on delivering smaller, completed sections of the project at each milestone. For instance, in the mixed-use development project, the initial increment might focus on finalizing the layout of major structures and road networks. Subsequent increments could then detail individual buildings, followed by utility systems like the storm sewer network, and finally, finer details such as landscaping and exterior aesthetics. This method allows for managing the project's complexity by breaking it down into more manageable, discrete parts, each with its own set of deliverables and review stages. A visual representation of these iterative and incremental methods is shown in Figure 6.5.

Many projects will use both iterative and incremental practices because each one encourages continuous feedback that helps achieve project success. While continuous feedback promotes engagement and maintains the trajectory of the design work, this methodology has some challenges. The iterative and incremental deliverable scope must be clearly defined and communicated. Does the 50% set include half of the design data of all site

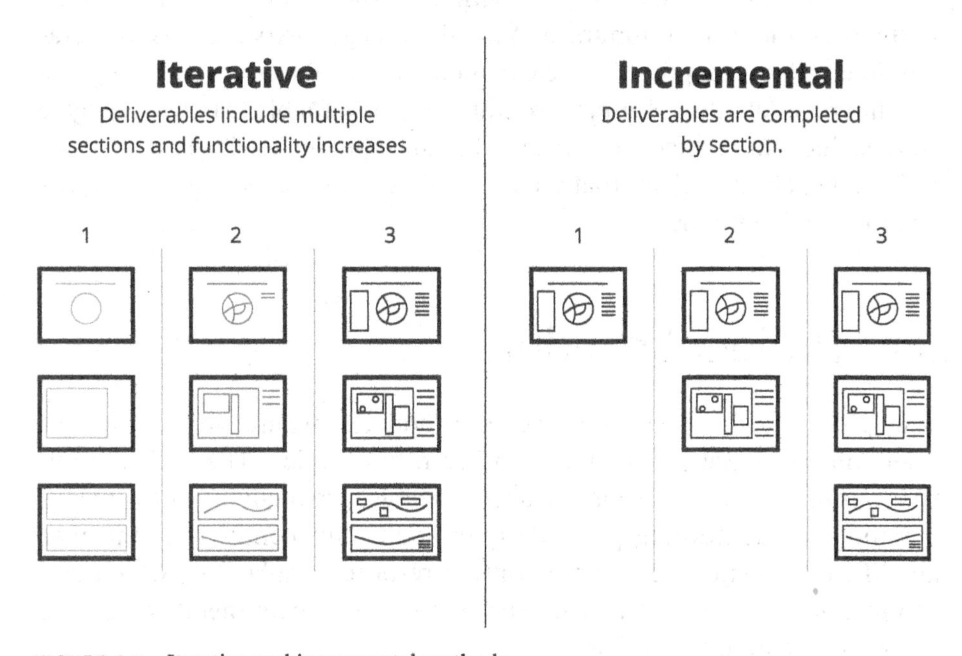

FIGURE 6.5 Iterative and incremental methods.

plan sheets or only half of the plan sheets? Which half is important to see early, and to what level of detail?

The design team should understand the client's expected scope for each iteration to provide the highest value information early. The value of the design elements is established by the client but often focuses more on the broader scope items and less on the technical details. Narratives, shop drawings, and tables of computations are necessary for design but are unlikely to provide value in early project phases. Similarly, quality control personnel need to understand the scope of each iteration (to avoid confusion regarding omitted content). For example, suppose the first design iterations only provide a basic network diagram of the storm sewer utilities and approximate shapes of the stormwater management system. In that case, there is little benefit in a detailed quality control review of hydrologic computations.

Adaptive life cycles represent a significant shift from predictive methodologies, offering a responsive and flexible approach that is advantageous in environments where change is expected. By embracing iterative and incremental practices, these life cycles allow for continuous stakeholder engagement, adaptability to evolving project needs, and an opportunity for ongoing refinement of project deliverables.

While adaptive practices come with their own set of challenges, such as the need for clear communication and managing expectations for iterative deliverables, adaptive life cycles benefit project objectives and recognize the inherent volatility of project conditions. They enable teams to navigate complexities more effectively, respond to stakeholder feedback faster, and ultimately deliver projects that are more aligned with evolving client needs and market conditions.

6.4 Project Schedules

Managing time in professional services projects, where progress hinges solely on personnel production, is inherently complex. The variability in tasks, coupled with personnel's unpredictable productivity rates, makes planning and scheduling particularly difficult. This complexity is further amplified in environments with shifting resource availability, which is a common scenario in matrix team structures where team members may be shared across multiple projects.

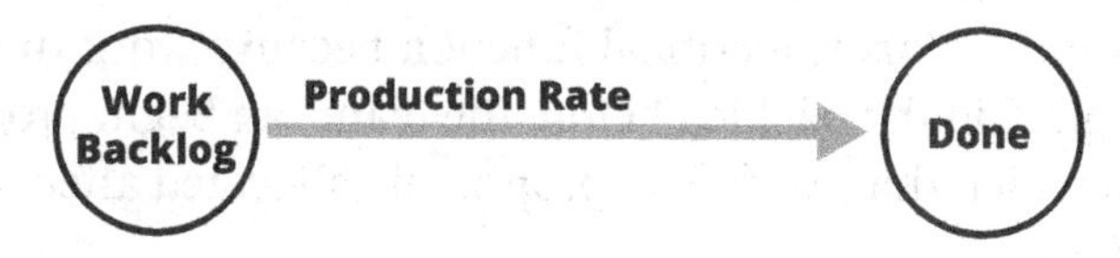

FIGURE 6.6 Simple workflow model.

One critical aspect often overlooked in simple scheduling models is the multitude of steps involved in completing a task. A simple model may inaccurately estimate the time needed for a single task, focusing on meeting established deadlines without accounting for the intricacies of the workflow. Successfully addressing this is not as simple as allocating time for the task itself; considerations must include planning, coordination among team members, potential revisions, quality control process, corrective actions, and unforeseen delays. Each of these elements can significantly impact the overall time required for project completion.

For instance, a simple workflow model likely considers linear progression: a backlog of work to be done, followed by the execution of those tasks, and ultimately, the compilation of completed work delivered by an established deadline. Figure 6.6 provides a basic visual representation of this simple model.

However, real-world scenarios demand a more intricate model. Work isn't truly "done" until it passes the rigors of quality management processes and is validated by the client. A client may not validate that the work meets project objectives as initially established with the project scope. Additionally, quality control may identify the need for revisions, augmenting the initial backlog. This cyclical flow – backlog, production, quality control, then returning to the backlog – better represents the realities of project schedule management.

> Personnel production rates are not discrete.

Another consideration is that personnel production rates are not discrete; they are shaped by factors like the inherent fluctuations in daily performance, the changes to team size, the complexity of tasks, communication requirements, and ramp-up for new team members. A small, consistent team will be more efficient than a larger team with fluctuations in assigned personnel. Additionally, the project manager plays a pivotal role in designating tasks based on team members' expertise and assessing task interdependencies. This is a critical function because work must be organized and assigned in the backlog before the team can begin production, and the degree to which that work is appropriately allocated affects overall project productivity.

These concepts were explored by Kenneth G. Cooper in "Four Failures in Project Management". Cooper identified the inherent variability of production rates by professional services, especially for architecture and engineering. The variability of the project's work production rate comes from resource (personnel) availability, the potential for undiscovered rework, and the influence of overworked and overburdened team members on productivity rates. Figure 6.7 offers a more comprehensive view of this expanded model.

The model in Figure 6.7 is expanded with a few more elements beyond the simple model of Figure 6.6 to include additional considerations:

- Before the work is considered done, the work must pass through a quality control gate. Any work that does not pass will generate rework and reenter the backlog.
- The production rate of work is variable. It depends on who is available and who is assigned to the work.
- Scope change, whether in predictive or adaptive life cycles, will also increase the work backlog and priorities as the client generates change orders.

The model in Figure 6.7 could be expanded further, recognizing that resource demands likely compete with other projects, meaning that resource availability is dynamic. Additionally, overburdened staff are more likely to

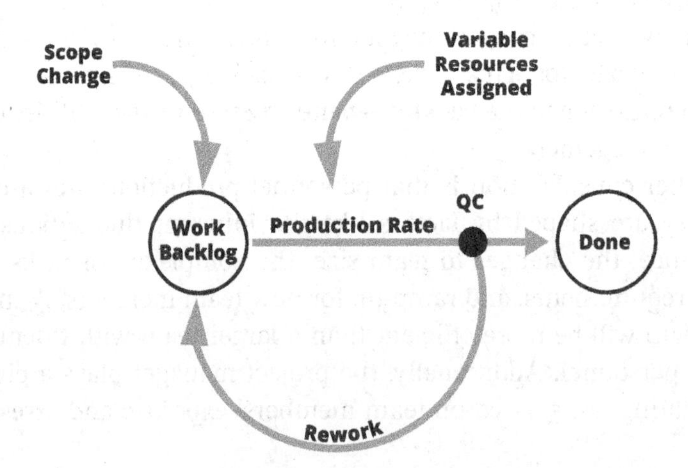

FIGURE 6.7 Expanded workflow model to consider production variables.

make mistakes and increase the number of issues identified through quality control. Larger teams require more communication effort, late changes require even more work, and many projects experience changes in scope that lead to more undiscovered rework. To establish realistic schedules and deadlines, project managers must account for these dynamics, even if only conceptually.

Many scheduling estimation techniques mirror those used for cost estimation, as detailed in Chapter 7 (Section 7.5). This close alignment between schedule and budget estimates is particularly relevant in professional engineering services, where the bulk of project expenses are tied to personnel time. However, schedule estimates are generally less tolerant of errors than budget estimates due to the inherent rigidity of time. While budgetary overruns or underruns can often be adjusted by reallocating funds across different tasks, any delay in a critical task can cause setbacks for the entire project timeline. For more information on the impact of delays on project scheduling, refer to Section 6.4.3, which describes the critical path methodology. The risk associated with project schedules is described in Chapter 8 (Section 8.5.2). Reviewing work and discovering cases for rework is a legitimate schedule operation, which begins with the quality management practice of self-certification (described in Chapter 5, Section 5.7). Reviewing work and discovering cases for rework must be accounted for with the schedule, often requiring work to be completed several days or weeks in advance of a deadline so there is an appropriate amount of time for quality control. Some of these operations are outside the purview of the project team members and must be monitored and communicated by the project manager.

Project managers must also recognize that scheduled tasks are interdependent, so it is critical to evaluate task dependencies. For example, the site-civil team may have completed the demolition plans for a project and categorized the work as "done"; however, late changes from the landscape design team would force rework of the site-civil demolition scope. Modifications in one task can necessitate revisions in another, and project managers must consider these dependencies when establishing the project schedule.

6.4.1 Schedule Dependencies

Many project tasks will depend on other tasks for initiation or completion. These relationships describe a dependency between project tasks and are categorized as discretionary or mandatory. A *discretionary* dependency represents a best practice that one task should be completed before another

A discretionary dependency represents a best practice that one task should be completed before another task. Mandatory dependencies describe a required order, which could be based on contract requirements or limitations with the nature of work.

task (or tasks). *Mandatory* dependencies describe a required order, which could be based on contract requirements or limitations with the nature of work.

For example, planting the site's landscape vegetation has a discretionary dependency on constructing a new office building. It is possible to plant the turf and trees before the office building is complete, but the construction activities could likely damage the new vegetation and require replacement and repairs. In this case, the office building *should* be complete (or near completion) before planting activities, but there is no mandatory requirement for the workflow.

A mandatory dependency would be the order of operations for road construction – there is no physical way to change the order of constructing the subbase, base, and top coat for a road. Similarly, a contract may stipulate that each phase of design work be submitted in a predefined order, establishing a mandatory dependency. Each project task (established from a work breakdown structure, as noted in Chapter 5, Section 5.5) can have a dependency on other tasks as established by logical relationships.

6.4.2 Logical Relationships

Dependencies between tasks can have various logical relationships that include finish-to-start (FS), start-to-start (SS), and finish-to-finish (FF). In each case, predecessor (Task A) and successor (Task B) activities are linked by a condition that governs when each task can start or finish. This relationship means that a task's start or finish relies on actions from a different task (or multiple tasks). For example, construction work cannot start until the design and permitting tasks are finished.

However, the scheduled date of a task does not imply a relationship between tasks. Many tasks can be scheduled early, potentially because of resource availability or a discretionary dependency, but the later tasks may not rely on early work. For example, a traffic impact study could be scheduled to start before the architectural design of a building, but there is no direct relationship or dependency between those tasks.

When a logical relationship is present, it must be identified and graphically depicted in schedule charts. This way, the project schedule describes the order of tasks and the critical relationships necessary to progress through the project work. Figure 6.8 provides a graphic representation of the most common logical relationships and a line connecting the start or finish relationship between tasks.

- Finish-to-Start
    - The most common is finish-to-start, where a successor activity (B) can start only after the predecessor (A) has finished.
    - Example: The asphalt pavement topcoat (B) can start once the base course (A) has been completed.
- Start-to-Start
    - A start-to-start relationship means that a successor activity (B) can start only after the predecessor activity (A) has started.
    - Example: The streetscape design can start only after the roadway design has started (which establishes the geometry). In this case, the tasks will likely require some concurrent effort once both have started.
- Finish-to-Finish
    - Finish-to-finish relationships will likely require similar concurrent task effort as start-to-start relationships. In this case, the successor activity (B) can finish only when the processor task (A) has finished.
    - Example: The quality control work (B) can start at different times during the design phase but can only be considered finished once the design is complete (A).

The logical relationships between tasks primarily define how efforts are interconnected within a project schedule. However, the timing between

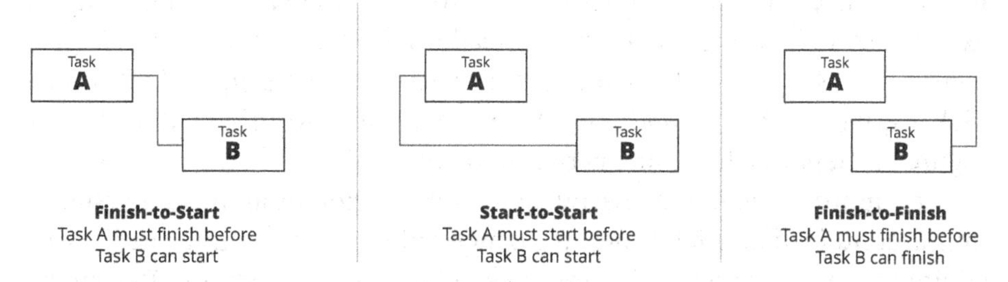
Task
A
Task
B
Finish-to-Start
Task A must finish before
Task B can start
Task
A
Task
B
Start-to-Start
Task A must start before
Task B can start
Task
A
Task
B
Finish-to-Finish
Task A must finish before
Task B can finish

FIGURE 6.8 Logical relationships in completing tasks.

these tasks also plays a crucial role. Temporal relationships, represented by *lead* and *lag* times, are essential for a nuanced understanding and management of the project timeline.

- Lag
  - The intentional delay inserted between tasks, such as waiting a few days after one task finishes before starting the next. This delay accommodates necessary pauses in work or administrative processes.
- Lead
  - Allows a successor task to start before the predecessor task has completely finished, effectively overlapping tasks to compress the project schedule.

As an example of lag time, consider a scenario where two tasks are connected by a finish-to-start relationship, indicating that the second task (Task B) can only start after the first task (Task A) finishes. If a waiting period needs to be introduced before starting Task B, this is accounted for as lag time. This concept is illustrated in Figure 6.8, showing that Task B does not immediately follow the completion of Task A, hence incorporating a necessary delay to accommodate processes.

Consider engineering design work (Task A) that precedes the issuance of a project construction permit (Task B). Completing the design work and submitting the engineering plans to a review agency will require a lag between finishing the design work and issuing a construction permit. The lag time reflects the time needed for the authority's review.

Conversely, lead time describes scenarios where the start or completion of a task is expedited relative to its linked task. Though less frequently utilized, this approach adjusts the tasks to allow overlaps, as might be seen in tasks with a finish-to-finish relationship. For example, in preparing construction documents, the engineering design plans and the project specifications have a finish-to-finish relationship, but the work can overlap to expedite completion. This overlap, or lead time, ensures that Task B begins before Task A concludes, facilitating a concurrent wrap-up of both tasks. The overlap of Task A and Task B in the finish-to-finish relationship of Figure 6.8 depicts a lead time between the two tasks.

To improve schedule clarity and address potential misunderstandings around lead and lag times, it's beneficial to delineate intermediary or additional tasks explicitly. For instance, adding a new task for "Review by

External Agency" accounts for the plan review period, avoiding the implicit assumption of lag time. This method enhances schedule transparency and identifies sources of potential delays and risks, particularly for tasks outside the direct control of the project team. By making these temporal relationships explicit, project managers can better navigate the complexities of project scheduling, ensuring that all stakeholders have a clear understanding of the project timeline and its dependencies.

6.4.3 Critical Path

The *critical path* is a fundamental scheduling concept that refers to the sequence of tasks that establishes the earliest completion of the project. Projects will have concurrent tasks with varying dependencies, but not all tasks will exist along the critical path. If a task's start or end date is flexible, the task is not on the critical path. This extra time is called *float* (or slack). Tasks without float exist along the critical path and are considered critical tasks. Identifying the critical path can be complex and best solved with scheduling software, but it generally requires the following steps:

1. Identify all tasks and estimate the duration to complete each task.
2. Establish the order of tasks, best represented through diagrams, based on logical relationships.
3. Starting with the first task, determine the earliest date each task can start. Each task should have an early start (ES) date. Add the duration to the ES to establish the early finish (EF) date.
4. Working from the last task to be completed, determine the latest that each task can end, the late finish (LF). Subtract the duration from the LF to determine the late start (LS).
5. Calculate the float for each task by finding the difference as LF minus ES.
6. The critical path is identified by the sequence of tasks with zero float (the critical tasks).

Task dependencies are crucial in this process as they determine whether a task can start or finish based on its relationship with other tasks. This process must also be mindful of working days (not calendar days). Figure 6.9 provides an example of sequential tasks on the critical path and one task (D) that has float.

Referencing Figure 6.9, suppose Task A, B, and C are sequential tasks with finish-to-start relationships where the width of each rectangle represents the work duration. Any change in the start or finish date of these tasks would change the start or finish date of the project. Assume that Task D can begin after Task A and must be completed before starting Task C but has a shorter duration than Task B. The float is the difference between the allowable earliest start (directly after Task A finishes) and the latest start time (finishing at the same time as Task B). When a task has zero float, it exists on the critical path (Tasks A, B, and C in Figure 6.9).

Tasks along the critical path may not be initially apparent, as even the seemingly trivial tasks can influence the project completion. Project managers should evaluate project schedules to understand the logical relationships, task durations, and then identify the critical path to know which tasks will influence the project completion date. Project managers must also monitor and update the critical path, as a delay or acceleration of tasks can shift the critical path.

Identifying the critical path is a pivotal step in project management, as it allows the project team to prioritize resource allocation strategically, thereby minimizing the risk of project delays. Critical tasks have a direct and often cascading impact on numerous other project operations. Therefore, it is

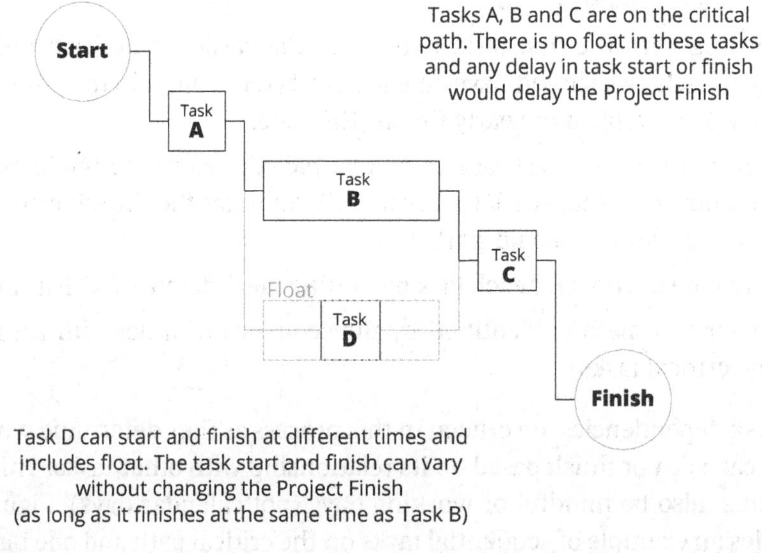

FIGURE 6.9 Sequential tasks located on the critical path (A–C), with one task that has float (D).

essential to clearly communicate the details of these critical tasks to the entire project team and ensure they are supported by reliable and consistent resources, particularly in terms of personnel assignments.

Conversely, understanding which tasks lie outside the critical path is equally important. This knowledge indicates where there is flexibility in terms of starting or completing tasks, providing valuable insights for effective resource management. In scenarios where resource reallocation is necessary, tasks off the critical path can often accommodate shifts in resource distribution without jeopardizing the overall project timeline. Additionally, this approach assists in managing project risks by identifying potential buffer zones where delays are less likely to affect the project's critical milestones.

As indicated in this section, resource allocation is crucial to project planning. While project schedules primarily focus on task dependencies and durations, these tasks require resources to complete. With limited resources, allocation must be considered when establishing a realistic schedule and subsequently monitoring and controlling the schedule. Some scheduling diagrams are designed to identify the critical paths and float (such as precedence network diagrams), or the critical path can be identified through other diagrams, such as Gantt charts.

6.4.4 Schedule Diagrams

The most common schedule diagram in the engineering industry is a *Gantt chart, which* serves as an intuitive visualization for reviewing tasks, progress, and schedule relationships. Using the work breakdown structure (WBS, as noted in Chapter 5, Section 5.5), each row of the chart represents a project task. The vertical columns represent the time scale, often in increments of days or weeks. The time scale represents workdays, removing time off for weekends, holidays, and potential weather delays (for weather-dependent activities like site investigations). The schedule of a task is represented as a horizontal bar in the row of the associated task.

Note the distinction between duration and scheduled work – *duration* is the amount of time necessary to complete the work, whereas the *schedule* identifies when the work can be completed. For example, the design of a site plan may require 40 hours of work (the duration) that may incorrectly assume the work can be completed in one 40-hour week; however, task

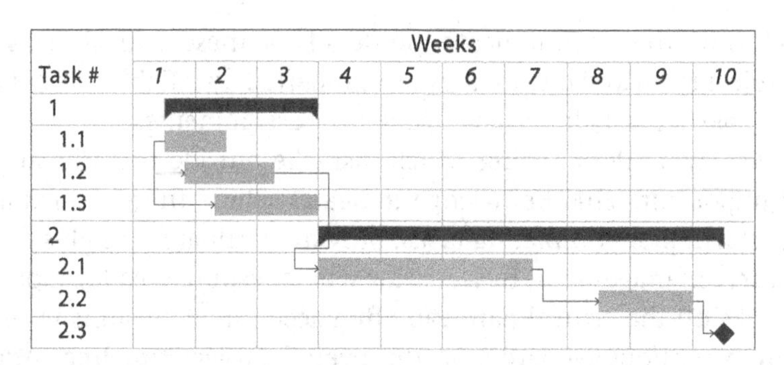

FIGURE 6.10 Sample format of a Gantt chart.

dependencies and resource availability will determine when the work can be completed and could require spending a few hours across multiple weeks. Gantt charts typically include lines that connect tasks to describe dependencies and identify milestones with symbols that make it easy to find critical dates. A vertical line will appear with the current date and scheduled activities as the project progresses through the schedule. Figure 6.10 provides a sample Gantt chart for reference.

For large projects, Gantt charts can become complex and challenging to comprehend, so most charts will include some hierarchy to focus on different tasks. Finding the right level of detail will benefit from referencing the hierarchy of the WBS, where the referenced tasks are meant to be informative without becoming too intricate.

Resource requirements can be identified with Gantt charts to track the potential overdemand of a resource or identify the level of effort required to complete a task. For example, each task could reference the personnel (resource) assigned to the work, such that the sum of each time step (day or week) could calculate the required hours per person. If the demand hours exceed the availability, the schedule will need to accommodate the adjustment or additional resources will be required.

Resource Loading

As the project schedule is developed, it is crucial to allocate resources to tasks. An initial draft of the schedule may overcommit one resource (person, equipment, material) to several concurrent tasks. *Resource loading* is a term used to describe the calculation of resources required to complete a task with a set duration. If a task is estimated to require three weeks of effort when using one resource and the work is scheduled for one week, then resource loading is used to determine the requisite number of resources to meet the deadline.

For instance, consider a task estimated to require three weeks of effort (i.e., 120 hours) using a single resource but is scheduled to be completed in one week. Resource loading comes into play to determine how many additional resources are needed to meet this one-week schedule. This involves making a detailed analysis of the work hours required for the task and then correlating them with each time increment. By assessing the total staff hours assigned to the task and summing up the overall hours needed, it is possible to evaluate the number of resources required to adhere to the schedule. In the example, it would require approximately three staff to complete 120 hours of work in one week (ignoring the production penalties of large teams). This approach ensures that the task is realistically planned, considering the limitations and availability of resources.

Figure 6.11 depicts an example of one format, which shows sub-levels of the requisite hours of work for each week of a task's duration (with variable weekly production needs), and a sum of total personnel hours at the bottom.

The example in Figure 6.11 identifies the total hours of production required for each week of a task; however, resource loading may estimate productivity rates based on how the organization has classified personnel (e.g., Engineer-1, Engineer-2, Engineer-3, etc.), often based on experience. It's essential to note that project tasks are multifaceted and require unique skills that not all available team members possess. Also, it is unlikely that an organization's classification of personnel is directly correlated to their

	Weeks									
Task #	1	2	3	4	5	6	7	8	9	10
1	120 Staff Hours									
Req. Hours	*12*	*36*	*48*	*24*						
2			60 Staff Hours							
Req. Hours			*15*	*15*	*30*					
3					150 Staff Hours					
Req. Hours					*75*	*30*	*30*	*15*		
4						200 Staff Hours				
Req. Hours						*15*	*30*	*60*	*40*	*40*
Σ Hours	*12*	*36*	*63*	*39*	*105*	*60*	*60*	*75*	*40*	*40*

FIGURE 6.11 Resource loading in a Gantt chart.

It is unlikely that an organization's classification of personnel is directly correlated to their efficiency and capabilities.

efficiency and capabilities. For example, all engineers on the team may be proficient in computer-aided design (CAD), but each team member could have a specialty skillset for specific production tasks. This means that only some team members can complete certain tasks, and some team members may be more efficient in production than others. A project manager should recognize these conditions and use expert judgment when estimating resource requirements to meet the project deadlines with consideration for unique personnel traits.

Considering these unique skills and efficiencies, the Gantt chart in Figure 6.11 can be expanded to include additional rows detailing the time required by each resource category per task. For instance, the first week of Task 1, requiring 12 hours of work, can be further broken down to show the hours needed from Engineer-1, Engineer-2, and the project manager. This level of detail in resource loading helps identify peak production periods (e.g., Week 5 in the example), which might not be immediately evident.

The complexity of resource loading varies depending on the scenario. In cases where resources and production rates are directly proportional (like machinery), the calculation is straightforward. However, it becomes more intricate in professional services, where personnel are the primary resources. For example, the 120 hours required for Task 1 cannot be evenly distributed among Engineer-1, Engineer-2, and the project manager due to differing productivity levels and availability. Engineer-1 is unlikely to perform any work required of the project manager, and the project manager may not be able to perform the duties of Engineer-2. Additionally, larger teams may face communication and coordination challenges, impacting individual efficiency, as discussed in Section 6.2.1.

Resource loading is instrumental in identifying scheduling constraints and potential resource shortages. It may reveal if the necessary number of resources are unavailable or if it's impractical to allocate excessive resources to a single task due to limited divisibility. In terms of personnel assignments, resource loading can identify if an individual or team is overextended, such as being scheduled for double shifts or assigned multiple concurrent tasks. For instance, the aggregated hours at the bottom of Figure 6.11 could include additional rows to sum the hours required of each team member and indicate if a team member is allocated more than forty hours in a

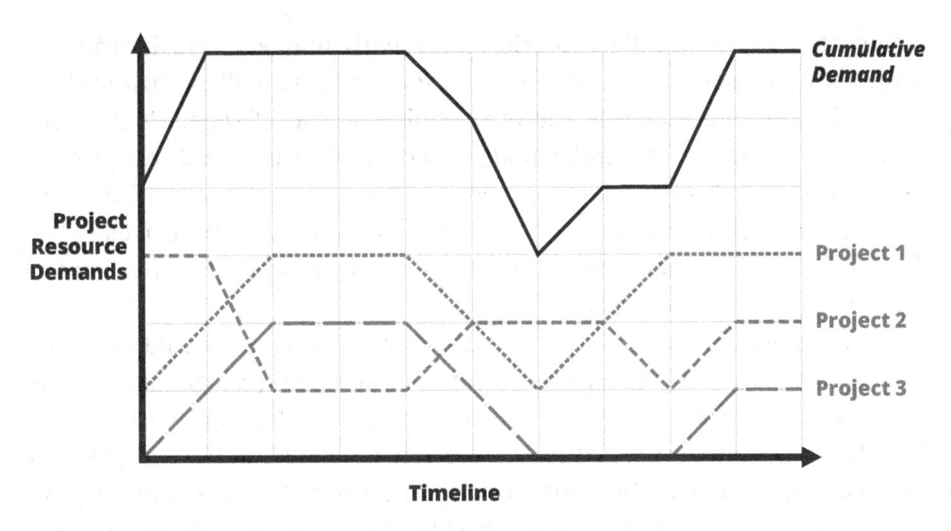

FIGURE 6.12 Variation of project resource demands across time.

week. Recognizing these conditions, the project schedule might need to be extended, or additional resources may need to be applied.

When engineering teams are engaged in multiple projects (e.g., operating within a matrix project team structure), managing resource allocation is more challenging. This is because individual project managers may narrowly focus on their respective projects and not realize the competing resource demands across different projects. As a result, managing resource allocation becomes complex and can lead to project-related difficulties. The team members may also experience stress as they find themselves in a position where they are left guessing which project work to prioritize for which manager. Figure 6.12 depicts how multiple project resource demands can change over time through cycles of plan submissions, reviews, changes, etc. Project managers must recognize this cumulative demand load in addition to project-specific demands, and establish methods to avoid undue strain on project resources (i.e., the people).

6.4.5 Measuring Progress

A task must meet the definition of done to be considered complete. *Done* means all work requirements have been completed and validated (as previously defined in Chapter 5, Section 5.6.2). Throughout the life cycle, the project manager will monitor the work progression to determine the status of deliverables. This metric, along with budgets, will inform the project's performance and how the completed work will be invoiced. Is the project behind schedule and over budget, ahead of schedule but over budget, or on track and under budget? The project expenditures provide insight into the effort and budget spent but do not provide any reference to how much work has been completed. To track progress and correlate the work effort with project costs, it is important to have an effective way to measure the completed work.

For engineering design, incremental work progression could be measured by the number of completed sheets in a site plan (as compared to the total expected number), by the linear feet of infrastructure designed, or by other metrics. However, this system can be challenging to implement because not all deliverable units (like the number of plan sheets) require similar effort. For example, the sheet with the legend will require far less effort than a page of hydraulic computations.

Early completion of defined increments, such as the 30% deliverable, can inform the effort required for subsequent phases, so long as the increments have been well-defined and represent their numeric title. This would indicate that the 30% deliverable requires 30% of the time and budget, which is not always true. Additionally, these metrics must consider when the complete work can be considered "done" or if there are underlying issues with deliverables. For example, submitting "complete" plans for jurisdictional review and approval may realistically represent 95% completion, as there could be changes requested by the reviewing agency that require refinements before the plan is approved and complete. Still, supplemental work, such as permit documentation, compilation of bid documents, and coordination between clients and engineers, is additional work that must be completed. To mitigate some of these challenges, adaptive methodology provides a metric that facilitates monitoring the project's progression, as described in this section.

Story Points

Adaptive methodology applies story points to tasks to quantify the effort associated with project work. In this method, the story point system follows a Fibonacci sequence (1, 2, 3, 5, 8, . . .) to establish a relative scale for units of work for different tasks (or *user stories*, as described in Chapter 5, Section 5.8.1). The point system does not correlate directly to a unit of time.

Instead, the point system is meant to be an abstract measure of effort such that it acknowledges the uncertainty of effort in professional services. The assignment of points requires expert judgment from team members and managers who understand task complexity and team efficiency. Expert judgment will recognize that some tasks can only be completed by certain personnel, and the experience of the personnel may determine how quickly the work can be done. Consequently, story points are tailored to the unique dynamics of each project team.

Expert judgment will recognize that some tasks can only be completed by certain personnel, and the experience of the personnel may determine how quickly the work can be done.

This relative weight system of story points overcomes basic metrics (such as the number of plan sheets) by considering how similarly sized deliverables can have varying levels of effort. For example, instead of measuring progress by the number of completed design plan sheets, each section of the design document could have an assigned set of points. As the site plan section is completed, it indicates the team's production rate and how much work is left to complete. The general notes sheets could be allocated as one point, demolition as two, road design as five, stormwater as five, etc. This point assignment ignores the number of parts (sheets) delivered and instead focuses on the team's effort to complete incremental deliverables.

The assignment of story points is placed on tasks that have an established team assigned to complete the work. This way, the project manager is aware of the capabilities of the team and the potential velocity of work. The *velocity* refers to the rate at which the team completes story points. At a set time interval (often the *sprint* for an agile project, or about every two to four weeks), the number of completed story points is recorded. Initially, the completed points per sprint can be volatile as the team becomes familiar with the project and their roles in completing work. More information on sprints and Scrum methodology is included in Chapter 5 (Section 5.8.2).

Over time, the goal is to stabilize the number of story points achieved during the time interval. The stabilization occurs as the project manager is more accurate in assigning story points, and the team becomes more productive. In this way, any deviation in production rate (such as completing fewer story points) signals the team that there is an obstruction in progress. This method also informs the team (and stakeholders) regarding how much work can be completed per unit of time. Figure 6.13 depicts how the amount

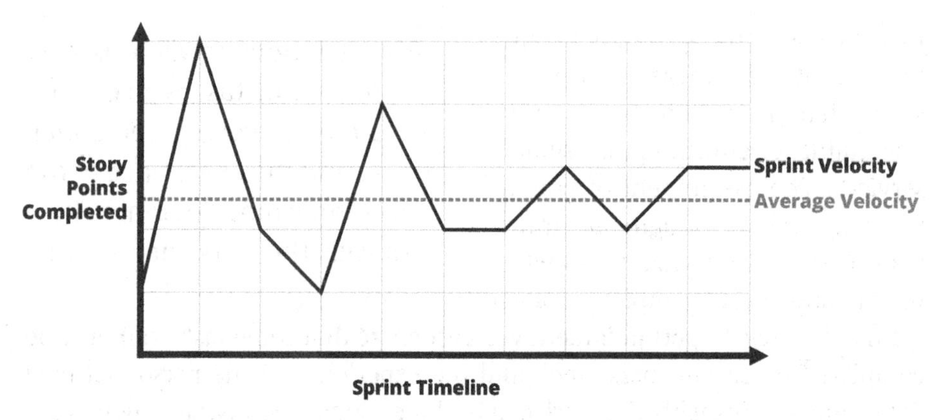

FIGURE 6.13 Completed story points across time.

of work completed (measured by story points) varies across a timeline for each sprint with expectations that the completion rate will become constant as productivity becomes more efficient and story point assignment becomes more accurate.

6.4.6 Schedule Modifications

Inevitably, project progress will succumb to disruptions, and the schedule must be modified – or the resources will need to be reallocated to hold a constant deadline. Modifications in scope, unforeseen conditions, variability in personnel production rates, and other factors can influence the team's ability to adhere to the project's original schedule. The challenges associated with establishing a schedule during project planning mean that the schedule is often initially inaccurate. A project that is "behind schedule" may actually indicate that the schedule was not accurately established or that other undocumented changes have affected the schedule, such as increasing scope or restricting personnel production. Recognizing that schedules often need modifications, PMI identifies methods for resource optimization and schedule compression used to monitor and control the project schedule.

A project that is "behind schedule" may actually indicate that the schedule was not accurately established or that other undocumented changes have affected the schedule.

Resource Optimization

Resource optimization typically comes into play after an initial scheduling draft is created for a project. A first draft of the schedule focuses on task dependencies, determining when each task should occur based on the project's workflow. However, the draft might not fully account for the availability and workload of the resources (like personnel or equipment) required for these tasks. This is where resource loading becomes crucial, as it helps uncover additional constraints that weren't initially apparent. For instance, consider a case where two different design tasks are scheduled to occur at the same time, but both require the expertise of the same individual. In this case, despite the task dependencies allowing for simultaneous scheduling, the limited availability of the required resource (the expert) necessitates a schedule change. The tasks must then be rescheduled to occur sequentially instead of concurrently, ensuring that the expert's time is allocated effectively without overburdening.

Resource optimization in project management involves two key techniques: resource leveling and resource smoothing. These techniques are essential for managing schedules effectively, particularly when multiple tasks simultaneously place excessive demands on resources. By applying resource leveling and smoothing techniques, project managers can adjust the schedule to better align with resource availability, ensuring that tasks are completed efficiently without overextending the project's resources.

Resource optimization in project management involves two key techniques: resource leveling and resource smoothing.

Resource leveling requires a schedule extension to balance (level) the resources based on available capacity. This process should be evaluated during early drafts of the schedule to avoid late modifications but may also be required if resource availability is disrupted. For example, if the original schedule would require personnel to work more time than is possible (or reasonable), the resources would be leveled based on a set threshold (e.g., 40-hour workweek), which could require a schedule extension. Figure 6.14 depicts how overburdened resources are leveled based on a threshold, which requires an extension to the timeline.

Alternatively, *resource smoothing* is used when there is a need to maintain the project deadline. This implies that the schedule has advanced beyond an initial draft and the project manager has evaluated resource availability.

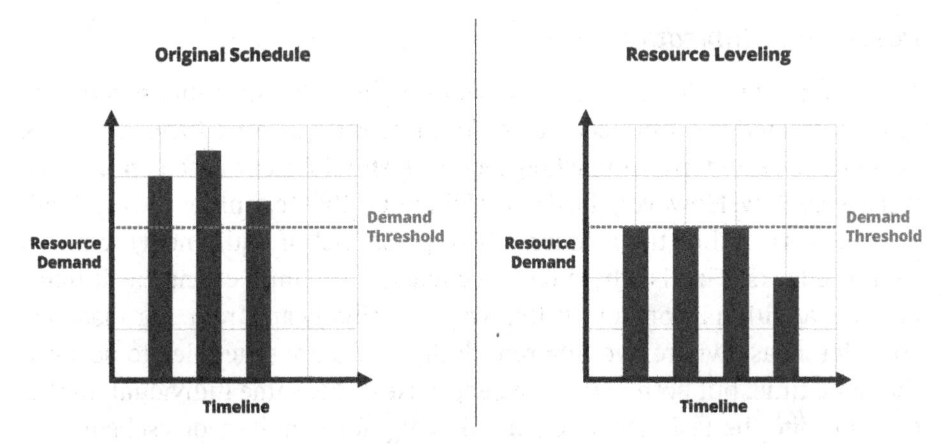

FIGURE 6.14 Resource optimization by leveling, which may extend the timeline.

Smoothing aims to avoid resource demand peaks and valleys that may occur across the project life cycle. Resource demand may exceed desired conditions (overtime) but would be smoothed to accommodate pre-defined limits.

For example, if an employee typically works a 40-hour workweek but the schedule dictates that some weeks will require 60 hours while others require 20 hours, the intent is to establish an acceptable average. This approach requires forecasting the efforts of resources and communicating the project needs so that early adjustments are made. This might include moderate overtime for several weeks to prevent severe overtime conditions during peak production time. Figure 6.15 depicts how variable resource

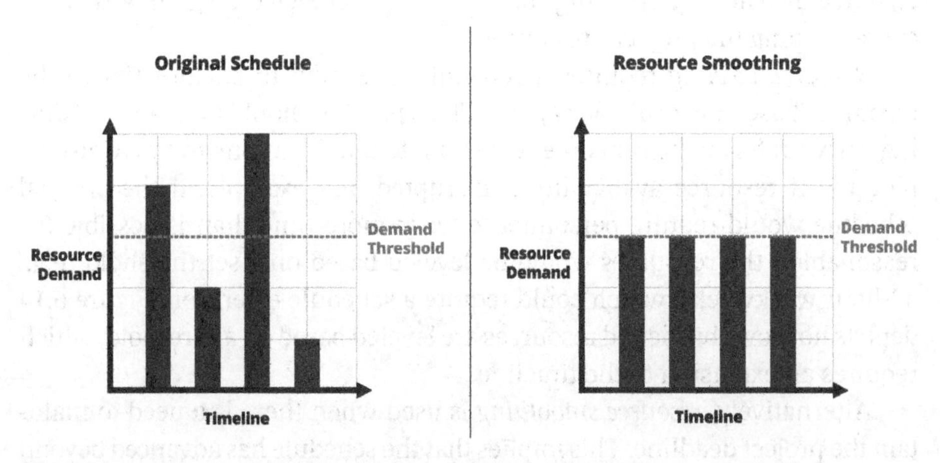

FIGURE 6.15 Resource optimization by smoothing, seeks to balance workloads.

demands in the original schedule would overburden or underwork the personnel each week, which could then be smoothed to provide a constant resource load.

Schedule Compression Techniques

When a project schedule requires acceleration, PMI identifies two techniques for compressing a schedule: fast-tracking and crashing. Schedule acceleration has a cost from increased risk and the need for additional resources.

Schedule acceleration has a cost from increased risk and the need for additional resources.

Fast-tracking is a schedule compression technique used in project management to reduce overall project duration. It involves overlapping tasks that would typically be done sequentially. This method is applicable when the logical relationships between tasks are discretionary rather than mandatory, allowing for the transition from sequential to concurrent execution.

While fast-tracking can expedite project completion, it comes with a trade-off: accepting higher risks associated with overlapping tasks that are usually performed sequentially. The risk from this approach includes a potential for errors and the need for rework, as tasks that are usually dependent on the completion of preceding tasks are now being done in parallel.

For instance, consider a scenario involving utility design (Task A) and a landscape planting plan (Task B) for a site development project. The best practice would be for the landscape architecture team to wait until the civil design has nearly finalized the utility design before beginning their work (completing Task A and then Task B). However, to fast-track the project, the project manager might decide to start the landscape design concurrent with the utility design.

This example highlights the requirement for closer coordination between the landscape and civil design teams – the risk is that concurrent tasks with discretionary dependencies may lead to potential rework and additional communication costs. This rework, along with the need for increased coordination, could extend the duration of individual tasks. However, since the overall goal is to shorten the total project timeline, it is achieved through this trade-off (more effort for individual tasks but faster completion of all work). Figure 6.16 depicts how two tasks (A and B) with a discretionary dependency can be fast-tracked with concurrent operations to reduce the total schedule.

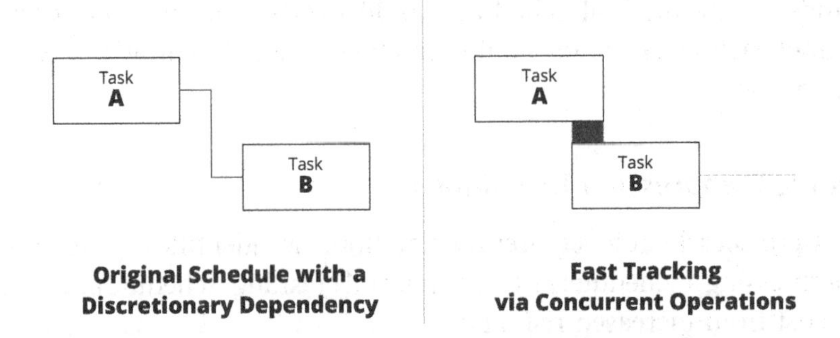
Task
A
Task
B
Original Schedule with a
Discretionary Dependency
Task
A
Task
B
Fast Tracking
via Concurrent Operations
The original discretionary dependency is abandoned for Task A and Task B in favor of concurrent work. This introduces risk but accelerates the schedule.

FIGURE 6.16 Discretionary dependency fast tracked with concurrent operations.

The other common compression method, *schedule crashing*, is when additional resources are assigned to the tasks to accelerate production and reduce task duration. Compared to fast-tracking, each task's logical relationships and dependencies could remain the same, but the task duration is diminished because of additional resources. This may require overtime work, more personnel, or an increase in the production rate (perhaps by using more personnel at a higher skill level and cost). This approach isn't linear and isn't always beneficial. The crashing technique can be applied when the project manager acknowledges the limitations and operating penalties.

Most tasks have a resource limitation with a diminishing return based on limited divisibility of work.

Most tasks have a resource limitation with a diminishing return based on limited divisibility of work. Even a single task could have multiple steps with mandatory dependencies and require a logical order of completing work, which means that more resources would not increase the production rate as expected. For example, suppose the original schedule assigned one engineer to design the stormwater system for a site with an estimated duration of five weeks. In that case, it does not mean assigning five people to the task will complete it in just a week (or that 25 engineers could complete it in one day). The task has a procedural method for stepping through subtasks, which must be completed in a logical order (such as establishing drainage divides, checking inlet capture efficiency, sizing conveyance systems, and then sizing the stormwater management system).

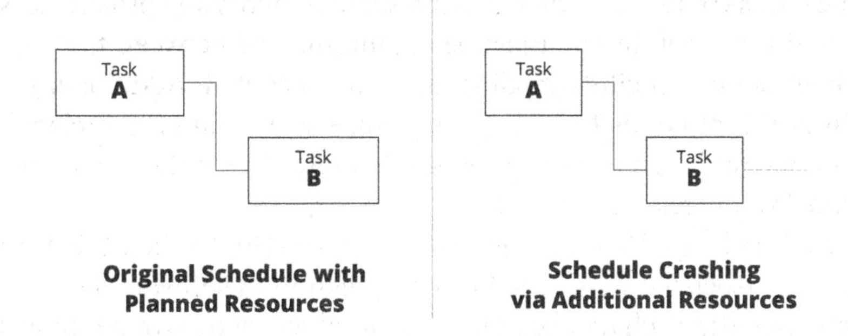
Task
A
Task
B
Original Schedule with Planned Resources
Task
A
Task
B
Schedule Crashing via Additional Resources
The dependency for Task A and Task B is maintained but additional resources are assigned to the tasks to reduce the duration.

FIGURE 6.17 Schedule crashing to accelerated production by assigning more resources.

The other condition associated with schedule crashing is operational penalties for adding additional resources. For design work, resources refer to personnel. Additional team members on a single task require further coordination and communication channels, which can be less efficient. The penalties can be severe depending on when new resources are added to the project. If schedule crashing is attempted only late in the project life cycle, and new personnel are added, they do not have the background knowledge of the context and processes as they begin project work (and therefore require ramp-up).

For example, suppose that two weeks before the deadline, the client directs the project manager to finish the plans a week early. Assuming the resources are available, adding twice as many personnel to the project will not double the production rate – it will likely slow production down as new personnel become familiar with the project operations.

Schedule crashing is more successful if planned early. Still, additional resources will not linearly improve production; it will likely decrease production initially and, therefore, increase project costs (refer back to challenges in Section 6.2.1). Figure 6.17 depicts how adding more resources could reduce the duration of Tasks A and B to improve the schedule.

Challenges with Schedule Modifications

When a client requests an aggressive schedule or an acceleration to the original schedule, the engineering team often feels obligated to affirm that the new deadlines can be met. Project managers must recognize that schedule modifications come with new challenges, often increasing project risks and costs, and those conditions must be communicated between the engineer and client before schedule modifications are accepted. Additionally, there are physical limitations to how quickly some work can be completed, and it's possible that some tasks cannot be compressed regardless of how many additional resources are assigned.

Fred Brooks (author of *The Mythical Man-Month*) identified several challenges associated with schedule compression in software development projects, but these challenges are also observed across other industries. Many of these challenges are not intuitive, as some managers or clients may inadvertently delay projects or increase costs when the intent is to expedite productivity. Brooks' law addresses the most common mistake, which is that when a project schedule is stressed, the reaction of many project managers is to add more resources. In practice, this only delays a project further as production penalties emerge, such as ramp-up time for new staff, the need for additional communication channels, and the fact that many tasks must follow a logical dependency with limited divisibility of work. These concepts are summarized in Figure 6.18.

Brook's Law
Adding new staff to a late project will only delay it further.

Ramp-Up Time
New team members require time to learn and contribute to the project.

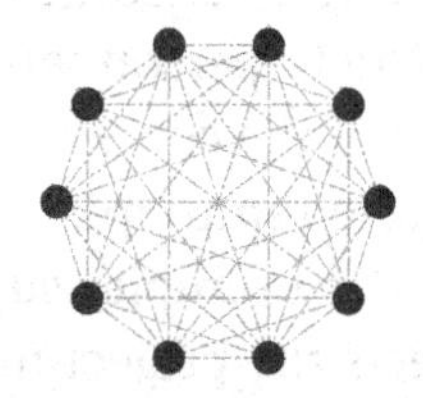

Complex Comms
More team members exponentially increase communication complexity.

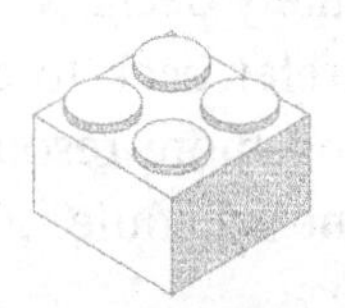

Limited Divisibility
There is a limit to how much a task can be split among multiple personnel.

Quality Risk
Rushed work can compromise the project quality.

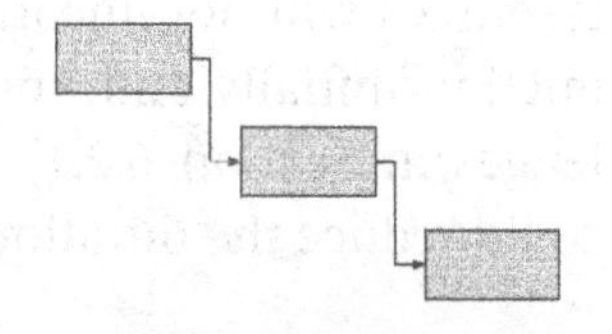

Task Dependencies
Some tasks have mandatory dependencies and cannot be expedited.

FIGURE 6.18 Summary of challenges in schedule modifications.

Clients and project managers should recognize that professional services, which rely on personnel, cannot be amplified using assumptions about mechanical production rates, as seen with manufacturing, where more machines result in a linear scale of production. The project team members are people, not machines; each person will bring unique personalities, skills, traits, and human complexities to their work.

The project team members are people, not machines; each person will bring unique personalities, skills, traits, and human complexities to their work.

Large teams require complex communication and coordination processes, and individual personalities and working relationships must be considered. With large teams, many personnel will focus more on their respective tasks, budgets, and schedules and lose sight of the larger project goals. For this reason, the team structure, RACI chart (described in Section 6.5), and communication procedures must be established to mitigate the challenges of applying schedule modifications. The recommendation is to establish a schedule based on an ideal and small-sized team. Adaptive methodologies suggest this is optimized with a team size of between three and nine people.

Large projects require multiple levels (sets of teams) where major tasks are assigned to disparate teams. For example, a small site design project would work well with a team of three to nine people comprised of multidisciplinary team members. For a larger development project, multiple teams would focus on different sections (by discipline or geographic region of the project), and higher levels would be required to coordinate and then communicate back to the sets of teams.

Enforcing (or suggesting) overtime is another tactic used for schedule modifications, which could remove some of the challenges of adding new team members. Still, overtime work brings different challenges associated with personnel limitations. In "Four Failures in Project Management," Cooper identifies some challenges that are specific to the AEC industry. Many team members will not be as productive after their normal (e.g., 40-hour) workweek, so the production rates would diminish and could lead to errors and then rework. This rework ends up delaying the project schedule instead of accelerating it. Sustained overtime can also lead to burnout, with a significant drop in production and project interest and, eventually, staff attrition. While overtime can provide a quick solution to improving the schedules, the required efforts, duration, necessity, and purpose should be communicated and agreed to by the project team.

Coordinated Schedules

In understanding the broader context of project operations, it is important to recognize that many activities extend beyond civil engineering tasks. The client's schedule, for instance, begins much earlier with initial steps like feasibility studies, agreements, financial arrangements, and other development-related tasks, all preceding the design work. The nature of the project determines the interdependencies of different schedules.

For example, the architectural design schedule for a commercial development might run concurrently with civil design operations but with different delivery dates for building plans and permits. These activities are interconnected, particularly in aspects like building location and utility connections, which are integral to both building and site design. Additionally, the client's timeline doesn't end with the design phase; it continues through construction and into the occupancy (or infrastructure use) and facilities management. This extended timeline underscores the necessity for coordinated scheduling across different project phases.

Challenges arise when task schedules are managed separately by different teams, leading to potential misalignments. For instance, a site-civil engineering team might be directed to start and complete their work before the building designs are complete, driven by the developer's eagerness to initiate site clearing and grading activities. While this approach accelerates the construction start date, it can introduce risks to the site-civil work if the building design continues to change. Consequently, the site-civil design might need revisions to accommodate updates in the building elements, such as door locations, utility connections, and others. Similarly, the building design process might face changes prompted by the client's evolving needs, such as tenant fit-out requirements.

Figure 6.19 illustrates these ostensibly independent yet interconnected schedules, highlighting the complexity of managing different phases of a project.

The engineering project manager is responsible for overseeing their specific scheduled operations, but it is crucial to identify and communicate any external dependencies to the broader project team. This includes highlighting potential risks to the project's schedule, scope, and cost that arise from the timing of input from other stakeholders. This responsibility extends to external stakeholders, like jurisdictional agencies, who have their own timelines for plan reviews and permit issuance. Beyond what is depicted in

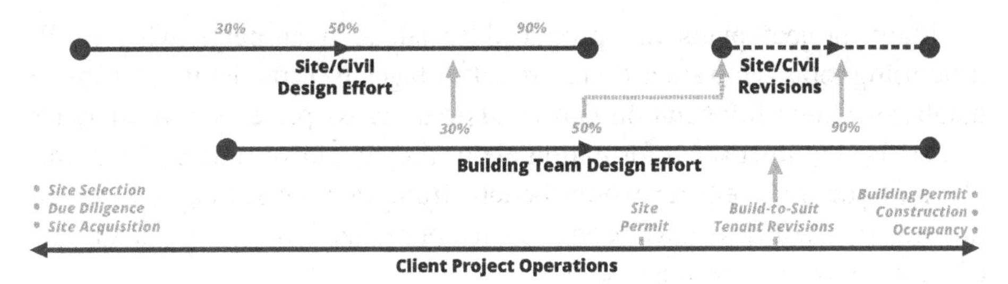

FIGURE 6.19 Managing coordinated schedules of a project.

Figure 6.19, numerous other interdependent schedules, including financing, construction, contracts, and more, must also be considered for comprehensive project management.

6.5 Project Plan Outline

The project plan is authored by the project manager while collaborating with the project team and is often based on templates with guidance from other organizational process assets (including prior similar projects). Although the project manager is responsible for authoring the project plan, the entire project team should be involved in developing the plans and see value in returning to the document throughout the project life cycle. This requires project plan content to be clear, concise, and organized for quick reference (with detailed content in appendices).

The client should also be involved in the development or review of the project plan, although some content may be reserved for internal use (such as team charters, as described in Chapter 10, Section 10.6). For example, some budget details or identified risks may contain propriety information and are not meant for external use. Still, the client can inform project risks based on their perspective or could clarify priority deliverables and confirm the schedule meets their needs.

PMI includes an outline for common project plan topics (as referenced in this text) and notes the importance of tailoring these project plans based on project size, team, industry standards, corporate policy, and other factors. For example, a small project may only require a consolidated project plan that is a few pages long, whereas complex projects could contain several volumes.

Many project plans are produced as siloed documents with word-processing tools. However, there are advantages to establishing corporate databases where information can be shared across projects and analytics can be used to uncover information about project performance. For example, a project risk register would benefit from viewing similar projects to understand what risks have been expected or realized in prior work and the effectiveness of risk responses.

The project plan is continuously updated throughout the project life cycle to benefit the team.

The project plan is continuously updated throughout the project life cycle to benefit the team and serve as a central resource for sharing and analyzing project risks, schedules, budgets, stakeholders, etc. The project manager and project team will determine the appropriate frequency of updates, but monthly updates based on invoices, the schedule, and progress are often appropriate. These project plans must be uniquely tailored to each project and can include a variety of content. A sample table outline of a project plan is included:

1. Project Overview
2. Scope
3. Communication
4. Schedule
5. Budget
6. Resources
7. Risks
8. Stakeholders
9. Quality
10. Change Management

More detail on these topics is provided throughout this book and many tools, processes, and documentation practices identified will benefit multiple sections. A brief description of content that could be included in each section of a project plan is described here.

1. Project Overview

The project overview should be brief and fact-oriented. The detailed project requirements will be provided in other supporting documents, so this section should summarize key elements and any highlights of the other topics. Some recommended information includes:

a. Project Name and Number
   - This includes the official project name from the contract or procurement documents, which often includes a client or engineering firm contract number. If there are accepted abbreviations or aliases for the project, these should be included. While this seems intuitive, project team members may otherwise establish their own conventions that confuse official correspondence.

b. Project Manager and Client Information
   - Contact information for the project manager and the point of contact of the client or project owner.

c. Project Team Overview
   - Key project team names and contact information, as well as their roles on the project. This is not intended as a comprehensive list in the Project Overview section (which could be covered in the Stakeholder section). It could be a tabular format with a short narrative about key personnel roles.

d. Scope Summary
   - A summary of the scope of work, limited to just a few sentences with reference to other sections of the project plan for more information.

e. Budget Summary
   - Total project budget, with reference to budget allocation for sub-consultants.

f. Revision Log
   - Number and brief description of the revisions to the project plan, recommended to be provided in tabular format.

2. Scope

This section describes the project requirements, often with reference to the contract (which could be attached as an appendix). For internal purposes, this section may add more (technical) detail or context compared to the scope defined in the contract so the project team better understands the tasks and approach. This could also reference the project's unique characteristics to highlight what is different from many other projects with a similar scope, which allows the project team to identify key requirements.

a. Project Objectives

- Begin with the objectives to remind the project team of the goals and intent of the project. This is often sourced from the original request for proposal (RFP) if one was issued. It includes any notes from early communication between the client and the design team. Refer to Chapter 2 for more information on defining objectives.

b. Phases and Tasks

- A hierarchy of project phases along with the main tasks, which could follow the project's WBS, as described in Chapter 5 (Section 5.5). This content should be updated as change requests are approved to reflect the current requirements. Phases are often sequential and coordinated to milestones, such as conceptual, preliminary, and final design or transitions between design and construction. This list should include references to the required deliverables.

c. Deliverables

- Many tasks will include a deliverable (e.g., plans, reports, etc.), which should be identified to indicate what is expected. Many task definitions will indicate the expected deliverable; however, some of the deliverables from some tasks may not be explicit or could benefit from clarifying the quantities, format, and recipients of deliverables.

d. Change Log

- A list of all formal change requests that have modified the phases and tasks of the project. This should prompt a change to the defined tasks (and a change to objectives, if applicable).

e. Definition of Done and Closing

- This section should describe the effort of transitioning and closing the project. While often not a significant effort relative to the project, resources and time are required to close and document the project efforts. This process can also document if and why a project was canceled.

3. Communication

This section ensures that all stakeholders are kept informed and that there is a clear understanding of the project's progress, issues, and decisions.

a. Communication Responsibilities

- This includes identifying the point of contact (for the client, design team, contractor, and other key stakeholders). This could include guidance to the project team on how communication should be managed with key stakeholders, often referencing who is responsible and how information should be disseminated.

b. Communication Process

- Expectations on meeting occurrences and format should be documented here. Any client requirement (or expectations) on communication frequency and format should be identified.

c. Record Keeping

- Many communication records are managed by the organization, but this section could include additional requirements about how communication content should be documented in the project files.

4. Schedule

The project schedule information is often depicted as a Gantt chart or similar scheduling diagram that identifies milestones, listed tasks, and task dependencies (as described in Section 6.4.4). The schedule should also include a reference to the required resources (at a high level), such as identifying the task duration with a note about which personnel will be assigned.

a. Schedule Parameters

- An introductory section could reference any assumptions or parameters that should be considered by stakeholders. This might include notes about when work should be finalized for quality control prior to a deadline (e.g., deliverables should begin quality control review at least one week prior to the submission deadline). Other high-level conditions should be listed, such as expectations about overtime or dependencies on external stakeholders.

b. Milestone Summary

- It is beneficial for quick reference to include a list of summary milestones for reference by team members (e.g., a bullet list of plan submission deadlines).

c. Schedule

- The project schedule is often shown as a Gantt chart; however, Gantt charts can become complex for large projects, so it's important to consider whether a condensed version is warranted in the body of the project plan with reference to the full schedule in an appendix.

5. Budget

This section includes the cost information for tasks (assigned to cost codes). For professional services, the budget will be closely correlated to scheduled tasks. Some budget information may be reserved for internal coordination and not shared between the client and engineer; alternatively, some clients (often public agencies) require transparency of project cost details. Some standard content is referenced here.

a. Budget Parameters

- A summary of personnel, their hours, and rates should be indicated to adhere to the budgeted amount. For example, how much of the budget is dedicated to project engineers, managers, or subject matter experts? Assumed reserve, mark-up rates, and other direct costs (ODC) should be indicated. Additionally, any special invoicing conditions can be noted in this section.

b. Task Budgets

- A list of project tasks with their assigned budget, which may be organized according to contract tasks, subconsultant fees, or internal operations. For example, the internal task budget information could include a breakdown by department (if more than one is assigned).

c. Burndown Chart

- A burndown chart (depicting the available budget compared to the schedule progress), or similar data visualization of project expenditures, can provide a quick reference to the project cost performance condition, which is best monitored with delivered work (value). These graphics are most useful when updated frequently (e.g., monthly) to depict current cost conditions.

6. Resources

The section on project resources generally includes requirements for people, facilities, and equipment. For civil engineering design projects, this will mostly focus on the people and skills required to complete the work. Information on resources should be referenced with budget and schedule data (e.g., which personnel are needed at what time for what cost?).

a. Team Personnel Roles

- The required skills and roles for the project should be documented for each project team member. This section could include an expanded organizational chart (beyond what was included in the Project Overview section). For example, the need for traffic engineers, hydrologists, and people with geotechnical and other skills should be listed. Additionally, external resources (such as subconsultants) should also be noted as project resources (with assigned responsibilities).

b. RACI Chart

- The resource information benefits from a RACI chart, which identifies team members by their roles and if they are responsible, accountable, consulted, or informed (RACI, pronounced as *racy*) for each task. The RACI chart provides transparency for team member roles. For example, many projects require expertise from multiple disciplines, with one project manager who approves (signs and seals) each section of the plan. Standard practice expects that only one person is accountable for each task (to sign and seal the work). Figure 6.20 provides a sample RACI chart with multiple tasks and engineer personnel.

c. RACI Chart for Teaming Partners

- Similar to a RACI chart tailored to project team members within the organization, it is necessary (and perhaps more critical) that teaming partners are aware of their responsibilities. This information will be defined in the contracts, but a RACI provides a visual summary of which team is accountable for which tasks. For example, if a

	Task 1	Task 2	Task 3	Task 4
Engineer A	A	A	I	I
Engineer B	R	R		
Engineer C	C		A	A
Engineer D			R	R

R: Responsible
A: Accountable
C: Consulted
I: Informed

FIGURE 6.20 Sample RACI chart.

landscape architect is responsible for designing the plazas, does that design include the topographic grading, or is that the civil engineer's responsibility? Does the geotechnical engineer write the earthwork specifications, or does the site-civil engineer? This content will mitigate any conflicts or confusion in the assigned scope between different teaming partners that likely have overlap.

7. Risks

A project plan includes a risk register of identified and prioritized risks along with any risk analyses (more information is included in Chapter 8). There are many tools available to identify, analyze, monitor, and control project risk, so this section is often carefully refined to each project based on stakeholder risk perceptions and current external environmental factors (economics, natural environmental, sociopolitical, etc.).

a. Risk Management Strategy

- Any initial references to risk management policies and procedures from project stakeholders. This might include standard practices for risk management and analysis to define detailed methodology, while the project plan focuses most on the outputs from risk analysis.

b. Risk Identification (Risk Register)

- A list of identified risks, often developed through prompts (described in Chapter 8, Section 8.3.1), which should be prioritized based on stakeholder input regarding the impact, probability, and timing of the identified risks. The register may also include summary content of other risk sections, such as a column for risk impact, probability, responsibility, and mitigation measures.

c. Risk Assessment

- A summary risk impact-probability chart is a standard way of depicting the magnitude of risks and identifying those with the most significant influence on project operations. The results may also be shown in tabular format with each identified risk.

d. Risk RACI Chart

- The risk section should also include a RACI chart applicable to risk management (similar to the one established for resources used with project tasks), which will establish who is accountable for monitoring the risks and implementing risk response initiatives.

e. Risk Response Strategies

- For those risks identified as high-priority for the project, this section would include the applicable risk response initiatives, such as implemented processes to mitigate, transfer, or avoid risks.

8. Stakeholders

This section includes information about stakeholders, such as including the project team, the client's team, and external stakeholders. More information on stakeholder management is included in Chapter 10 (Section 10.7).

a. Stakeholder Engagement Strategy

- This section describes how stakeholders will be engaged, including methods of communication, frequency of updates, and strategies for managing differing stakeholder needs and expectations. For most projects, this engagement strategy will include details only for key stakeholders.

b. Stakeholder Identification (Stakeholder Register)

- This section includes a list of the individuals, groups, and organizations relevant to the project, often prioritized by key stakeholders. This could be formatted as a table with interest, power, responsibility, and engagement strategy columns.

c. Stakeholder Power-Interest Chart

- This stakeholder assessment tool can depict where stakeholders are charted based on their power and interest in the project, which will vary throughout the project life cycle. A sample power-interest chart is shown in Chapter 10 (Section 10.7.1).

9. Quality

The quality section will reference a firm's quality management system (QMS) and the project's adherence to requisite standards. Resources such as the International Organization of Standards (ISO) include guidance for standards, and the project plan may include a matrix implementation.

a. Quality Objectives

- A summary of the project team's standards and objectives, including applicable policies that must be implemented and how quality will be controlled and reported for the project.

b. Quality Manager Role

- Identification of the individual responsible for the quality management (not the project manager) who will initiate and control quality management processes.

c. Quality Assurance (QA) Process

- A description of the systematic activities implemented within the quality system to provide confidence that the project will fulfill quality requirements. This might involve process audits, quality reviews, and continuous improvement processes.

d. Quality Control (QC) Process

- An outline of the procedures for monitoring and controlling the quality of work produced. This includes review processes to ensure the deliverables meet the requirements (and defined quality criteria). This might include referencing the design review checklists, notes about review certifications, documentation requirements, and others.

e. Quality Metrics

- As established by organizational policy or direction of the quality assurance team leader, this section could include performance indicators to help monitor quality throughout the project. This could include the number of design errors, comments, or other performance parameters (cost and schedule conditions) – these may best reference expectations or industry standards when available. For example, including a list of how many errors are identified compared to what is typically seen for similar projects.

f. Continuous Improvement

- Throughout the project, this section can serve as a resource for lessons learned (good or bad), root causes, or other feedback about how the project's quality and performance could be improved (or what new implementations have worked well).

10. Change Management

This section describes the project change management processes. When changes are requested, whether scope, budget, or schedule, this section establishes the process of evaluating, approving (or disapproving), and implementing the change. The content in this section should identify who has the authority and responsibility to review and approve changes and how a change request should be communicated. This section may also include a register of each change that has prompted a revision to the project plan (if not already noted in the Project Overview section).

In Appendix A, Part 6 of the Millbrook Logistics Park, the client identifies a desire to accelerate the project timeline, particularly the transition to final engineering and site permit phases. This suggestion prompts a strategic dialogue between the team members, focusing on the feasibility and implications of such an acceleration. A concurrent design strategy emerges that seeks to balance the desire for expedience against the risks of premature work and the potential need for rework. The narrative explores the dynamics of project management under pressure, the challenges of aligning ambitious timelines with practical project execution, and the collaborative effort to navigate these complexities while managing stakeholder expectations and contractual realities.

CHAPTER 7

Cost Management and Monitoring

Value determines price.

CHAPTER OUTLINE

Management Essentials for Civil Engineers: A Practical Guide to Business, Communication, Ethics, and Risk, First Edition. Cody A. Pennetti, C. Kat Grimsley, and Brian M. Grindall.

This chapter describes methods of project cost management from the perspective of a civil engineering firm. Beginning with the foundations of cost metrics and value, this chapter includes information about cost parameters, project fee structures, and budget estimates. Methods of monitoring and controlling project costs, as well as forecasting costs, are also included.

7.1 Introduction

For project managers in civil engineering, cost management represents a vital and intricate task. It necessitates diligent oversight and a deep understanding of cost-related project performance metrics. This chapter examines the nuanced elements of project financials, considering the perspectives of clients, project investors, end users, and engineering firms. The chapter emphasizes managing the engineering firm's design costs; however, it is important to note that effective cost management aims to align the financial objectives of the engineering firm and the client.

> Project managers must understand the relationship between cost and value.

Project managers must understand the relationship between cost and value in civil engineering projects. From a client's perspective, value is not solely determined by the engineering firm with the lowest cost but rather by a team's balance of cost-efficiency, quality, and timely delivery. Clients perceive value in the degree to which the project meets their requirements and expectations, often including on-time delivery, lasting durability, functionality, aesthetic appeal, marketability, and the fulfillment of project objectives. For engineers, the focus should be on managing costs to avoid financial losses and establishing a price best representing the value of their work. This involves understanding the client's definition of value and aligning the project deliverables accordingly.

Cost management is more than just keeping direct expenses within a budget. It encompasses a range of financial variables, including employee compensation, project fee structures, budget estimates, and monitoring costs of subconsultants and multiple business units. This involves understanding wage structures and considering corporate overhead expenses and profitability goals. Translating these costs into project budgets and client invoices is critical for maintaining financial stability and effective cash flow management.

Proper cost management ensures client satisfaction by delivering value while simultaneously achieving the business success of the engineering firm. A comprehensive understanding of these components and their interplay is essential for any civil engineering professional engaged in project management.

7.2 Cost Metrics

Effective cost management begins with the understanding of cost metrics and the variables that influence the financial performance of a project. In the initial project planning and proposal stages, a project's budget is formulated, informed by the project schedule and scope (the triple constraints, as described in Chapter 1, Section 1.3.2). For civil engineering design projects, labor costs associated with the project team constitute most of the expenses. The time spent by the project manager and team members working on the project translates directly into labor costs charged to the project and invoiced to a client.

One of the core responsibilities of a project manager in civil engineering projects is to create a budget that comprehensively covers the estimated costs of material and labor required for the project while achieving a level of profitability. The estimation of costs typically forms the basis of the project budget and directly influences the fee presented to the client. However, it's crucial to recognize that in some scenarios, the value of the engineering work may surpass the calculated labor costs. In this way, the price of the engineering services could be greater than the estimated direct costs. This value can arise from various factors, such as the firm's investment in innovative technologies that enhance production efficiency or from the unique combination of institutional knowledge, established professional relationships, and expert skill sets that the engineering team brings to the project.

The value of the engineering work may surpass the calculated labor costs.

Value-added elements can justify a higher project fee, reflecting the premium nature of the services provided. Therefore, while the initial step in budget formulation is understanding and estimating the labor costs (time) involved, it is equally important to consider the additional value the project delivers to the client. This holistic approach ensures that the budget covers the cost of labor and adequately represents the quality and uniqueness of the engineering work. Once this comprehensive budget is developed, factoring in both cost and value, it is articulated to the client through a detailed cost proposal. This proposal sets the stage for establishing the project fee in the ensuing contract, providing clarity and justification for the costs involved, thereby aligning the client's expectations with the project's financial and value propositions. More information about value concepts is included in Section 7.2.1.

While an organization's financial structure can be intricate, this chapter focuses on the core metrics that a project manager can monitor and control (to some degree). The concepts in this chapter include both project revenue and profitability. *Revenue* is a standard metric based on the project's paid invoices – a project's revenue represents how much the client has paid for work performed and invoiced.

Invoice amounts are determined by the amount of work completed (or value delivered). An invoice for a fixed-price contract will indicate that a certain percentage of work has been completed and invoice for that amount. For example, if 50% of the work has been completed for a $100,000 project, the invoice would indicate $50,000. Alternatively, some work may be billed as time and materials (T&M) efforts, in which case each hour spent by each team member is invoiced with the determined billing rates based on categorizations of personnel (e.g., Engineer-1, Engineer-2, etc.).

While revenue identifies the inflow of money for project efforts, it does not consider the costs to perform the work. The revenue generated from project work is distributed to various expenses associated with the project and the engineering firm's operations, including employee compensation and the engineering firm's overhead. This means an organization could receive revenue but lose money because of high costs.

For instance, consider a scenario where the budgeted labor cost and contracted project budget were $50,000. Unfortunately, the expenses escalated to $60,000 due to errors and unforeseen rework. At the end of the project, when the engineering firm receives a payment from the client, it is per the agreed budget of $50,000, which falls short of covering the costs. Despite generating positive revenue (the engineering firm has received a $50,000 payment), the project incurs a financial loss because the revenue does not cover the costs incurred.

Net *profit* is measured as revenue minus the costs of goods sold, such as operating expenses, interest, taxes, etc. In most cases, a project manager can determine the project's revenue based on the invoices of work billed; however, profit is more complex and is likely abstracted based on metrics established by an organization. This is because the project manager cannot calculate cost components such as the organization's taxes or depreciation to determine net profits. For this reason, a metric of earnings before interest, taxes, depreciation, and amortization (referenced as the acronym EBITDA) is more frequently used to consider the revenue and operating costs visible from the managerial level.

While EBITDA has the advantage of being a more tangible measure of profitability, it can be somewhat misleading as a business measure of financial performance. Still, these metrics are frequently referenced to establish a macro performance measure. In most cases, project-level cost metrics will consider project profitability based on the revenue received minus the direct costs (primarily the labor cost for professional services) and predefined indirect costs such as overhead. Standard terminology associated with cost measures includes *unloaded rate, loaded rate, fully loaded rate,* and *billing rate*, defined as:

- Unloaded Rate
  - Unloaded rates focus solely on the labor cost of an engineer, excluding any overhead or indirect expenses. Essentially, how much does the employee get paid for their time?
  - These rates are used for estimating the direct labor cost component of a project without considering the broader financial implications associated with overhead and administrative expenses.
- Loaded Rate
  - Loaded rates include the combination of an engineer's unloaded rate and the employee's benefits, such as paid time off, vacation days, insurance, and others.
- Fully Loaded Rate
  - The fully loaded rate adds to the loaded rate by including a portion of the company's indirect costs or overhead expenses, such as office space lease costs, taxes, administrative staff salaries and benefits, nonbillable time, and equipment.
  - These rates provide a comprehensive view of the actual cost incurred by the engineering firm for an engineer's time and effort, ensuring that projects cover both direct and indirect costs.

- Billing Rate
  - The billing rate is designed to cover all costs associated with the fully loaded rate and adds profit. This rate is often pre-established in tiers based on employee categories (Engineer-1, Engineer-2, etc.) as client-facing pricing information.

This terminology can vary by engineering firms, clients, and cost management software – the terms referenced in this book follow those published in the Bureau of Labor Statistics (BLS) to provide general applicability. The formulaic relationship between these rates is referenced in Table 7.1.

More information is provided in this chapter for the formulas and terms listed in Table 7.1, but the table provides an initial reference to the relationship between the rate categories. The unloaded rate (W) is more predictable because it could be reasonably estimated based on employee wages (although compensation adjustments must be considered for long-term projects). Benefits are often calculated as a percentage of the employee wages because many costs are correlated to the wages (such as payment for time off). Overhead may be calculated as a percentage of the loaded rate, but considering it as a fixed price per billable hour worked is often simpler.

The fully loaded rate is variable because external environmental factors influence overhead costs. For example, changes in the engineering firm's office lease costs, interest rates, organizational investments, company-wide utilization, corporate profitability, and others will influence overhead costs.

TABLE 7.1 Relationship between variables in rate calculations

Rate Category	Variable	Formula
Wages (unloaded rate) as cost per hour	W	
Benefits percentage, as a percentage of wages	B_P	
Benefits, cost per hour	B	$B = W \times B_p$
Loaded rate, cost per hour	L	$L = W + B$
Overhead, cost per Hour	O	
Fully loaded rate, cost per hour	F	$F = W + B + O$
Profit per hour	P	
Billing rate, price per hour	R	$R = W + B + O + P$

An organization manages these costs, but variability is still expected across months and years, which impacts profitability. For these reasons, both unloaded and fully loaded rates are used in cost management, while billing rates are used with external-facing project pricing.

7.2.1 Value

It is essential to understand the distinction between cost, value, and price. Cost is the financial expenditure associated with designing and developing a project, encompassing wages, materials, and overhead expenses. While cost is a tangible and measurable aspect of project planning, value is often a less tangible metric. Value encompasses the benefit, quality, and functionality that the project will deliver to its stakeholders. In the design phase, value is derived from making informed decisions that optimize project performance, durability, sustainability, and client and user satisfaction, all while adhering to scope, cost, and schedule constraints.

It is essential to understand the distinction between cost, value, and price.

Balancing cost and value is challenging because it requires engineers to identify opportunities for cost savings without compromising the project's overall quality, safety, and long-term viability. Effectively managing this balance is the project manager's responsibility as they coordinate with the client and end users to ensure investments translate into enduring and beneficial infrastructure or new buildings for the community.

Engineers must understand that value and cost are not always correlated. Sometimes, the value delivered by a project is greater than its cost, creating a positive return on investment for the engineering firm, client, and the broader community. Conversely, the cost of a project may surpass its perceived value, leading to questions about the viability of the undertaking. All aspects of project management are critical in delivering the highest value for the lowest cost.

The pricing of a project, set through either competitive bidding or negotiated contracts, typically exceeds the actual costs incurred. However, price is not required to match these costs precisely, except in cases where the contract's payment terms specifically mandate such alignment. Instead, pricing

The value of the work delivered determines the price of the services rendered.

should be based on the project's value to stakeholders. Some project fee structures, such as fixed-price, can capture added value for the engineering firm. A client may agree to a price exceeding the actual cost of the engineering work because the client has established a greater value of the product independent of the engineering firm's cost. The value of the work delivered determines the price of the services rendered.

This valuation considers how implementing an effective project management approach will benefit project delivery and mitigate risks. For example, suppose an engineering firm has invested heavily in market research, technology, and training to deliver project results more efficiently. With these investments, the engineering team can deliver more value at a lower (project) cost. Additionally, value can be provided with novel design solutions that can provide benefits, such as construction efficiency, enhanced quality of life, reduced environmental impact, and increased safety. Still, it is the engineers' responsibility to communicate and quantify the intrinsic value of services and products.

The investments made by engineers in their professional growth, such as acquiring specialized skills, securing industry awards, and gaining recognition for innovative designs, contribute to their ability to deliver high-value projects. These investments benefit the reputation of the engineers and their firm and enhance the quality of projects. Similarly, clients who demonstrate efficient and effective implementation of services and products will foster a collaborative client-engineer relationship that attracts and encourages innovation. Clients are active participants in a cycle of excellence, driving demand for high-value services and products to attract quality engineering teams.

The relationship between cost and value in civil engineering projects is dynamic, and pricing should reflect a project's holistic value. As noted in other chapters, a client's perceived value can prioritize a variety of project attributes (e.g., schedule, quality, safety, etc.) and must be understood by the project manager. This nuanced understanding of price and value is crucial for achieving successful project outcomes and fostering a culture of continuous improvement in civil engineering.

7.3 Project Costs

The cost of time for engineers, attorneys, architects, planners, and other project team members is likely the most substantial part of the project design budget for civil infrastructure and real estate development projects. It is necessary to understand that project personnel costs include more than just an employee's wages. There are many indirect costs driven by investment decisions related to training, technology, quality management, and consistent efficiency of delivered work. This context is the foundation for estimating project design budgets and forecasting costs while considering equitable and competitive compensation for team members.

From a client's perspective, a civil engineering project has no value until the engineering work is complete, the product is delivered, and the system is operational. Decisions on cost management at the micro level of project-related work, such as the cost of labor during the design phase, must consider this perspective. While clients do not usually disclose all financing details to engineers, the value of work can be implied by the project scope. For example, a transportation agency will not recognize the value of the new highway until after construction is complete and the system is in operation. Similarly, a private commercial developer will have significant investments in the project during the planning, design, and construction phases but the value is not realized until the project is built and leased, and rent is collected. Project managers should recognize these implications (even if only through general awareness) when deciding on project operations, design solutions, processes, schedules, and resource allocation. This section contains principles for cost management based on the primary project costs.

From a client's perspective, a civil engineering project has no value until the engineering work is complete, the product is delivered, and the system is operational.

7.3.1 Employee Wage Structures

The two main categories for employee wages are salary and hourly. The exact payment terms will vary with each employee's contract. Either compensation method can be full-time, usually expecting the employee to work 40 hours per week.

With *salaried* employees, there is a set annual pay based on 40 hours a week for 52 weeks of the year (2,080 *hours*). The alternate *hourly* classification does not imply a part-time employee status. Most design engineers will be full-time, even if paid hourly, with an expectation that they work 2,080 hours per year. The distinction is that a salaried employee has an established compensation rate, which pays the same regardless of how many hours they work (even if they work above the minimum 40-hour requirement). If a salaried employee is required to work overtime to meet deadlines, the employee's pay remains constant. Alternatively, hourly employees operate with a different pay structure and are paid for each hour they work. Hourly employees who work 45 hours during one week are paid for each of the 45 hours. In many cases, working overtime is paid at the same rate as the first 40 hours of work, or in some cases, it may include a higher rate (such as 1.5 times more than the standard rate), depending on pay types and employee agreements.

This distinction between hourly and salary is vital for project financials and employees considering job opportunities and the expectations associated with performance. A salaried position may have higher compensation but expects frequent overtime, which effectively diminishes the hourly paid rate of the employee (i.e., they make less per hour the more hours they work). Conversely, an hourly pay rate may benefit an employee who is expected (and willing) to work additional hours beyond the standard 40-hour work week. For example, if an hourly employee works 4 extra hours each week, it would be nearly a 10% higher salary. However, these extra overtime hours would not change the compensation for a salaried employee. Many organizations will offer a mix of pay structures based on employee role or position. They may include other incentives such as benefits, paid time off, bonuses, and profit sharing.

The calculated unloaded rate (or raw labor costs) for the employee is often referred to as a direct cost for the project budget. This is computed as an employee's hourly rate, or the annual salary divided by the 40-hour weekly schedule (2,080 hours per year). For example, an employee making $100,000 per year has an unloaded cost of $48/hour, which is what the employee sees reflected on a pay stub.

Most organizations offer benefits to employees, such as vacation time, sick leave, and holidays. This time is paid to the employee, but an engineering firm does not generate revenue during this time. For example, out of the 2,080 possible work hours each year, many organizations might offer 9 or 10 holidays and several weeks of vacation, which account for almost 10%

of potential billable time (resulting in about 1,872 billable hours). The organization must allocate funds to pay the employees for this time off. Other benefits, such as tuition reimbursement, retirement plans, and insurance, will also have associated costs to the engineering firm, which must be accounted for. For context, the US Bureau of Labor Statistics estimates that the total cost to a company for an employee is approximately 140% of the salary, where the additional 40% accounting for employee benefits.

The wages and the benefits provided to the employee account for the loaded rate. While the labor cost is a significant component of project costs, labor is only one component of the fully loaded rate, which includes overhead.

7.3.2 Overhead Costs

While labor is a primary cost, an organization will incur other costs for business operations. Office space must be leased, and equipment (computers, survey equipment, furniture) must be purchased or rented. Other common costs include software purchases, corporate insurance, and professional license reimbursements. Additionally, engineering firms depend on staff that maintain company operations and support production personnel, such as legal teams, human resources, accountants, marketing teams, communication departments, corporate executives, and IT support. While these corporate teams do not bill projects and, therefore, do not directly generate revenue, they are critical to the success of the project teams and the organization. These costs are often referred to as indirect costs. Figure 7.1 shows a schematic representation of how revenue received by the organization is distributed.

While labor is a primary cost, an organization will incur other costs for business operations.

As depicted in Figure 7.1, overhead is the most significant cost factor for revenue distribution. Engineers may only see their employee wages, but this is a fraction of the revenue distribution. Revenue must cover employee benefits and organizational operations (overhead) that support and enable project work.

Employees are not always working on projects, and therefore, not all time is billed to generate revenue. One component of overhead costs is *utilization* (primarily applicable to the private sector). This term often refers to how much billable time an employee has worked compared to the time

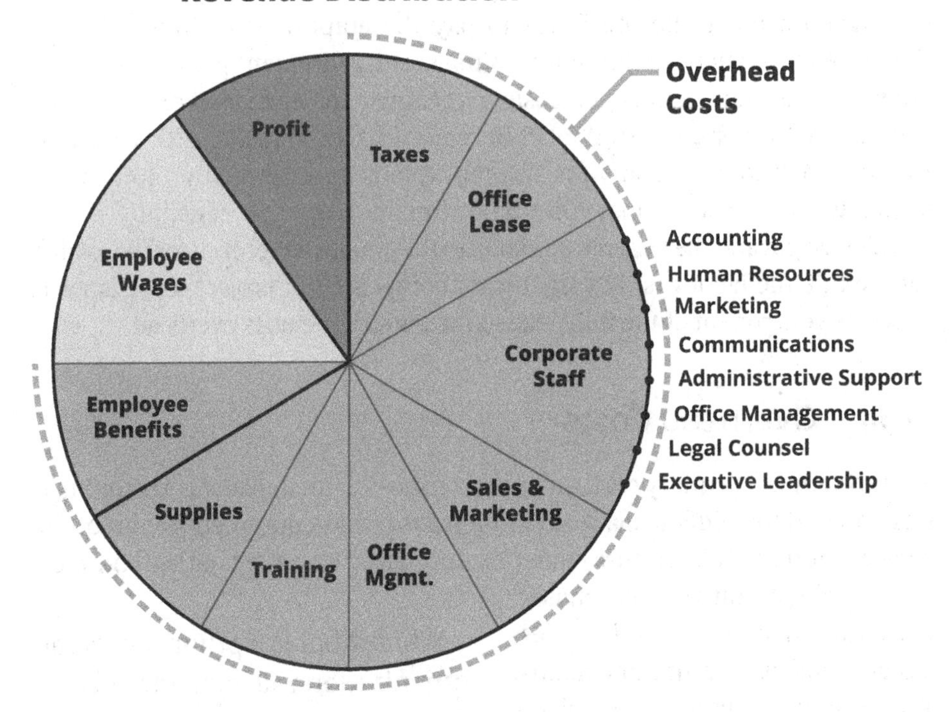

FIGURE 7.1 Revenue distribution example

not associated with billable work. Some employee work is not billable to clients but necessary for business operations, such as writing proposals to win work, internal meetings, marketing events, corporate and industry conferences, and training.

Many organizations will set utilization goals, often correlated to an employee's position. Entry-level employees usually focus on engineering and production work, staying 95% utilized. This means that, on average, the employee will spend 38 hours each week working on billable projects and about 2 hours on tasks that are not billable to a project. Comparably, a project manager might be closer to 75% utilized (30 out of 40 hours billable to projects) and otherwise spend a quarter of their time each week on proposals, marketing, office administration, or internal operations meetings. At the higher tiers, with corporate executives, it's expected that there is almost no utilization, meaning that all their time is nonbillable and does not directly generate revenue (but their time is spent supporting the engineering firm operations and personnel working on projects).

Overhead Cost Calculations

The overhead costs will vary by organization, but the cumulative costs of overhead (equipment, leases, insurance, corporate staff, etc.) must be considered. The total company overhead cost is distributed across all employees and the potential billable labor hours to determine a fully loaded rate. Note the distinction for *billable* labor hours; only hours considered as utilized time could be billed to generate revenue. For example, an organization of 500 people may only have 400 that are production staff (engineers and managers), with each of those 400 having a range of utilization that equates to about 600,000 billable labor hours (as compared to all hours from all employees at 500 × 2,080 = 1,040,000 paid hours). The cumulative costs are then divided across the billable labor hours. All overhead costs might equate to $45 million and are then allocated across the 600,000 direct hours to determine the $75/hour overhead cost. Based on these conditions, an organization will closely monitor both utilization (which affects the total labor hours) and indirect costs to make sure the overhead costs are managed.

Overhead is sometimes reported as a percentage associated with direct labor costs. The direct labor costs are the sum of loaded rates (wages and benefits) for all billable hours. The calculation follows a similar procedure as defined for a fixed hourly overhead rate but measures overhead costs as a percentage of direct labor costs in lieu of billable hours. Continuing the prior example, if the average loaded rate of the 400 production staff is $70, then the total direct labor costs are $42 million ($70 × 600,000 billable hours). From there, the overhead percentage is computed as the total of indirect costs divided by direct costs. In this case, the $45 million is divided by $42 million, which is an overhead rate of 107%. At the project level, the total loaded costs are evaluated with the overhead percentage before applying a fixed-fee profit for a cost-plus-fixed-fee (CPFF) structure. Essentially, the project must generate enough revenue to cover the wages plus an additional percentage (e.g., 107%) to break even.

Some overhead costs vary across an organization's geographic market or business unit. For example, a downtown office likely incurs higher office lease costs than the organization's office in a suburban or rural setting. Similarly, the cost of living and the compensation can vary by geographic region. With a significant fluctuation in cost, an organization might choose to establish variable billing rates for different geographies. This is used sparingly to avoid confusion for clients who might work across different geographies and expect the same billing rates for the same tiered staff, but it remains an option.

7.3.3 Billing Rates and Multipliers

When wages, benefits, overhead, and profits are summed, the result is a *billing* rate. For example, if an employee makes $50/hour and receives benefits worth an equivalent of 30% (an added $15/hour), and the organization has an overhead rate of $75/hour with a goal of about 10% profitability, the marked-up rate is the sum of these costs at $154/hour. From this calculation, an appropriate billing rate might be around $160/hour. Figure 7.2 provides a summary of the components of unloaded loaded rates that are included in the billing rate.

An employee classification usually establishes billing rates and is tiered in round numbers (e.g., $160, $180, $200/hour). For example, an Engineer-1 could represent someone with up to three years of experience and might have a billing rate of $160/hour, and then an Engineer-2 billing rate might include engineers that have three to six years of experience for $180/hour, and so on up to the highest billing rate tier. These tiers and billing rates will likely have associated qualifications, such as professional licensure, advanced degrees, and industry certifications.

These rate tables are often published to clients (although usually without a detailed list of qualifications for each billing category). Some public agencies will define consultant billing rates based on industry-standard criteria, which could differ substantially (higher or lower) from an organization's billing rate classification. An example of billing rate tiers and qualifications is shown in Table 7.2.

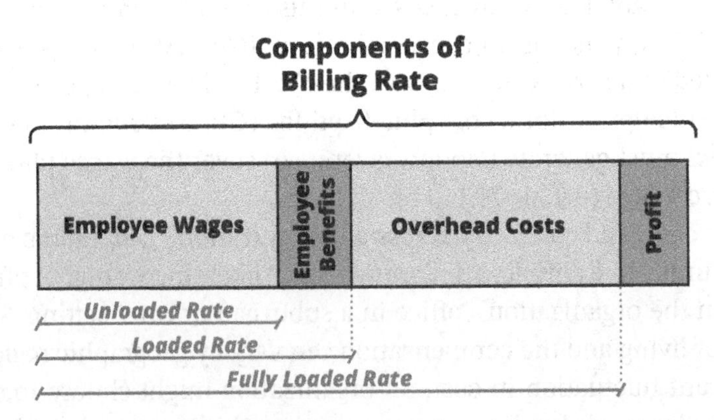

FIGURE 7.2 Components of billing rates.

TABLE 7.2 Rate tables

Billing Category	Employee Qualifications	Hourly Billing Rate
Engineer-1	0–3 years	$160
Engineer-2	3–6 years	$180
Engineer-3	6–10 years, PE license	$200
Project manager	10–15 years, PE, PMP	$240

Each billing rate has a range of experience, which means employee compensation varies within each tier. For example, for the Engineer-1 tier, an entry-level engineer is likely paid less than the engineer who has worked for three years. Still, the entry-level and the three-year engineer's billing rates would be the same because they share the same billing category (Engineer-1). An engineering firm may also have different billing rates across geographies or services. For this reason, many cost metrics will monitor budgets with both the unloaded rates and billing rates based on the cost performance metric.

Cost Multiplier

A *cost multiplier* is a common cost performance metric, which compares the billing rate (or invoiced amount) to the costs associated with the unloaded rate. For an initial example, this is calculated by dividing the billing rate by the unloaded rate. From the prior example, a billing rate of $160/hour and an unloaded rate of $50/hour would equate to a multiplier of 3.2 (160/50). This multiplier is often compared to an engineering firm's breakeven multiplier to determine if the project is profitable. The breakeven rate is computed as the fully loaded rate divided by the unloaded rate. In the example, the organization must bill at least $140 every hour the employee works to break even on the project ($50/hour labor + $15/hour benefits + $75/hour overhead), which equates to a break-even multiplier of 2.8 (140/50). A larger multiplier indicates a better ratio of revenue to cost.

Engineering firms do not usually disclose these break-even rates and multipliers, as the values reflect the operating costs and efficiency of the organization. These fully loaded rates are a calculated number reserved for internal accounting purposes (except when it must be disclosed with cost-plus billing types, described in Section 7.4.3).

The multiplier can vary even by staff that have the same billing rate. Continuing the prior example (with benefits at 30% and overhead at $75/hour), suppose an entry-level engineer earns $45/hour with a fully loaded rate of $134, whereas the three-year engineer earns $57/hour, equating to a fully loaded rate of $149. Both are billed at the Engineer-1 tier of $160/hour, but their multipliers differ (3.6 and 2.8, respectively). The higher multiplier of the entry-level engineer indicates higher profitability per hour billed for the organization compared to the three-year engineer. However, there is an expectation that a three-year engineer will have increased productivity, higher-quality deliverables, and be less dependent on hands-on instruction from peers, which may yield higher profitability on fixed-price fee structures (defined in Section 7.4.1).

An organization may not have a requirement to correlate employee compensation with billing rates, meaning that a well-qualified engineer could satisfy the criteria for a higher billing rate but have relatively low wages (which results in a higher multiplier). Conversely, an engineer could be well-compensated but does not meet standard criteria for a higher billing rate, achieving a lower multiplier.

The prior examples of multipliers pertain primarily to hourly rate structures, but cost multipliers are also used to monitor performance on projects with a fixed price. For example, suppose a project has invoiced $12,000 of work and the sum unloaded costs from all project personnel is $4,000; then the project multiplier is 3.0. This project multiplier is compared to a firm's breakeven multiplier to determine if the project is profitable. If the sum of unloaded costs were $6,000, then the multiplier would be 2.0. This lower multiplier is likely less than the break-even multiplier and indicates the project is not profitable. More information on monitoring project budgets is included in Section 7.7.

One exception to the mechanics of billing rates and multipliers pertains to cost-plus-fixed-fee (CPFF) contracts. These are often established by public agencies and set a limit on the profit, which requires the engineering firm to disclose their overhead costs and then select an allowable billing rate per employee based on the employee's wages, benefits, overhead, and a fixed profit percentage.

Several nuances must be considered to understand how to monitor a project's financial performance, including how overhead is calculated and how overtime affects profitability.

Considerations for Overtime

Recognizing that overhead costs and the breakeven multipliers are a function of billable labor hours, the overhead is "paid" once employees have met their utilization goal (e.g., once they have worked for 38 hours if expected to be 95% utilized). The cost of overtime depends on whether an employee is paid salary or hourly, but indirect costs will also influence overtime revenue.

This way, overtime could be considered more profitable compared to standard working hours. With the original salary example, the billing rate for the employee was $160/hour, which includes a $75/hour overhead and indirect cost; however, with overtime, the $75 is resolved (i.e., the overhead is accounted for), but the billing rate remains at $160/hour. Each hour billed to a client (as time and materials) can now realize the $75 as additional profit. Note that some overhead costs could carry through overtime if there are material or utility costs for working, but there is generally a substantial financial benefit.

Additional profitability can also be gained if the employee is salaried compared to hourly. Continuing with the original example rates, the employee unloaded labor costs are $50/hour and would not be incurred for a salaried employee (who does not receive overtime pay). Meanwhile, the engineering firm can charge the standard billing rate for each hour worked by the employee. This is a substantial financial benefit often recognized by an organization, which will encourage (or require) overtime work. For example, a salaried employee with a 95% utilization rate (38 hours of billable labor) could generate $570 weekly profit for the organization if the billing rate includes a typical $15/hour profit. After working just 10% more each week, the employee has more than doubled that weekly profit to $1,210 (at $160 per hour, with 4 additional hours per week, adding about $640), which substantially differs from the expected financial operating conditions.

While the financials of overtime show benefits, this practice should be used with consideration for the project team members. Recognizing that engineering is a professional service that relies on the attentive expertise of people, there is a diminishing return on productivity when an employee works severe overtime. This means completing a normal hour's work could take more than an hour of overtime. More importantly, employees may feel exhausted, frustrated,

There is a diminishing return on productivity when an employee works severe overtime.

and disenfranchised when working long hours, which could lead to staff attrition (resignations). It's necessary for project managers to recognize these effects on the project team and consider opportunities to refine scheduling and cost management as needed. When overtime is necessary, it should be communicated with the project team to manage expectations.

7.3.4 Calculating Labor Costs

For many organizations, the direct labor costs are assigned to a project, and the individual tasks are tracked using timesheets to best calculate, monitor, and control these costs. Each employee will document how much time was spent on a project-related task during the days and weeks of the project. Figure 7.3 shows a sample timesheet, where the first column is the project identification number, the second column is a unique identification for a project task (i.e., task code), and the remaining five columns identify the days of the week and how many hours have been allocated to each task.

Each project's task codes are based on internal financial operations and contract requirements. A contract may dictate that different project phases or tasks have independent budgets, and some tasks may have different billing types. For example, the contract may require that meeting time is billed as time and materials (billed per hour worked), whereas site plan design work is a fixed price (paid based on the earned value of work). More information on fee structures is included in Section 7.4.

The budget and the fee structure are associated with each task so the project manager can monitor work performance based on the associated scope. The task codes may also be organized for internal accounting purposes. For example, one task may require work from two departments in the company, so the budget is split for each department with an assigned code, even if the task only has one line item on the invoice.

A project manager can calculate the budget spent based on the hours worked by the employee. Suppose an employee has a billing rate of $160/hour and an unloaded rate of $50/hour. After the week is complete, a project

Project	Cost Code	M	T	W	R	F
PROJECT #0005	SITEPLAN	6	8	6	7	8
PROJECT #0005	EXHIBITS	2			1	

FIGURE 7.3 Sample timesheet.

manager will often have access to a report of hours worked by employees for each project task. For the example in Figure 7.3, the employee has spent 35 hours on the SITEPLAN task for an unloaded labor cost of $1,750 (35 hours × $50/hour), with a marked-up value of $5,600 (35 hours × $160/hour). The timesheet shows three hours for the EXHIBIT task, which would have an unloaded labor cost of $150 (3 hours × $50/hour) with a marked-up value of $480 (3 hours × $160/hour). These initial calculations will inform the project manager of costs and potential revenue; however, the invoiced amount will vary based on the work completed during these hours and the project fee structures, as described in section 7.3.5.

7.3.5 Invoices

Invoices are sent to clients to collect payment for work performed. Before the invoice is developed, cost performance reports provide information on the project expenditures based on employee timesheets. In some cases, these performance reports are compiled each week as the timesheets are complete and later summarized (often monthly) to report the amount spent and remaining budgets. Based on different fee structures, the project managers use this information to identify how much to invoice a client.

Figure 7.4 provides a sample invoice proof. The fixed-fee tasks (SITEPLAN) are shown with information regarding the contracted work amount, the current percentage of work completed (determined by the project manager), the total amount earned (percentage completed of the contract amount), how much was previously billed, and the difference between earned and billed as the current amount owed. The sample T&M tasks (MEETINGS) provide a breakdown of which personnel billing rates have worked on the project, the number of hours worked, their billing rates, and the total amounts owed. The total of the fixed-fee billings and the T&M sum to the total invoice due for this period of work (e.g., monthly invoice).

While engineers are paid within a week or two of their time spent on a project, an engineering firm's revenue collection is delayed by several months. This occurs because the project costs must be allocated, and invoices must be sent before payment is collected. After a summary of monthly costs, the project manager will work with an accounting team to develop the invoice (usually a few weeks after the end of the month). The invoice is sent to the client, who could take 30–60 days to pay (if paid on time), and then the payment is collected by the engineering firm. This means that revenue for time worked

INVOICE PROOF

Bill To:	________	Invoice #:	________
	________	Invoice Date:	________
	________	Due Date:	________
	________	Client #:	________
		Contract #:	________

FIXED PRICE BILLING

Task Description	ID	Contract Amount	Percent Complete	Amount Earned	Previously Billed	Current Amount
SITEPLAN	L001	7,500	80	6,000	2,500	3,500

TIME & MATERIALS BILLING

Task Description	ID	Personnel	Hours	Rate	Amount
MEETINGS	T001				
		Engineer 1	3	160	480
		Engineer 3	5	200	1,000
				Total For T001:	1,480

TOTAL INVOICE DUE: 4,980

FIGURE 7.4 Sample invoice.

(and value delivered) is not collected until several months after the work has been performed (and the engineer has been paid by their employer).

It is crucial to distinguish between the actual costs incurred on a project and what has been invoiced to the client, as these figures often do not align in each billing period. This discrepancy arises due to the fee structure and the timing of invoice issuance. In some cases, project work has started but is incomplete and cannot be billed to a client. For instance, if a project is billed by milestones, the work can only be invoiced to the client once the milestone is complete. Until that time, costs may accumulate for weeks or months before they are visible to the client or generate revenue. This is sometimes referred to as *work in progress* (with the acronym WIP), representing accrual of costs for work that is not complete and therefore not billed.

To maintain project financial stability, managers must accurately track costs and anticipate the timing of revenue from invoices. This responsibility includes monitoring billings and payments and (as needed) diplomatically reminding clients of due payments to prevent any potential interruptions in project progress.

One method to address potential discrepancies between completed and invoiced work and collection of payment is through retainage and retainers. *Retainage* refers to a portion of the payment, often a percentage of the total fee, withheld by the client until the work is satisfactorily completed. For instance, on a $100,000 project, a client might hold back 10% ($10,000) as retainage. Although the engineer considers the project design done and invoices the client 100%, the client may hold a portion of payment until certain conditions are met, such as design approval and permit issuance (thereby assuring the work is actually done).

Conversely, a *retainer* is a sum paid upfront by the client to secure the services of an engineer and mitigate the risk of nonpayment. Typically, an engineering firm might request a 10% retainer from a new or high-risk client to safeguard against situations where the client fails to pay after the project has commenced. This advance helps cover costs that accrue early in the project, reducing the firm's exposure to potential losses.

Project fee structures will govern when and how invoices are sent. More details on fee structures are provided in Section 7.4, but it is essential to understand the invoice process in order to consider project cash flow operations.

7.4 Project Fee Structures

The project fee structures (or billing types) can be established in different ways. A project may include several fee structures assigned to each task or phase of the project. The choice of billing types does not relate to employee

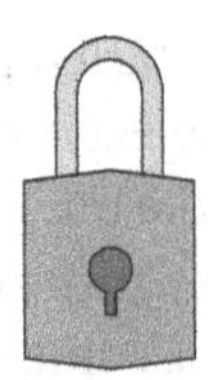

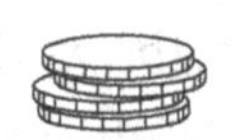

Fixed-Price

- A predetermined price is established for the scope.
- Invoices are based on percent of completed work.
- Best for well-defined scope with is primarily controlled by the bidder.

Time and Materials (T&M)

- Each hour of personnel time on a project is billed based on published standard rates.
- Usually includes a budget estimate for total likely expenses.
- Best for scope with potential variability, such as meeting time.

Cost-Plus

- A formulated billing rate based on published labor costs and overhead.
- Often includes maximum allowable profit percents, such as cost plus *X*% profit.
- Primarily used by government agencies.

FIGURE 7.5 Fixed-price, hourly, and cost-plus billing.

compensation methods, meaning that the fee structures for project tasks should not be confused with compensation structures for employees. The three most common billing types are fixed-price (sometimes referred to as lump sum), time and materials (T&M, or sometimes referred to as *hourly*), and cost-plus. Each of these is summarized in Figure 7.5.

The choice of billing type is either pre-set by the client or recommended by the engineering firm and then refined through negotiation before being finalized in the project contract. Ultimately, the contract terms and conditions will establish the enforceable conditions of the fee structure. Each fee structure presents unique benefits and risks for both the client and the engineering firm, as described in this section.

7.4.1 Fixed-Price

The client will pay a predetermined amount for the work performed with a *fixed-price* billing type. Civil infrastructure and real estate projects span several months and years, so these fixed price amounts are paid in increments during the project. For instance, if the work to complete a site plan has an established price of $300,000 for the scope of services, then the engineering firm will send invoices to the client based on the incremental amount of work completed. The total amount could be invoiced each month based on the current percentage of work completed, such as $30,000 after completing 10% and $150,000 after completing 50%. Alternatively, invoiced amounts may be associated with predefined project milestones, such as conceptual design, preliminary engineering, final engineering, and construction documents.

The project manager evaluates the cost of work and the amount of completed work, which informs the amount invoiced to the client; however, there may be some disparity between the cost of work and the amount of completed work. For example, the project manager could be concerned if the project has only met the requirements of the 50% deliverable but has spent 60% of the budget. In most cases, there is no reconciliation for the engineering firm if the costs exceed the fixed-price budget. Conversely, if the project has completed 100% of the scope of work but only spent 85% of the budget, this difference is gained by the engineering firm as additional profit.

Fixed-fee price structures serve as a low-risk condition for the client because the costs are established; however, fixed-price can represent a high-risk condition to the engineering firm because the work efforts may exceed the budget estimates. For this reason, a fixed-price structure is generally reserved for a well-defined scope of work that the engineering firm controls. For example, an engineer is likely comfortable estimating the effort required to design a site plan with a familiar site program and could reasonably establish a fair budget for a fixed price.

Even with a familiar scope of work, the project manager must closely monitor the project to mitigate the potential for scope creep, which would reduce the profitability of a fixed-fee task (refer to Chapter 5, Section 5.2 for more information about challenges with scope). When the scope cannot be well-defined or a service is meant to accommodate client needs (like meeting time), these tasks are better established as time and materials (hourly) billing structures.

7.4.2 Time and Materials

With the *time and materials* billing type, each hour worked by the design team is invoiced based on the established billing rate. Like a fixed-price structure, an initial budget is established upfront; however, the T&M budget is often considered an estimate instead of a strict limit because the scope may be difficult to define. For example, T&M billing is frequently used for meetings and coordination time. The client and engineering firm acknowledges that the meeting frequency and duration are primarily outside the control of the engineers (although engineers should strive for efficient and value-driven meetings). Similarly, a T&M budget may be established for work where a third party, such as a jurisdictional plan and permit processing effort, can influence the scope and effort.

While the T&M budgets might be considered initial estimates, the project manager should keep the client apprised of the funds spent and the remaining funds so the client understands when the budget will be depleted. For example, when the costs have reached a certain threshold (e.g., 80%) or if there is an early indication that the forecasted costs will exceed the initial budget, the project manager and client should document the condition and modify the budgets (or scope) accordingly.

T&M billing may include a not-to-exceed (NTE) provision, which sets an agreed limit to the budgeted amount. This may be used when the client seeks to achieve the best condition of a fixed-price structure (with a known limit of spending) but only pays for the time spent by the project team. For example, if the project has completed 100% of the scope of work but only spent 85% of the budget, the client only pays for the 85%. If the project is complete but the team has spent 110% of the budget, the client still only pays 100% of the fee.

There tends to be a moderate risk for the client and the engineering firm with the T&M billing type (although the risk is greater for the engineering firm when an NTE provision is applied). The engineering firm will make a predetermined profit based on the labor cost and the established billing rate. For example, the scope of work may estimate that the project design team will meet with the client once every two weeks for one year, and each meeting will last an average of two hours, so a billing rate of $200/hour would establish the budget of about $10,400 (assuming only one participant from the design team). Each hour worked by the engineer is billed at a rate of $200/hour, and is then invoiced based on the number of hours and the established rate.

The T&M structure can accommodate budget adjustments by following the appropriate change management processes (see Chapter 5, Section 5.6.1). Suppose at the halfway point of the project, the team realizes that the meetings tend to go long (as they tend to do). In this case, the project manager may determine that they've already spent 75% of the budget and should submit a notification to the client with justification for a budget increase to cover the remaining project meeting time. However, if an NTE provision was included, the implication is that there is no tolerance for fee increases until a formal change order (CO) is authorized.

Ethics in Time and Materials Billings

Project managers should recognize the unique ethical considerations associated with T&M billing contracts. Engineers and their firms must provide their clients with professional engineering services, including exercising appropriate judgment on allocating team members' time to complete the project work. For a T&M task, the engineering obligation extends to controlling how much the client will be billed based on personnel's time spent completing the work.

Engineers are ethically obligated to prioritize client service by operating with maximum efficiency, even if it results in reduced revenue for their firm. T&M tasks are structured so that the engineering firm receives revenue only when its staff are actively engaged in the project work. With T&M, each hour worked is an hour billed to generate revenue. Greater efficiency, which translates to less time spent, naturally leads to lower immediate revenue. However, demonstrated efficiency benefits clients by providing fair and competitive pricing and faster delivery of work, enhancing the firm's value and reputation in the long run. This approach not only upholds professional integrity but also fosters trust and long-term relationships with clients.

Engineers are ethically obligated to prioritize client service by operating with maximum efficiency, even if it results in reduced revenue for their firm.

These practices are crucial as clients typically rely on the engineering firm's expertise in resource allocation, placing the firm in a position where it must carefully balance ethical considerations and risk management. For example, consider a scenario where an engineering firm is providing services on a T&M fee structure, such as meeting and coordination efforts. In this situation, the firm is presented with a two-fold, and potentially contradictory, responsibility: (1) to limit staff attendance in meetings with the aim of conservative billing, thereby helping the client's project budget stretch further, and (2) to ensure the right personnel attendance in these meetings when necessary, to comprehensively address all project aspects, recognizing the resultant increase in costs and resource usage. The potential conflict between these competing imperatives can introduce an ethical dilemma.

As with all ethical scenarios, an engineer's first task is to recognize an ethical situation and then evaluate the circumstances to determine how best to proceed. In the case of T&M billing, examples of ethical dilemmas typically surround the engineer's decision for how many team members – and therefore how many billable hours – are asked to attend meetings or work on a particular assignment. For example, for meeting attendance, sending multiple engineers to the same meeting will increase the billable cost of the meeting and consume more of the client's project budget than sending fewer team members. Therefore, rather than simply sending multiple engineers to each meeting to increase billing for the firm, there is often an ethical dimension to deciding who should attend different meetings under T&M contracts.

The following are a few examples of potential advantages to larger meeting teams that may justify the additional expense to the client:

- When the topic of discussion requires all parties to coordinate clearly, this can most effectively and efficiently be done in person.
- The client, team, or other project stakeholders must resolve an ongoing dispute or agree over a contentious topic that cannot be accurately discussed through other forms of communication.
- If specialized input is needed, the meeting discussion will be more productive if regular team members and specialized engineers attend.
- If there is an especially complex outstanding technical problem, the project would benefit from the brainstorming efforts of multiple engineers looking for innovative solutions.

Examples of potential disadvantages to larger meeting teams that may suggest the additional billing is unjustified:

- Having multiple engineers in the same meeting may lead to an unnecessary debate that does not advance the project and may create confusion.
- When team members with no substantive role in the meeting feel they need or want to participate, they use meeting time to make nonmaterial comments.
- Senior management attends and causes the meeting to pivot away from technical discussions and focus elsewhere (such as business development and marketing with the client).

Examples of team participation scenarios that may have less clear ethical implications:

- A junior engineer would benefit from experiencing a client meeting, and the project manager is obligated to provide professional development opportunities for junior team members; however, the junior engineer may not materially add value to the meeting outcome.
- An engineer worked on a prior phase of a multiyear, multiphase project but has not been assigned to work on the current phase; including this engineer in the project meetings may – or equally may not – potentially provide insight into issues that the current team will face.

Ethical decisions often require a nuanced approach, balancing between various perspectives and strategies rather than a simple "either-or" choice. Instead, it requires a consideration of different strategies and viewpoints. Open and clear communication between engineers and clients about the requirements of each task is essential to dispel any notions of inefficiency or negligence. Additionally, engineers have an ethical obligation to manage resources effectively. This safeguards the client's interests and protects the engineering firm from potential accusations of substandard service or underperformance. Effective resource management in this context is about finding the right equilibrium that serves both the client's needs and the firm's reputation for quality and integrity.

The project manager should be transparent with the client about who is needed for a meeting and be efficient with the organization and execution of the meeting. More information about communication practices relevant to meetings is included in Chapter 9 (Section 9.5).

7.4.3 Cost-Plus

The cost-plus fee structure is often seen with public agencies and can be used to establish transparency in consultant profitability on a project. This model is similar to T&M and fixed-price conditions, but instead of using an organization's preset billing rate, the client will request backup financial data that document employees' direct and indirect costs, which establishes the fully loaded rate (wages, benefits, and overhead). Then, a profit percentage limit is applied to that fully loaded rate to establish an allowable cost-plus billing rate.

For example, with a $50/hour unloaded rate, $15/hour for benefits, and $75/hour overhead ($140/hour fully loaded rate), the client may limit profit to 7% for a billing rate of $150/hour. This calculated rate would be used for invoiced work and documented in the calculations for the budgeted fixed-price amount. For instance, if the engineering firm has proposed a fixed-price budget of $12,000, then the client would expect to see some calculations that identify an effort of 80 hours of work (at the $150/hour rate).

7.4.4 Alternative Fee Structures

While the most common fee structures are fixed-price, T&M, and cost-plus, the project team can agree to other payment terms. A few other options are:

- Cost-Plus Performance
  - This is a modification of the standard cost-plus, except that it includes a performance fee determined by the client. For example, a developer may accept that the project will take 12 months to design, but they establish a performance fee if it is completed sooner. This could encourage the design team to work overtime or assign additional personnel. This option is also common for construction services.
- Cost-Plus Incentive Fee
  - An option for both fixed-price and the standard cost-plus structures is adding an incentive fee as a shared benefit for the client and engineering firm with a predetermined split rate (e.g., 80/20). For example, a developer may realize additional lease revenue from a new project if the work is completed ahead of schedule and may split that financial benefit with the design and construction team.
- Fixed-Price with Economic Price Adjustment
  - The lengthy time of civil engineering projects means that the financials are vulnerable to economic shifts. Increased material costs, fuel costs, and labor costs can all influence how profitable a project is by the time all the work is done, compared to the initial estimates at the start of the project. The economic price adjustment mitigates that risk by establishing parameters for the engineering firm to adjust its price based on documented economic changes.

- Unit Price
  - When associated with labor effort, the unit price structure is an extension of fixed-price but typically assigned to multiple, small, repeatable tasks. For example, a neighborhood development may have 100 residential lots, and the engineer could establish a unit price to complete the lot grading plans for each (unit).

Each of these alternative fee structures (and many others) offers unique considerations for project performance and delivery. Clear, detailed contractual agreements are essential to define the terms, conditions, and criteria for any adjustments or incentives, ensuring mutual understanding and alignment of interests between clients and engineering firms.

7.4.5 Other Direct Costs (ODC)

Project budgets have two primary cost types: labor and *other direct costs* (ODC). While generally minimal relative to the labor costs, the ODC budgets cover other project-related resource costs. This includes printing, travel, permit fees (if elected to be paid by the engineering firm), lodging, presentation materials, and others. Federal policies often dictate the costs associated with travel and lodging (e.g., cost per mile of travel in a personal vehicle). Printing and material costs often have a standard rate (cost per page) or could be invoiced with receipt of costs incurred. Some clients may include provisions regarding reimbursement amounts or expectations (e.g., only reimbursing economy-class flight expenses or basic hotels).

ODCs often include subcontracted services that are crucial to a project. When engaging subconsultants, their expenses are also typically classified as ODCs because these costs, directly linked to the project, are passed on to the client. Contract terms regarding these expenses can vary; in many instances, the prime consultant is not required to itemize the billing details of subconsultants to the client but rather includes a consolidated fee covering all ODCs. Subconsultants should understand that invoicing for these costs falls under the prime consultant's purview. Furthermore, it's common for contracts to specify that subconsultants receive payment only after the client has paid the prime consultant.

In addition to the direct costs of subcontracted services, contracts often specify a markup on these costs, typically ranging from 5–15%.

This markup is designed to cover the engineering firm's overheads related to processing and managing these ODCs, including administrative, finance, accounting, and project management efforts. This practice not only ensures that the firm's additional resource usage is compensated but also maintains project profitability.

For effective client communication and financial planning, the engineering firm is responsible for documenting all required ODCs during the budgeting phase. This documentation should provide a detailed and transparent view of the project's anticipated costs, including both direct expenses and associated markups.

7.5 Budget Estimates

With knowledge of employee compensation, overhead, and project fee structures, project managers can develop budget estimates for the project. The estimate's accuracy is correlated with the time necessary to develop the estimate and the available information on prior project cost performance. Note that estimating methodologies used with budgets are frequently used with schedule estimates. More complex estimating methods tend to be more accurate but rely on more historical cost data and require more time.

More complex estimating methods tend to be more accurate but rely on more historical cost data and require more time.

Budget estimating techniques include analogous, bottom-up or top-down, and parametric estimation. Analogous cost estimates provide quick reference without needing a lot of historical cost data but do not contain the details of other methods. Top-down and bottom-up methods require a scope breakdown to assign costs to disparate tasks, which takes more time. Parametric estimating goes a step further than task breakdowns and considers additional variables that could influence costs for project tasks.

Many projects will use a mix of estimating methods based on the complexity of tasks, risk considerations, prior experiences, or the need to allocate fees or provide various levels of detail during early planning and feasibility studies.

7.5.1 Analogous and PERT Estimates

Analogous estimating informs the budget by considering the costs of prior similar projects. A new cost estimate can be established by reviewing prior project costs and comparing similarities in project scope.

Suppose a previous residential project required the design of 100 homes and had a site design cost of $100,000 – from analogous estimating, a new project that involves the design of 200 homes would theoretically cost about twice as much. However, projects are unique, and it is not advisable to assume all cost variables can be directly correlated or consistent with similar projects. Analogous estimating is quick, but best serves as a starting point. Expert judgment is required as external environmental factors, economic conditions, team members clients, and other conditions will influence the project's performance.

An effective analogous estimating technique requires organized records of prior project financial performance. For example, if an engineering firm is developing a cost estimate for the design of a new retail development project, the team would start by gathering budget data from prior retail projects. The data gathering extends beyond the original cost proposal, and the final cost performance of the project must be investigated. Was the project profitable in the end, given the scope and budget? Was there a design change order that requested additional costs? Is the current project team and client the same? Ideally, analogous estimates are informed by recent projects with similar geographies, customers, and market conditions and performed by similar project teams.

A large sample size of budgets will be more effective in informing cost ranges, and details about the associated tasks provide a better reference for tailoring the current cost estimate. For example, some costs may scale with the scope (meetings, quality control review, design work time, etc.). In contrast, other costs are generally uniform across project types (time to prepare permitting forms, project file setup, etc.). For these reasons, analogous estimating is best for internal discussions and initial project planning considerations.

As a high-level cost estimating technique, analogous estimates can benefit from weighted averages or *Program Evaluation and Review Technique* (PERT) methods. The team can develop optimistic, pessimistic, and most likely estimates based on project uncertainty. For example, suppose there is a set of similar project types with variable performance. In that case, the historical data might reference how some projects faced challenges (pessimistic), some were highly efficient (optimistic), and many were standard conditions (most likely). Based on this distribution, a cost estimate can be established that averages the optimistic, pessimistic, and most likely – alternatively, a weighted average can be used with the most likely conditions (often by a factor of four) to emphasize the expected conditions. While PERT is often associated with analogous estimating, the method can also be used for other estimating techniques.

7.5.2 Bottom-Up and Top-Down Estimates

With more detail and effort, bottom-up and top-down estimates establish costs for individual tasks. *Bottom-up* is one of the more time-intensive and accurate estimates, often using the work breakdown structure (WBS) that has identified the tasks required to complete the scope of work (refer to Chapter 5, Section 5.5 for more information on WBS). Each WBS component has a cost estimate that is then assimilated into a total project cost.

For example, when preparing a budget for a road design, a WBS is created for various components, such as plan production for road alignment, utilities, stormwater management, and landscaping. Each of these tasks—or their sub-tasks—is assigned an estimated cost for the required work. This method is helpful for contracts that will list distinct budget lines for each task, especially in complex or unfamiliar project types. As budgets are set for each task within the WBS, this detailed financial information can be itemized in project proposals and listed with invoices. Proposal itemization can aid discussions between the client and engineer about budget adjustments in case of scope changes or scope eliminations of certain tasks.

While a detailed WBS is essential for precise cost estimation, excessive itemization in proposals and contracts should be avoided. Overly specific cost breakdowns can complicate time tracking for team members, making it difficult to list their hours against an array of tasks accurately. For example, while it is reasonable to itemize the stormwater management work as a distinct cost in a proposal, breaking it down further into individual components like each pond, pipe, and catch basin in an invoice could lead to confusion. Therefore, although the WBS can include detailed tasks for estimation purposes, the cost codes and the structure of invoices should reflect a more general categorization.

Alternatively, *top-down estimation* begins with the total estimated cost and then assigns the costs to various tasks. This occurs when the project budget has already been established, and the team must understand how to allocate funds for each task or team member. For example, if the client has an established construction budget, they might have an assigned allocation for design costs (e.g., 8% of the construction cost), establishing the team's design budget. This budget sets the top estimate, which is then carried down to the tasks and teams. Financial records and expert judgment may facilitate a formulated approach to assigning budgets to various tasks this way.

7.5.3 Parametric Budget Estimates

Following some of the same principles as analogous estimates, *parametric estimates* provide a more formulated approach to budget estimates. For parametric estimates, unit prices and project quantities are used for a systematic calculation to determine the costs with an added parameter. Parametric estimates are commonly seen with construction projects, where project materials, such as square yards of pavement, could be quantified for costs of materials, labor, and management. As with other methods, costs are best informed by recent and local information.

Parameters can take on many forms beyond material quantities by using expert judgment. For example, the project could include design costs based on the roadway centerline distance, number and size of residential lots, distance of utility corridor, number of reports and permits expected, and others. While this provides a quantifiable baseline, it is beneficial to represent a parameter for the complexity or by considering the capability of the project team. A qualitative scale system could inform complexity (e.g., less to more complex). Similarly, based on the project schedule, is a small or large team required to complete the work (expecting that larger teams require additional coordination and time)? These qualitative parameters can be translated to a numeric representation to scale costs, such as adding 0–10% based on complexity. The formula then could include these parameters, backed by relevant historical data, such as:

$$\text{Road design cost} = (\text{Total lane miles}) \times (\text{Cost/Lane mile}) \times P$$

Where:

$$P = [1 + (\text{Complexity parameter}) + (\text{Team size parameter})]$$

With this example equation, a baseline estimate is established from prior projects (and expert judgment) to determine the design cost per distance of the road. For example, several prior project design budgets could be normalized as design cost per mile. From there, variables are referenced to consider how a more complex project or a larger team (perhaps to expedite the production) increases design cost per distance of road. For example, a baseline design cost of $100,000 per mile of roadway, with a complexity parameter of 5% and a large team parameter of 3%, would yield an estimate of $108,000 per mile. While these estimates use numeric methods and historical data, they serve primarily to inform the project team of cost estimates and are not intended as explicit budget values.

7.5.4 Multiple Estimating Methods

Establishing the project budget will benefit from multiple cost estimation methods. Early considerations during proposal planning stages can evaluate a quick understanding of cost, using analogous estimating, to consider expectations on labor hours and the appropriate pursuit budgets. Additional details can be used before the cost proposal, working toward bottom-up or parametric estimating.

Establishing the project budget will benefit from multiple cost estimation methods.

Many projects require estimates from functional managers across multiple disciplines that are compiled into the total budget estimate. Any combination of estimating methods may be used for disparate project budgets, and the project manager is responsible for coordinating these costs.

There is a benefit to documenting how budgets are established, with relevant calculations referencing scope, time, and staffing (rates) used to establish the budgets. This documentation also improves historical data (for reference on subsequent projects). Budgets should be established independently across the team and compared to evaluate relative costs for the scope. For example, when comparing budgets, it would be unusual to see the cost of land survey services exceed the design budget for a new road project. Like the methods of analogous estimating, historical data can inform the expected relative weights associated with each team's budget estimate. For instance, building design fees would be much larger than site design, and surveying and geotechnical services would likely be less than site design fees. Prior projects may have summarized the typical allocation of fees across each team.

7.6 Costs Across Multiple Business Units

In addition to understanding project-level financials, there are several considerations for how the associated work is executed in the context of more extensive programs and corporate structures. Many projects will require resources from multiple departments of an engineering firm and from other companies. For example, an engineering firm could have several business units (i.e., engineering departments) that include specialties in transportation, stormwater management, landscape architecture, geotechnical, and others. The project work associated with a transportation project would engage the personnel of other departments, likely operated by other functional managers. In this case, the transportation project manager would coordinate with other functional managers across multiple departments to understand budgets and monitor project costs. Similarly, a large project might require staff from multiple geographic offices in transportation departments but with disparate functional managers. This format requires the project manager to consider a few additional cost components.

7.6.1 Budget Allocation

To establish the initial budget, the project manager will coordinate with other business units to understand the scope and fee estimates for the required work. That budget effectively becomes the allowance for the other business units, which the project manager monitors. Still, the accountability of cost performance is typically assigned to the functional manager. The project manager and functional manager must coordinate the work efforts, as decisions by the project manager could affect the scope, schedule, and costs of other business units.

The work effort and responsibilities are expected to shift (or settle) during the project as the scope is refined. In some cases, a business unit may offer to support other teams to achieve a challenging schedule. While the overall project cost might remain the same, these shifts in responsibility would influence the financial performance of each business unit. For instance, if the landscape architecture business unit decides to spend more time on the project to support some production tasks for the transportation engineers, the landscape team would likely overrun its initial budget. The total project cost remains the same, but the landscape budget would be overrun while the transportation budget might be underrun. These conditions would show adverse performance for one department while strong performance for the other, which could influence the perceptions of each team. Each organization may approach these conditions differently, and the project manager may need to reallocate budgets to best represent the effort of each business unit.

7.7 Monitoring Project Budgets

As budgets are estimated, established, and allocated, there is a significant effort required to monitor the budgets throughout the project life cycle. Many financial systems will include processes to report weekly or monthly expenditures linked to project cost codes and personnel timesheets. These reports are often formatted to include cost metrics for task budget, amount spent, amount billed, and variances. In some cases, the unloaded costs (employee wages), fully loaded costs, and multipliers are all reported.

7.7.1 Cost Reports

Most organizations will publish internal *cost reports*, which enable project managers to monitor project financials. These reports are often automatically generated through financial management systems that include information about the project budget and the time spent by project team members. A project manager can use the cost report to determine the appropriate invoicing, which often occurs monthly. Financial reporting systems often require information from the project manager to input work performed and forecasts for project expenses, such as the percentage of work completed, to calculate the invoiced amount for a fixed-price contract. The format varies, but this section provides some of the typical information in the report:

- Project Name and Task Codes
  - Cost reports will have a hierarchy by project and tasks with the associated task cost codes and a summary by project. A project manager will identify the project's financial performance and each task at these levels.

- Budget
  - With each task, the report will include data on the original budget established. Budgets may be shown as the price and contracted amount or as an unloaded budget.
- Cost
  - Each task will also list the cost to date, determined by how many hours of an employee's time have been charged to the task. Similar to the budget, this may be shown as the sum of hours by billing rate or by the unloaded rate.
- Variance
  - Budget minus cost provides the variance. When the variance is positive, it indicates a budget is still available for the task and project. This number provides a quick reference to budgets but does not consider the value of work completed compared to the costs, which is identified with a cost performance index (CPI).
- Cost Performance Index (CPI)
  - The CPI is a ratio of earned value (EV) to actual cost (AC) that provides a quick reference to financial performance by considering how much work has been completed. This is calculated as EV/AC. The earned value is input from the project manager or as determined by other contract metrics. For example, based on the effort of work completed, the project manager would use their judgment to document the percentage completed, multiplied by the task budget, to establish the EV. Or the contract may stipulate that a percentage of payment is released after a certain milestone has been completed. The actual cost is the budget spent as determined by the billing rates and invoiced amount.
- Cost Multipliers
  - As noted in Section 7.3.3, cost multipliers represent a ratio of unloaded rates and fully loaded rates, or billing rates. An organization will have a calculated breakeven multiplier and a target multiplier. The cost report will provide the current project multiplier based on budgets and actual expenses to indicate financial performance. This multiplier is computed similarly to the CPI but uses the unloaded rate to establish the cost.

Project	Task	Budget	Costs	Variance	% Spent	% Complete	CPI
SITE PLAN							
	ROAD DESIGN	240,000	142,000	98,000	59%	50%	0.84
	LANDSCAPE	90,000	35,000	65,000	39%	55%	1.41
	STORMWATER	120,000	80,000	40,000	67%	85%	1.27
	PERMITS	6,000	6,500	(500)	108%	100%	0.92
	TOTALS	456,000	263,500	202,500	58%	61%	1.05

Variance:	Budget - Costs
% Spent:	Cost / Budget
% Complete:	*From Project Manager*
CPI:	% Complete / % Spent

FIGURE 7.6 Sample cost report.

The example cost report in Figure 7.6 shows the financial metrics for a project that includes four tasks. This level of detail is often shown as part of a monthly cost report before invoices. Each task is listed with the budget and the costs to date. When project managers have more than one project, these cost reports often summarize the information for each project. Calculations are shown for each task's variance (budget minus cost) and the percent spent (cost/budget).

The column for percent complete is often input by the project manager based on the earned value of completed work. The CPI is then calculated as a quick reference to the financial status of each task (EV/AC, or % Complete / % Spent in this sample). A value of one (1.0) indicates the job is progressing as expected, greater than one (1.0) is a positive condition (for the engineering firm), and less than one (1.0) indicates poor financial performance. In this example, the Road Design and Permits task has a CPI of less than one (0.84), indicating that work performed for these tasks has cost more than the value of work created. The next two tasks have a CPI greater than one (1.41 and 1.27, respectively), indicating the opposite condition. From the totals shown, the project has a CPI just above one (1.05), which is acceptable, but with about one-third of the project left to complete (based on % Complete), this value will likely change and should be monitored.

This example includes budgets and costs based on a multiplier. Some cost reports will include the unloaded rate calculated with the job's multiplier to determine the scaled budget and cost. This is included because there is a variance between billing rates and the effective multiplier for each employees based on compensation (as noted in Section 7.3.3). Employees may bill the same amount for each hour of work performed, but the unloaded labor costs differ if their compensation is different. Additionally, a project manager may compare this value to a target multiplier, which corresponds to the desired profitability for the project. For example, if the organization has assigned a target multiplier of 3.0 for the project, then the $120,000 budget for Stormwater would have a calculated unloaded labor budget of $40,000.

Figure 7.7 shows how unloaded budgets and costs determine the effective multiplier (the column on the far right). Continuing from prior examples, if the target multiplier is 3.0, the unloaded budget is calculated by dividing the marked-up budget by the unloaded rate. Direct labor costs are summed by the hours billed based on an employee's unloaded rate. Finally, the effective multiplier is calculated as the amount invoiced divided by the unloaded cost. The invoiced amount will often match the billing rate (for a T&M price structure), or it will be based on the percentage of work completed (for a fixed-price structure).

Comparing Figure 7.6 to Figure 7.7, the performance measured by the CPI shares similarities to the cost performance measured by the cost multiplier. There may be cases where a task's loaded budget has a good CPI but could still reflect adverse financial performance (multipliers below 3.0) based on the high labor costs of the employees compared to the billing rate. Conversely, a budget may be exceeded when calculating loaded billing rates but may perform well with the unloaded cost (as demonstrated in the Permits task). These disparities between CPI and the multiplier come from the variability of an employee's effective multiplier (as noted in Section 7.3.3).

Project	Task	Budget	Invoiced	Unloaded Budget (3x)	Unloaded Costs	Unloaded Variance	% Complete	Eff. Multipl.
SITE PLAN								
	ROAD DESIGN	240,000	120,000	80,000	50,700	29,300	50%	2.4
	LANDSCAPE	90,000	49,500	30,000	13,000	17,000	55%	3.8
	STORMWATER	120,000	102,000	40,000	25,400	14,600	85%	4.0
	PERMITS	6,000	6,000	2,000	1,950	50	100%	3.1
	TOTALS	456,000	277,500	152,000	91,050	60,950	67%	3.0

Invoiced:	*% Complete x Budget (Fixed Price); or Hourly Billing*
Unloaded Budget:	Budget / Multiplier (3x)
Unloaded Costs:	*Based on unloaded labor rate and time*
Unloaded Variance:	Unloaded Budget - Unloaded Costs
% Complete:	*From Project Manager*
Effective Multiplier:	Invoiced / Unloaded Costs

FIGURE 7.7 Sample cost report with the effective multiplier shown.

Project managers may wish to communicate project budgets to their team members by referencing the expected hours of work instead of the dollar value. This is often a more tangible metric for team members who are evaluating the scope and the schedule. The calculation is performed by dividing the task budget by the billing rate, which is often the most visible set of metrics for personnel (alternatively, a time budget can be computed by dividing the direct labor budget by the unloaded rate). This calculates the number of hours allocated to the task. This is best performed when the billing rates and employee wages are similar for staff assigned to a task.

Evaluating the budget in hours can facilitate cost monitoring by the project manager and the team members as they monitor the current hours billed (based on timesheet entries) for each week or day. For example, an employee with a billing rate of $160/hour working on a task with a marked-up budget of $72,000 would have about 450 hours (based on that employee's billing rate). In many cases, multiple staff with various billing rates will work on one task, so it's important to understand how the billing rates correspond to budgeted hours. This conversion of budget to time for various billing rates is shown in Figure 7.8.

The project manager must understand this calculation to determine the budgeted time for the project based on available resources. This information should also be communicated to the team by referencing the hours expected to complete a task or achieve a milestone. The project multiplier and profitability will decrease if the labor effort exceeds budgeted hours. If the budget is a fixed-price structure and the actual labor hours are less than the time

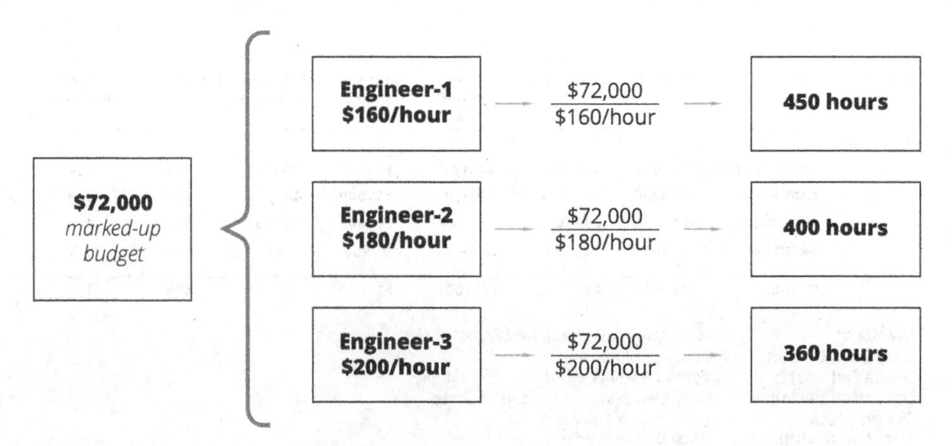

FIGURE 7.8 Conversion of budget to time.

budget at the project completion, then the task (or project) is more profitable. Given the long duration of many civil engineering projects, managers must consider that employees may see an increase in wages and billing rates, which would reduce the budgeted hours.

7.7.2 Cost Considerations for Change Orders

Engineers and clients must be comfortable with discussions regarding project changes, especially when modifications to scope or schedule warrant a cost-related adjustment. External environmental factors, unforeseen conditions, schedule changes, resource limitations, and other factors beyond the client and engineering firm's control will influence real estate and civil infrastructure project parameters and can lead to the need to make these changes. Ideally, a resulting change of scope is well-documented through formal change orders (as noted in Chapter 5, Section 5.6.1) so that there is a way to establish the appropriate financial change.

Engineers and clients must be comfortable with discussions regarding project changes, especially when modifications to scope or schedule warrant a cost-related adjustment.

A *change order* (often referred to by the initialism CO) is a document that captures the reason, scope, and cost and modifies the project contract to allow the engineering firm to work on the change and be paid for the unanticipated work. The CO typically begins with a *potential change order* (with the initialism PCO) that outlines the required scope or schedule change, a narrative regarding the financial impacts, and an associated cost of the proposed change. The PCO provides the client with an opportunity to understand the impact of a suggested or required change. Similar to the original contract, the cost identified in a PCO can be negotiated between the engineer and the client. However, the client likely has little choice in who can perform the work because the current engineer is uniquely knowledgeable about the project context and resources. Still, a client and the engineering firm can review the original contract fees to validate that PCO fees are representative of prior costs. Once the PCO has been reviewed, it is issued as a CO to contractually change the project's scope, schedule, and budget.

Late change requests are more challenging to estimate.

Late change requests are more challenging to estimate. A scope change is less expensive to implement early in the project; however, when the project is near completion, that same change will cause more rework and create severe challenges in tracking all affected project elements as the scope changes cascade through the project documents. Additionally, resource availability to accommodate late change requests may influence costs. For example, a road design task may have originally been completed by an Engineer-2 with a low billing rate. However, a late change may limit resource availability, so only an Engineer-5 can complete this work. While the Engineer-5 is likely more efficient, the engineer may be unfamiliar with the project and require time to learn the project requirements and status (ramp-up time). This condition would change the cost of the PCO. When a change request is received, the project manager should communicate the implications of these costs (and schedules) because the effort and impacts are likely unknown to the client.

7.8 Forecasting Costs

While monitoring the project budget, the project manager forecasts expenses and operations to anticipate the final financial performance. A well-established budget estimate (often from bottom-up estimating techniques that directly associate resources and costs to project tasks) can be used with progress reports that provide good metrics for the completed work. It is only practical to reference the percent complete metrics (and the associated CPI) if the project manager accurately estimates and monitors the project completion. The initial effort of estimating project costs based on tasks and resources should be reevaluated throughout the project to identify projections.

The *Estimate to Complete* (ETC) is a financial metric calculated after a project has commenced. It represents the expected cost required to complete the remaining project work.

If the project's performance deviates from the initial budgetary expectations or if performance is expected to change, the ETC requires a recalculation.

Under normal circumstances, the ETC is calculated as the difference between the original budget and the actual costs incurred so far. However, if the project's performance deviates from the initial budgetary expectations or if performance is expected to change, the ETC requires a recalculation. This recalculation involves a detailed examination of the remaining scope and associated costs, incorporating insights from recent project performance to predict future expenses more accurately. The goal of recalculating the ETC is to identify potential budget overruns early. If it is determined that a project may exceed its budget, the calculated ETC allows for the timely implementation of corrective measures to manage the financial impact. A few examples of disruptions that would warrant a recalculated ETC and methods to mitigate each are presented here.

- Personnel Availability
  - If the initial budget was established based on expectations that certain team members would be assigned to the work and those conditions are no longer possible, this would warrant recalculating the budget. For example, if an engineer proficient in an esoteric hydrologic modeling program was reassigned, a different engineer may need to learn the software and would be less efficient. These factors, plus differences in billing rates, would change the estimated budget.
  - *Mitigation:* If the project manager identifies financial hardship based on personnel availability, this should be discussed with a department manager or functional managers that determine how personnel time is allocated across various projects. In cases where the assigned personnel are not immediately qualified (e.g., untrained in software), it should be determined if the project or overhead should burden the cost of training. This is primarily a cost owned by the engineering firm.
- Rework and Quality Control
  - Based on quality control challenges or the project's inherent complexity, extensive rework would be a reason to recalculate the project budgets. For example, if internal quality control processes (or external jurisdictional reviews) prompt more comments than expected, this could indicate that the ETC should be recalculated.
  - *Mitigation:* Any future quality issues are mitigated by a root cause analysis, which ideally solves the source issue to improve future production. This analysis may determine if staff require training. Or, the cause may be external, where rework and late changes are prompted by communication challenges with the client or because task workflow is inefficient (e.g., a mandatory dependency that is missed).

- Scope and Schedule Changes
  - Whether proactive or reactive, a change in schedule or scope will likely affect the project costs. The scale and timing of the scope or schedule change will influence how the budget is modified, which should be identified with change orders that describe scope or schedule modifications. For example, adding a new project phase while accelerating the schedule will increase costs and prompt the need for a new ETC.
  - *Mitigation:* These cost modifications are established through change orders, which would document the increase or decrease to the project budget. The exception is when a scope item was unintentionally omitted, or the schedule was miscommunicated. A change order between the client and engineer would not exist in this case. Mitigation would require the project manager to evaluate why the scope was omitted, and if there are any budget reserves or opportunities to improve efficiency.
- Estimating Errors
  - The original budget may have erroneously estimated the project costs. This would be identified while reviewing the historical financial performance, indicating that the actual costs are misaligned with the earned value. The disparity may arise due to either a computational mistake or the complexities associated with estimating a unique or high-risk project, especially when the project scope is ambiguous or poorly defined.
  - *Mitigation:* If the error stems from an unclear scope item or a disagreement regarding exclusions, the engineer and client should openly discuss the source of the issue. This may necessitate a change order to adjust the project scope and budget accordingly. Ideally, the project budget should include contingencies or reserves, especially for high-risk projects that are more susceptible to estimating errors. These reserves act as a buffer to absorb unforeseen costs. When T&M fee structures contribute to estimating errors, the project manager should proactively communicate with the client. If, for example, meetings become more frequent or extend longer than initially anticipated, the client should be notified as soon as the pattern emerges. This allows for recalculating the budget estimate and ensures transparency in cost management.

- Project Delays
  - A delay in the project can affect the budget even if the scope remains constant. If a schedule delay occurs before the project starts, a cost impact can be caused by economic conditions (inflation) or personnel costs (salary adjustments) that were not accounted for. For example, if the project manager is awarded the project but the notice to proceed is delayed by a year, the project costs would likely differ. Similar to a delayed start date, a challenge could also come from a mid-project delay. In this case, economic inflation and salary compensation are still factors, but the cost associated with ramp-up is a new issue. Ramp-up cost (and time) is required because an extended departure from the project will erode knowledge and halt the momentum that the team had previously established.
  - *Mitigation:* Many contracts or proposals include provisions for inflation adjustments based on time limits. For example, a price proposal or the billing rates are updated annually and reissued to the client – or the client may establish allowable rate adjustments through an escalation clause (noting that the project fee is only viable until a specific date). A delay for an in-progress project would require a discussion between the engineer and client to develop a mutual understanding that this clause would apply. The team should follow a procedural documentation process to mitigate knowledge erosion during delays. These delays may also prompt other issues, such as personnel availability and rework caused by external factors (such as policy changes), which could prompt a formal change order.

As highlighted in this section, various disruptions – from changes in personnel availability to unexpected rework and scope alterations – necessitate a reassessment of the ETC. However, recalculating the ETC is not just a mechanical task; it is meant as a proactive measure that aids in identifying potential budget overruns early, thereby allowing timely corrective actions. The process of recalculating the ETC due to the disruptions listed in this section does not imply that a client is obligated to change the budget or compensate the engineer. Determining the ETC begins as a responsibility of the project

> Recalculating the ETC due to the disruptions listed in this section does not imply that a client is obligated to change the budget or compensate the engineer.

manager and, depending on the circumstances, could warrant a discussion with the client and potentially a change order. The practical application of these mitigation techniques requires a dynamic approach where project managers continuously evaluate project progress, scope, and schedule against the original estimates.

In Appendix A, Part 7 of the Millbrook Logistics Park, the engineering team evaluates the current project expenditures and the appropriate invoicing for the client. The concepts of cost and earned value are considered based on the estimated measure of completed work by referencing an example invoice proof. Based on the cost of completed work and a discussion of earned value, the engineering team establishes an appropriate billing amount for the client. The engineering team also forecasts costs for several tasks based on effort spent to date and estimations of remaining work.

CHAPTER 8

Risk Management

What could go wrong?

CHAPTER OUTLINE

Management Essentials for Civil Engineers: A Practical Guide to Business, Communication, Ethics, and Risk, First Edition. Cody A. Pennetti, C. Kat Grimsley, and Brian M. Grindall.

Risk management is interconnected with all other project management principles identified in this book. Risk affects planning, procurement, scope, quality, leadership, and schedules. This chapter introduces risk concepts and terminology before describing policies and strategies for risk management, risk assessment, and risk analysis. This content helps project managers to effectively identify, evaluate, and respond to risks.

8.1 Introduction

Risk is inherent to all projects and organizations. It is especially prevalent in civil infrastructure and real estate development projects due to the size, complexity, and unique parameters for each project site coupled with the dynamic nature of economic, sociocultural, and natural environments that can impact the project's development and its end users. Each street, structure, bridge, utility, wall, and element of the built and natural infrastructure has inherent risk that affects human life. The choices made during the planning and design of these civil engineering systems carry the weight of consequences – for better or worse – and can include environmental harm, monetary costs, or a direct impact on community health and human life.

In this way, risk is a broad concept underlying many of the challenges and opportunities encountered by projects and operations that require evaluation from multiple perspectives based on competing objectives. Should vehicle mobility be prioritized over pedestrian convenience? How will the project team's design choices impact the lives of the community members? When determining a risk-cost-benefit analysis, is it morally acceptable to assign a value to life? What benefit will the project provide to the client, to the users, and to other stakeholders? What are the trade-offs – when one stakeholder benefits, who is burdened? Ultimately, who is accountable for the consequences?

These rhetorical questions demonstrate that risk management, assessment, and analysis is a vast field applied to all project types and industries and even personal conditions. This chapter will use industry-standard terminology and describe methods tailored to civil engineering project management. Project managers must delve deeper into the extensive and multifaceted field of risk management beyond this initial review. The material presented in this chapter serves as an introductory overview, offering fundamental insights into risk management.

Ultimately, project teams must acknowledge the reality of limited resources in their projects and environments. Given these constraints, mitigating every conceivable risk is neither feasible nor practical. Instead, the primary role of risk assessment is to guide stakeholders in allocating limited resources toward risks that pose the greatest threat to project objectives. Additionally, project teams must recognize the dynamic nature of their operating environments. Continuous risk assessment is essential throughout the project life cycle to understand how emerging conditions might shift risk priorities. This ongoing process enables teams to reallocate resources effectively, ensuring they are directed towards newly prioritized risks in response to the evolving project landscape.

Mitigating every conceivable risk is neither feasible nor practical.

It is critical to note that, for the same project, clients will have different metrics and perceptions of risk than engineers. For example, a commercial developer will be concerned with a range of risks beyond the design process, which can impact the successful acquisition, financing, construction, leasing, operation, and cash flow of a commercial real estate asset. These various risks are often beyond the control of the developer but can, to some degree, be addressed through contract terms to allocate, schedule, or otherwise manage the risk and any associated liability. Chapter 4 covers key points on contracts and liability.

8.2 Principles of Risk Management

In the civil engineering context, risk manifests across the many multidisciplinary components of the project operations, including procurement, development, construction, management, finance, leasing, and related ownership settings. It is also necessary to recognize that risk influences projects and more extensive operations, such as programs and portfolios. Still, risk will influence the client's operations, the engineering firm, team members, and other stakeholders. Therefore, a comprehensive risk management strategy for a single project must also consider the holistic view. This approach should account for the collective risks that influence all participants and phases of the project life cycle. Further, risk management must be recognized as a continuous effort, carefully evaluated, and monitored throughout the project life cycle. By embracing this inclusive perspective, engineers and project managers can develop robust risk mitigation strategies that safeguard the interests of all involved parties and ensure project resilience.

> A comprehensive risk management strategy for a single project must also consider the holistic view.

Risk is a multifaceted concept, encapsulated by varying definitions from various sources:

- *Black's Law Dictionary* frames it as "the hazard of property loss . . . or the degree of such hazard," highlighting potential loss as a risk factor.
- *Barron's Dictionary of Finance and Investment Terms* defines it as a "measurable possibility of losing or not gaining value," distinguishing risk from the immeasurability of uncertainty.
- *Barron's Dictionary of Real Estate Terms* discusses risk as "Uncertainty or variability. The possibility that returns from an investment will be greater or less than forecast."
- Project Management Institute (PMI) provides a definition that perhaps resonates most closely with engineering \: "Risk is an uncertain condition that, if it occurs, has a positive or negative effect on one or more objectives."

Indeed, risk is the complex interplay of uncertainty, decision-making, trade-offs, and interdependencies, investigating how decisions today will influence future outcomes in a system. In project management, risk and uncertainty are distinct concepts. Risk refers to situations where the probability and impact of outcomes can be quantified, allowing for systematic assessment and management. By contrast, uncertainty arises when the probability and impact of outcomes are not known or cannot be accurately measured, complicating strategic planning and response. This does not imply that uncertainty is devoid of risk. Rather, uncertainty introduces scenarios where traditional risk metrics are ambiguous or unavailable.

Clients may not always be able to use a quantitative approach to risk discussions or mitigation across their larger project perspective. These interpretations underscore the multifarious impact of risk on projects. Identifying and comprehending risks is a critical step, yet it is merely the starting point in the comprehensive discipline of risk management.

Although risk is often associated with undesirable events, there are two types of risk: threats (adverse risk) and opportunities (positive risk). Adverse risk is an uncertain future outcome that can threaten loss or inhibit gain; it is something to avoid, reduce, eliminate, or mitigate, whereas positive risk may be pursued. Regardless of the type, when risk is embraced and encountered deliberately, businesses can avoid adverse consequences and exploit risk to promote opportunities.

There are two types of risk: threats (adverse risk) and opportunities (positive risk).

Risk management must extend beyond project-based risks and consider broader perspectives regarding the appropriate level of support and action for potential risks. It is a continuous effort required to adapt to the dynamic environments of the project life cycle. In this way, risk management aims to influence the probability and impact of risks, whether seeking to increase a positive outcome or reduce the magnitude of threats. With proper implementation, there is a better chance that the project, stakeholders, and the organization will be successful.

In a vacuum, evaluating probability without considering the associated impacts (or vice versa) can result in extraordinary risk conclusions. For example, imagine a meteor striking the earth precisely at the project site. Such an event would result in a significant impact warranting response actions; however, such an event is highly improbable (although statistically possible), and the project team would be well served to progress through the project as planned and not spend resources thinking of ways to avoid potential meteor impact zones.

8.2.1 Frameworks

Risk management encompasses the entire process of monitoring and controlling project risk. *Risk assessment* falls under the umbrella of risk management and is concerned with identifying and understanding risks. *Risk analysis* is a part of risk assessment, focusing specifically on evaluating the probability and impact of the identified risks. Together, these concepts create a framework for civil engineering projects to navigate the inherent complexity and uncertainties of projects. As established by Kaplan and Garrick, risk management begins with three questions:

- What can go wrong?
- What are the likelihoods?
- What are the consequences?

The answer to *What can go wrong?* helps inform risk identification, which can come from various catalysts or be experienced through non-event-based project conditions (such as inherent variability in project performance that can lead to errors). The questions regarding likelihood and consequences focus on the process of identifying and performing a qualitative and quantitative analysis of the risks. Likelihood establishes forecasts of the probability that a risk will occur, while consequence addresses the costs, whether monetary, sociocultural, environmental, or other, that impact the project objectives. However, these reflections fall short of a complete assessment. From here, Haimes establishes that risk management requires three additional questions:

- What can be done?
- What are the trade-offs?
- What are the impacts of the current decisions on future options?

Risk management considers what can be done for the risks identified through risk assessment, starting with the question: *What can be done?* This prompts an evaluation of the appropriate risk responses, such as mitigating, accepting, avoiding, or exploiting. It is essential to understand that project stakeholders have limited time and resources, so the risks must be prioritized, often leading to trade-offs. These trade-offs consider (a) what benefit is achieved at the cost of the risk response and (b) who benefits. The final question considers the impacts of the current decisions on future options, which acknowledges that project choices and implementation of risk responses can limit future options and introduce new constraints (or, potentially, new opportunities). The initial chronology of risk management and the integrated framework is shown in Figure 8.1.

This risk management framework is integrated into all project activities and must be monitored throughout the project life cycle. In general, the risk management process progresses once the risk management strategy has been established:

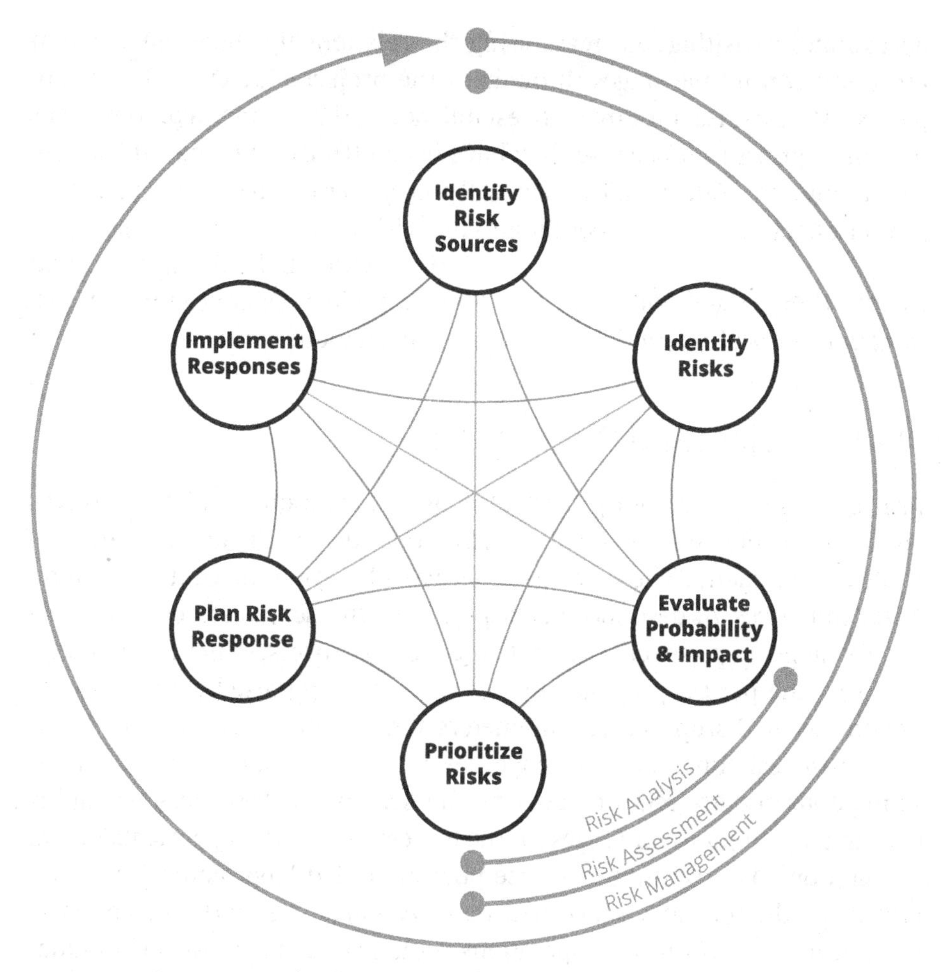

FIGURE 8.1 Interconnected risk management framework.

1. Identify the risk sources, often through prompt lists and organizational historical data.
2. Identify the risks based on the risk sources.
3. Analyze the risks to evaluate the probability and impact.
4. Prioritize which risks to focus on with risk response initiatives.
5. Plan the appropriate risk response initiative based on the priority risks.
6. Implement risk response initiatives.

While the risk management process starts as a sequential series of steps, it becomes a dynamic framework once the project begins. This approach necessitates revisiting and reassessing the relationships between different processes at multiple stages throughout the project's life cycle. For example, as risk response initiatives are established and implemented, some risks may be deprioritized because they have been effectively managed (but are still monitored). Additionally, risks will emerge based on external factors or project operations, and responding to one risk may uncover or create new risks. Ultimately, the team should seek project resilience through risk management strategies.

Responding to one risk may uncover or create new risks.

8.2.2 Project and Product Risk

Risk is considered across project objectives, operations, and the project's resultant deliverable and product (i.e., the constructed infrastructure and real estate development). These perspectives extend beyond the confines of individual initiatives. As described in Chapter 5, the two scope categories of a project also apply to perspectives of risk: the project processes and the product.

From the project process perspective, risk is appraised with respect to its potential to disrupt project parameters and initiatives. This includes risks that impact schedule, budget, scope, resources, and quality, among others. Ambiguous project requirements or the scarcity of resources exemplify risks that could pose challenges for the project team and impact stakeholder expectations. For example, a scarce pool of skilled labor could jeopardize project production and scheduled delivery dates. Similarly, mismanagement of resources or project operations could lead to poor cost performance and an unprofitable project for the engineering design firm.

Alternatively, from the product perspective, risk pertains to the operational performance of the constructed infrastructure or the real estate development. It plays a role in shaping the design and the scope of work. For instance, clients may opt for a design emphasizing resilience, such as advanced stormwater management systems or fortified structural components, to offset adverse environmental events. This choice underscores a preemptive stance against anticipated adversities to the constructed product.

This book acknowledges both product and project risks but emphasizes project risk more. This emphasis provides broader relevance across various engineering project management responsibilities and underpins the necessity for a robust risk management strategy that promotes project success despite the uncertainties.

8.2.3 Impact and Probability

Although it has many parameters, risk is most often measured by the impact of an event and the probability that it will occur. The impact of a risk is the consequence associated with a project operation, objective, or component. For example, if the event occurs, will there be an advantageous opportunity or an adverse effect on the schedule, quality, safety, or cost? In this way, risk impact is multifaceted, and the measured impact is influenced by the project objectives (such as a client's prioritization of minimizing cost, staying on schedule, or others). Probability is complex, both in mathematical application and in a qualitative representation, and is often categorized as classical, statistical, or subjective probability:

- Classical probability considers the potential for an outcome within a sample space, such as the likelihood of an event occurring several times through observed occurrences – a coin toss, for example.
- Statistical probability is based on recorded data (empirical) and used to predict the likelihood of a future event. For example, civil engineering hydrological models often use recurrence intervals, such as a 100-year storm referring to a rainfall event with a 1% chance of occurring in any given year (a 1 in 100 chance).
- Subjective probability is based more on judgment and is often reserved for conditions where there is no historical data, if information is not available or too complex to evaluate. For example, if an engineering firm estimates future revenue, it may assign a probability of winning certain projects.

In general, risk probability expresses the likelihood that an event will occur with an associated impact. Figure 8.2 provides an industry-standard classification of risks based on varying knowledge of risk impacts and the probability of occurrence, as classified by unknown-unknown, known-unknown, unknown-unknown, and known-known.

The top right of Figure 8.2 depicts the desired classification for risks, those that have a known consequence and probability of occurrence. Examples of risks associated with these classifications are as follows:

- A known-known risk is project knowledge, a documented fact about the condition of a project. For example, a project team knows that inclement weather will occur during construction and will disrupt activities.

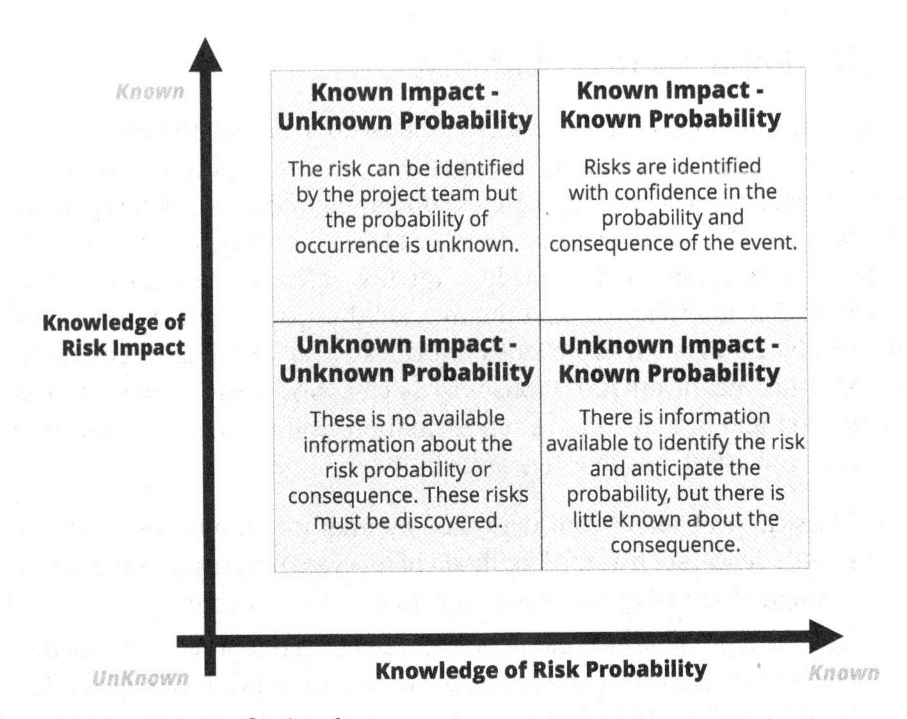

FIGURE 8.2 Risk classification chart.

This disruption can be measured (construction delays) and published information is available to the project team to understand the likely number of inclement weather days that can be expected each year.

- With an unknown-known condition, the probability of a risk can be measured, but there is little information on the impact. For example, severe weather events, such as 100-year storms, have a known probability (1% chance of occurrence each year), but if the event is not modeled, then there is no knowledge about the impact.
- A known-unknown risk has a known or expected impact, but there is no information about the probability of the occurrence. Suppose the project team has identified a risk that a critical material is available from only one supplier. In this case, the team can determine how losing that supplier would impact the project schedule and price, but the team would not know the probability of the supplier going out of business.
- Unknown-unknown conditions are best managed through organizational resilience because the team does not (and cannot) know the risk,

impact, or probability of the occurrence. For example, an earthquake in a region without any documented seismic activity could be catastrophic and not considered by the project team.

The project team, with support from the organization, should seek to move risks to the known-known classification. This effort of shifting risks into the known-known classification underscores the proactive nature of effective risk management. It involves strategies such as continuous monitoring, gathering expert opinions, conducting thorough research, and employing advanced predictive models. By doing so, project teams can convert unknowns into knowns, reducing uncertainty and enhancing their influence over project outcomes. However, this shift necessitates the allocation of resources and time, underscoring the need for strategic decision-making in determining which risks to prioritize for management.

It is important to note that at the larger development project level, clients will likely face conditions where subjective probability is paired with unknown-knows and known-unknowns. For example, in the public sector, the client may not know if budgets will be cut, especially after an election year, which can create problems without dedicated funding. In the private sector, a developer may anticipate that a planning committee will likely request changes to a proposed mixed-use commercial project in response to community feedback. While the client and engineering team can build relationships with these stakeholders and will typically be aware of points of contention, it is often impossible to know what changes will be requested, exactly how they will affect the project, and how long it will take to resolve them. Future economic and market trends can also create subjective risk for private sector clients, especially since development projects take years to deliver.

8.2.4 Magnitude

When evaluating risks, project managers must recognize the relationship between impact and probability, which is defined as *risk magnitude*. Risk magnitude is critical in risk management and assessment because impact and probability are not directly correlated; they can vary independently. Focusing solely on impact could lead to the oversight of less severe risks, which, over time, could have a cumulative effect just as detrimental as higher-impact risks. Certain risks might carry the potential for significant disruption, yet their low probability renders them less immediate concerns.

On the other hand, more common risks with predictable, manageable impacts may not seem pressing, but their frequent occurrence could pose a consistent threat to project stability and performance.

For example, consider the risk of a catastrophic failure of a significant piece of construction equipment, like a crane. While the impact of such an event would be substantial, potentially causing injury and significant delays, the probability could be very low (but still possible). Conversely, the risk of minor delays from supply chain issues might have a much higher probability but with a potentially lower impact, as these can often be anticipated and planned for with minimal disruption. Note that, depending on the commodity impacted and length of delay, supply chain disruptions can also have catastrophic consequences for a project if they adversely impact the critical path during construction and delay the building's delivery, or resulting in price spikes leading to material increases in overall project costs. The specific condition must be evaluated, and the probability and impacts will inform the appropriate risk response initiative.

By considering both dimensions – impact and probability – a more nuanced understanding of risk emerges. Risk magnitude explores how potential events will disrupt the project objectives. This understanding aids in allocating resources, ensuring that mitigation efforts are balanced and proportionate to the risk. It also underscores the importance of a dynamic risk management plan, which is sensitive to the evolving nature of risks throughout the project life cycle. Moreover, this dual assessment facilitates the development of a risk matrix that helps visualize where each risk falls in terms of priority, guiding decision-making processes (as shown in Section 8.4). Each risk is mapped to a matrix cell based on the associated probability and impact, which then informs the magnitude of the risk.

Emergent conditions can change the impact or probability of a risk event or uncover new risks throughout the project life cycle. External factors, such as technological advancements, regulatory changes, political shifts, sociocultural dynamics, and other sources, can influence risk magnitude because the impact or probability of a risk changes. While any single event may not severely disrupt the project, the aggregate effect could be substantial, mandating a risk-management strategy that includes adaptive scheduling and budgeting. It is also crucial to develop contingency plans for both high-impact with low-probability events and low-impact with high-probability events, as both can derail a project if not adequately managed.

8.2.5 Risk Parameters

Impact, probability, and the resultant magnitude are the standard parameters associated with risks; however, there are several other parameters that are commonly evaluated. Figure 8.3 provides a summary of these risk parameters.

Most notably, temporal domains play a critical role in risk management. The assessment and management of risk must consider when the risk will occur to determine the impact in relation to the current operating status. Additionally, the timing of a risk event can influence the available response and mitigation measures based on schedule, budget, and resource availability. Further, the occurrence of a risk or the risk management strategies may result in trade-offs for future events. Changes in scope are a good example of how temporal domains influence risk – a scope modification early in the project is easier to accommodate compared to a late change, which would

Temporal Domain
Recognizing how the timing of an event will influence the magnitude.

Perception
How risks are perceived by stakeholders, which may influence prioritization regardless of probability and impact.

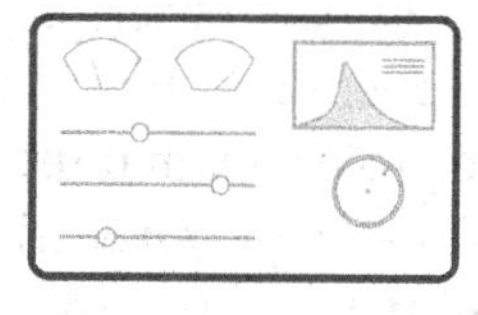

Control
The degree to which a risk can be managed or controlled, given the appropriate resources.

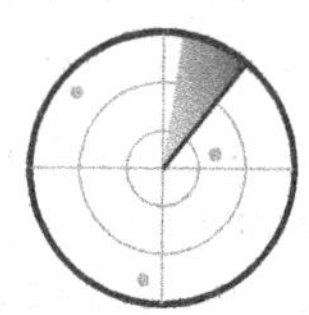

Detectability
The ability to observe the occurrence and impact of a risk event.

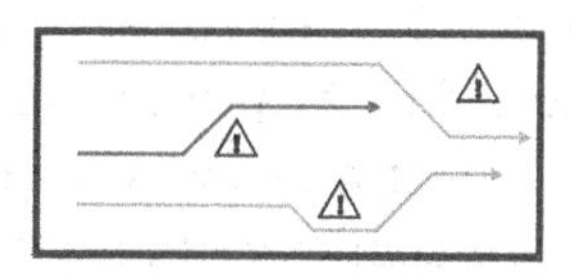

Strategic Objectives
The magnitude of the risk beyond the individual project as it relates enterprise operations.

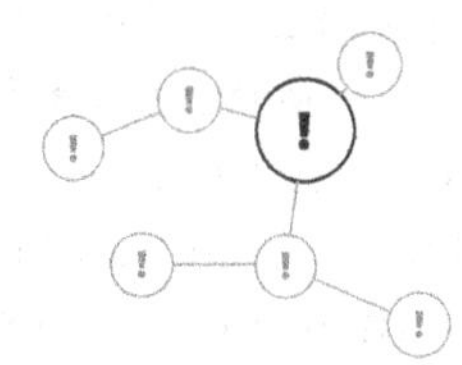

Connectivity
How one risk may influence the probability and impact of other risks.

FIGURE 8.3 Risk parameters.

require rework, compromise quality, and pressure the budget and schedule based on a limited choice of resources.

Understanding the perception of risk is important for managing stakeholder relationships. If there is a lack of knowledge (or confidence) of the risk impact or probability, stakeholders' perception of the risk could determine the prioritization and resource allocation. For example, if the community is unfamiliar with hydrologic modeling, there may be a perceived risk that a project will adversely affect flooding conditions, regardless of how the computational models estimate the impact. Often, these perceptions are shaped by the overarching goals of the engineering firm or client, which might not always be apparent to the project team.

It is crucial to recognize that a risk may have a different degree of influence on the broader strategic objectives than on day-to-day project operations. This could mean that while a design flaw (risk) might slightly delay the project schedule, it could significantly undermine the client's confidence (strategic objective) in the engineering team. It is also helpful to realize that a "slight delay in schedule" could represent a significant cost to the client and impact other project deadlines, depending on the specific context, so the client may have a different risk interpretation of the same event. Hence, the project team must assess their capabilities and authority to manage such risks – do they have the resources to manage project quality effectively? The ability to control risk is another aspect that extends beyond general awareness. The project team must evaluate whether they possess the capability and authority to manage the risk effectively.

Detectability is another critical consideration – what is the team's capability and velocity to identify a risk once it materializes? For example, is project data available to detect if tasks are being completed at the planned rate and within budget? The primary risk is that a project falls behind schedule and exceeds budget, but identifying the risk requires resources and performance data to be communicated to the project manager. This is particularly relevant for dormant risks, which may evolve over the project's life cycle, and their prompt detection is essential to mitigate their potential impact.

Risks do not exist in isolation; they are interrelated.

Finally, risks do not exist in isolation; they are interrelated, and the occurrence of one risk can create new vulnerabilities or exacerbate the probability or impact of others. For example, the failure to secure a construction permit on time may delay the project construction start and lead to cost escalations (risks), thereby damaging stakeholder relations (new vulnerabilities). This could ultimately increase the risk of contract termination (critical event). This interconnectedness and interdependency (connectivity) of risks must be carefully evaluated to understand the entire landscape of potential challenges and to develop a comprehensive risk management strategy.

8.2.6 Risk Appetite

The concept of risk appetite and tolerance plays a critical role in project operations. Organizations must define their willingness to bear risks and at what point they are deemed unacceptable. Should the project spend $4,000 to mitigate a risk with a 20% probability of causing $20,000 of monetary loss? This risk appetite will determine whether an organization or project team is willing to take on a risk. This determination directly influences decision-making, resource allocation, and the formulation of risk responses and contingency reserves.

A client's risk appetite directly influences the scale, scope, and ambition of the project. For instance, a client with a high-risk appetite might pursue an innovative building design with cutting-edge features, like a high rise with a complex structural system or a bridge using new materials. They are willing to accept the higher risk associated with such work in exchange for the potential rewards of efficiency, market recognition, and long-term cost savings. On the other hand, a client with a low-risk appetite would prefer more traditional, time-tested methods and designs that ensure predictability in costs, timelines, and outcomes.

From the civil engineer's perspective, an engineer with a high-risk appetite might be more inclined to recommend and design projects incorporating novel technologies or unconventional methods that have not been extensively tested. They are prepared to navigate the uncertainties for potential gains in efficiency, performance, or cost-effectiveness. For example, new design technologies might show promise, but are unfamiliar to the team and involve tools that are untested. Conversely, an engineering firm with a low-risk appetite will advocate for design methods based on established practices and standards. They prioritize reliability and are more likely to recommend conservative design margins and robust contingency plans to mitigate potential risks.

8.2.7 Responsibility of Risk Management

As with any project process, team members must be aware of their risk management responsibilities. Chapter 6 (Section 6.5) includes a description of how the responsibility of project tasks should be attributed to project team members by using a RACI chart (responsible, accountable, consulted, or informed) – this tool can also be used for risk sources and events.

For instance, if community opposition is identified as a low-priority but high-impact risk, the team may assign the responsibility of monitoring that risk to one team member. In this way, there is a clear direction for the expected action taken from team members to continuously monitor the risk and keep others informed if there are changes in risk parameters.

8.2.8 Personal Risk

This book focuses primarily on project-based risks, which can escalate to risks that impact the program, portfolio, or organization; however, engineers should be mindful of personal risks associated with project operations. Personal risks to individuals involved in a project, such as a project manager, extend beyond the project tasks and can impact their professional and personal well-being. For instance, a project manager may face safety risks, especially in overseeing construction sites or large-scale engineering projects, where they are regularly exposed to hazardous working conditions. Additionally, their professional reputation is at stake; mismanagement of a project, failure to adhere to ethical standards, or inability to meet deadlines can lead to long-term career repercussions. Ethical risks are a significant concern, as decisions made under pressure or without proper due diligence can lead to breaches of professional conduct, potentially resulting in legal consequences and damage to personal integrity.

These personal risks underscore the importance of comprehensive risk management strategies encompassing project-related and individual risks. It is imperative that engineers not only monitor these personal risks but also report and communicate them effectively. Such diligence in risk communication is essential

It is imperative that engineers not only monitor personal risks but also report and communicate them effectively.

for safeguarding both individual well-being and the success of the project, upholding professional standards, and maintaining personal integrity in the face of potential challenges.

8.3 Risk Assessment

Approaching risk deliberately with the intent to respond proactively requires a methodical process of identifying and correlating risks. *Risk assessment* includes identifying and correlating risks among various project activities such as procurement, design, construction, financing, and operation. In this way, risk assessment is identifying and evaluating events (i.e., possible threats and opportunities) that could affect the achievement of objectives and project success.

Risk assessment requires an understanding of likelihood across the timeline of a project. It considers past events that could influence current conditions, present factors actively shaping outcomes, and future occurrences yet to unfold. In this way, risk assessment considers how current actions will influence the operations of future conditions, often with varying trade-offs.

For instance, consider a situation where a client attempts to navigate ambiguous regulatory codes to minimize costs or expedite the design process, potentially compromising the project during quality control reviews in the design or construction phases. This could manifest as the client advocating for less stringent safety measures or materials that seem to meet minimum code requirements. The trade-off in this scenario might involve deciding between adhering to the highest safety standards, which could increase costs and extend design timelines, versus opting for a quicker, less costly approach that risks noncompliance or future operational issues. The project team must consider how current decisions impact the project's integrity or lead to costly rectifications in later stages, while still respecting the client's instructions. If an engineer believes a situation will potentially create ethical, safety (beyond minimum code requirements), or other problems for the project, the issue should be raised internally and may require formal discussions with the client. Chapter 9 includes best practices for how to clearly document and communicate difficult topics.

Risk assessment allows stakeholders to view uncertainties more deliberately and then respond to the threats and opportunities that may arise. The risk assessment process consists of three components:

1. *Identify the sources of risk.*
2. *Identify risks.*
3. *Evaluate probability and impact.*

The first two components consider the methods to identify various sources of risk. As risks are identified, the next step includes risk analysis, which involves an evaluation of the probability and the associated impacts of risks to understand the magnitude. In the context of risk management, the risk assessment must consider several risk parameters, including connectivity, timing, and controllability. Risk assessment is the starting point for risk management, leading to prioritization and risk response implementation. These processes are repeated through the project life cycle as risk parameters and project operations change.

8.3.1 Identifying Sources of Risk

Risk identification in civil engineering and real estate development begins with a broad exploration of potential sources of risk. These risks can arise from various aspects of a project, encompassing technical challenges, market fluctuations, inherent performance variabilities, legal issues, and more. Understanding these diverse sources is crucial for developing effective risk management strategies.

A standard strategy to identify and organize risks is with a risk breakdown structure (RBS), which shares similarities with a work breakdown structure (WBS) described in Chapter 5 (Section 5.5). Sources of risks can begin with high-level categories, often referenced through prompt lists (there are several common prompt lists used in the industry). For example, the prompt ELEMENTS references Economic, Legal, Environmental, Market, Enterprise, National Governance (political), Technical, and Sociocultural. The project team uses these high-level categories to facilitate risk identification practices, such as interviews, root cause analysis (described in Chapter 5, Section 5.7.3), brainstorming, and others. An example of the ELEMENTS prompt list is included:

- Enterprise
  - This considers risk sources from the organization, such as corporate management, processes, corporate strategy, project policies, and similar others. Enterprise risk sources can also include personnel risks based on skills and availability.

- Legal
  - Risks related to contracts, legal, and regulatory aspects, such as changes in laws, zoning requirements, contracts, and other regulatory frameworks. A common example includes modifications to building codes necessitating last-minute design alterations or the challenges posed by ambiguous scope that are subject to varying interpretations.
- Environmental
  - Pertains to risks from natural and environmental factors; the risks are distinct but tied to environmental regulations and legal risk sources. This could include the risk of discovering protected wetlands, evaluating floodplains, identifying endangered species, and the requisite legal conditions to protect these natural features.
- Market
  - Any market conditions that could influence project design or the anticipated performance of the product, such as demand, competition, and industry trends.
- Economics
  - The financial aspects or cost considerations, including project funding and cost performance, as well as larger national and global economic considerations (inflation, interest rates, exchange rates, etc.).
- National Governance
  - This source ranges from the local to national levels, including government policies, political stability, and public-sector initiatives. This is connected to legal sources, as laws can change with new elected officials.
- Technical
  - Technology, innovations, and processes could been see as a risk to the project. These could include new design technologies, contract requirements for specific design tools, or the processes applied to the work productions.
- Sociocultural
  - Encompasses risks from social and cultural aspects, which includes a focus on community relations and the project's social impact. For example, community perception of the project, media coverage, and others.

Prompt lists, such as ELEMENTS, help facilitate conversations that lead to risk identification through processes such as an RBS (an example is shown in Table 8.1). Further, many agencies or private firms will have risk management policies that identify their priority risk sources. For example, an organizational root cause analysis may document common project risks based on prior issues. Similarly, other projects may maintain a lessons-learned register with a retrospective of project challenges and opportunities realized throughout the project. Finally, the projects are likely part of programs and portfolios, which could have their own risk sources and risks identified. It is important to realize that a client's perception of the same ELEMENTS categories will be applied at the larger project level and will be different, although often overlapping, with an engineering assessment.

Each stakeholder group can influence, contribute to, or help manage project risks in different ways.

Once risks are identified, understanding the role of stakeholders becomes essential. Each stakeholder can influence, contribute to, or help manage project risks in different ways. Stakeholders in civil infrastructure and real estate projects span a wide range. The stakeholders contributing to a project's success often include sellers, buyers, government agencies, end-users, tenants, tenant customers, neighbors, investors, lenders, architects, engineers, and construction contractors. More information on stakeholder relationship management is included in Chapter 10 (Section 10.7).

The role of each stakeholder is a threshold consideration within the civil engineering risk assessment process. Stakeholders collectively include many participants with interests ranging from significant to trivial. For example, a project may require an easement across an adjoining property, which may be crucial to the project if the neighbor's willingness to grant an easement is required. However, if the project can be redesigned to avoid such an easement, the neighbor's participation may no longer be necessary for project success. In both cases, the neighbor is a stakeholder; however, the neighbor's actions carry a greater risk in the former compared to the latter scenario.

8.3.2 Identifying Risks

An event that initiates a risk is distinct from assessing the magnitude of the disruption caused by a risk. Recall that the accepted definitions of risk describe the *effect on objectives*. The risk itself is different from the event, or events, that could initiate a risk – many disparate events could generate similar risks. For example, the risk of delays in construction materials could be caused by events such as road closure, weather delays, delivery vehicle crashes, and manufacturing issues. With any of these events, the focus remains on the effect on project objectives. This understanding must also encompass non-event risks, like the variability in project operations or the ambiguity of stakeholder requirements, which can be just as influential as discrete events. Identifying risk events promotes essential discussions and actions in risk management; however, the emphasis remains on evaluating how a risk can disrupt priorities in project operations.

The emphasis remains on evaluating how a risk can disrupt priorities in project operations.

Critical events in risk management create risks that significantly derail a project from its objectives – or, in the case of opportunistic risk, the event may propel a project toward its objectives. It's essential to discern which events could initiate risks that severely impact the project's success so the project team can prioritize risk response strategies. For example, failing to pass a critical building inspection could halt the entire project. In contrast, minor cosmetic building flaws, though important, do not pose a severe threat to the project's timeline or financial outcomes. Prioritization helps focus risk management efforts on where they can have the most significant impact on maintaining project integrity.

Civil infrastructure and real estate development projects progress through distinct phases marked by a series of milestones and critical events. These encompass scheduled events, such as groundbreaking ceremonies or progress review meetings, as well as unforeseen occurrences, like an unexpected utility discovery during excavation. Project phases, each with a defined start and finish, range from pre-construction activities like obtaining permits to phases like site mobilization. While some phase transitions are fixed – for example, the completion of a building's foundation – others depend on variable factors, such as deciding the optimal time to initiate a property sale based on market conditions.

In this dynamic timeline, forecasting becomes a pivotal tool in risk assessment, differing from prediction by dealing in probabilities rather than certainties. In civil engineering, forecasts might include estimating the likelihood of cost overruns based on current design-phase performance, predicting construction delays due to historical weather patterns, or projecting market trends during the leasing phase. These forecasts, grounded in data and trends, help evaluate scenarios that feed into risk management. They guide the prioritization of project initiatives and strategies, taking into account the significance and impact of various risk events.

As the project life cycle progresses, the risk landscape evolves.

As the project life cycle progresses, the risk landscape evolves; early-stage risks like the accuracy of topographical surveys might transition to later concerns, such as compliance with updated building codes or integration of new technology. Recognizing and planning for these evolving risks is vital to maintaining project integrity and meeting the project's objectives.

Strategies to Identify Risks and Risk Events

As noted in Section 8.3.1, RBS is an effective method used to identify risks and risk events. Using a prompt list (such as ELEMENTS) as the RBS Level 1, each sublevel of the RBS would identify additional details on the types of risks associated with each category. This breakdown can continue until risk events are associated with each source. This hierarchical framework helps organize the sources of risks and risk events. In this way, the RBS prompts the project team to consider a comprehensive list. An initial RBS framework will likely come from organizational process assets, as the engineering firm and client have established a standard list based on project experiences and best practices. A sample RBS for the design work of a civil infrastructure project is outlined as shown in Table 8.1.

By using an RBS, the project team can create risk response strategies for each potential risk, as identified by risk sources and potential events, ensuring that risks are managed proactively throughout the execution of project work, such as during early design phases. The expectation is that this list of events is evaluated through risk management techniques to establish a targeted list, prioritized based on the magnitude of the risk and informed by other risk parameters. The RBS should be regularly reviewed and updated as necessary, ideally following a standard cadence, such as conducting a monthly audit.

TABLE 8.1 Sample RBS for the design of a civil infrastructure project.

RBS Level 1 Sources	RBS Level 2 Sources	Risk Events	Risk
Enterprise	Design	• Calculation errors or assumptions lead to inadequate system designs	Schedule delays; system failure from errors
	Technology	• Outdated design software leading to errors • Incompatibilities with software between design teams	Schedule delay;
	Survey	• Omissions in utility survey data leading to rework and construction conflicts	System failures from errors
Legal Risks	Contractual	• Unrealistic scope leading to client disputes • Noncompliance with schedule due to extended jurisdictional review times	Diminished client relationship; schedule delays
	Regularity compliance	• New environmental legislation that creates design noncompliance issues • Unforeseen code changes requiring rework	Schedule delays; System failures from errors;
Environmental	Weather	• Severe weather beyond estimates from typical conditions	Schedule delays
	Geotechnical	• Presences of large quantities of rock • Unsuitable soils, requiring replacement	Schedule delays; profitability
Market Risks	Economic changes	• Recession impacts the client's funding source and ability to pay • Inflation requires scope changes to adhere to the project budget	Profitability; schedule delays
	Supply chain	• Disruption leading to delays in material availability and requiring design alternatives	Scope changes; Schedule delays
	Client needs	• Shifts in client leadership or market conditions change design preferences and methods	Scope changes; schedule delays

8.4 Risk Analysis

Once project risks have been identified, the next step is risk analysis, which is the process of evaluating the probability and impact of risk. Engineers must contend with the reality of finite resources and limited time, making it impossible to address every potential risk. It is essential for the project team to have this mindset before investing in a risk management strategy – this condition necessitates a strategic approach where the selected set of risks (and the subsequent response initiatives) must be prioritized based on their potential influence on the project objectives.

Engineers must contend with the reality of finite resources and limited time, making it impossible to address every potential risk.

Risks must be analyzed based on the coordinated probability and impact (i.e., the risk magnitude). Critical risks that could significantly derail project objectives need immediate attention from risk response initiatives, while less critical risks are monitored for any changes in their status or impact. For instance, what is the probability that a risk will occur and cause the referenced severity of impact? Risk analysis seeks to prioritize those risks with the most significant magnitude considering the anticipated operations and the project dynamics.

Risk prioritization is a critical step that stems directly from risk analysis and is aimed at identifying which risks have the most significant influence on the project. This process involves assessing risks based on their potential impact, likelihood of occurrence, and relevance to the project's objectives to determine their overall threat level or opportunity potential. By doing so, it becomes clear which risks warrant the most attention. Subsequently, the focus shifts toward developing and prioritizing response strategies for these top-priority risks, as detailed in Section 8.6. These response strategies, which may include actions ranging from risk mitigation to outright acceptance, are carefully evaluated and selected based on how effectively they can address the risks, their practicality, and how well they align with the project's overarching goals.

8.4.1 Qualitative and Quantitative Analysis

Impact and probability are expressed qualitatively or quantitatively. Qualitative methods are subjective, as they rely mostly on expert judgment and experiences in lieu of numeric methods. The lack of quantification does not imply low confidence in the risk analysis – instead, it acknowledges the inherent complexity of many risks that cannot be expressed numerically. Historical data may influence the qualitative risk analysis, but it is not used to compute results. In this case, standard practices use measures such as low to high, minor to critical, rare to probable, and similar others.

Quantitative risk analysis is more data-driven, using historical data and statistical modeling to quantify the impact numerically (e.g., costs) and the probability as a percentage (e.g., 30% chance of occurrence). Some quantitative analyses may begin qualitatively based on the availability of information, often during early project phases. The assessment type may evolve as more data becomes available or the project team invests resources into a detailed analysis. For example, if regional geotechnical maps indicate the likely presence of rock on a site, qualitative methods may be used until geotechnical borings gather more information for a quantitative analysis. Most projects will rely on both methods based on applicability for the identified risks. Strategies for both assessment methods are included.

Strategies for Qualitative Analysis

Qualitative analysis is prone to subjective assessments influenced by the stakeholders' perceptions. For this reason, the project team benefits from identifying and documenting those parameters that contribute to the qualitative assessment. For example, a narrative that documents the stakeholder's prior experiences, assumptions, and other information relevant to the qualitative analysis could be included. Diverse perspectives will improve the robustness of the analysis. Many project team members will contribute expert knowledge in the project-specific operations. Still, a supporting facilitator (such as a supervisory engineer from another team) could inform the assessment based on different experiences.

Choosing a qualitative analysis is often based on risk complexity, data availability, resource requirements, and risk characteristics. Table 8.1 includes several risks listed in an RBS that are appropriate for a qualitative analysis. For example, the impact and probability of adverse public perception could not be quantified, so a qualitative analysis based on expert judgment would inform the project team of the likelihood of the event. Expert judgment could initially influence the assessment of adverse public perception, but because each project is unique, there is only an approximate analogous association. Still, qualitative analysis can inform the prioritization of risks to determine how resources should be allocated to the identified risks. The qualitative analysis associates the severity and likelihood of a risk, which is often depicted through probability-impact matrices, as shown in Figure 8.4.

The probability-impact matrix organizes how risks are categorized and then prioritized. Attention is focused on the darker regions of Figure 8.4 based on the magnitude of the risk, which could be a combination of high probability and high impact conditions. Priorities often begin with risks determined to be high probability and impact but extend to those with a high magnitude based on an independent qualitative assessment of probability and impact. For example, risks with high probability and reoccurrence, even with low impact, would also be highlighted with a probability-impact matrix.

While the analytics of a qualitative assessment will not typically inform quantitative project metrics (such as budgets or schedules), the qualitative analysis serves to identify the risk priorities. Again, the dynamic nature of projects requires that these assessments be performed regularly, specifically when there are changes in the project parameters. For example, if the risk for adverse public perception were deemed high probability and high impact, it would warrant a more thorough quantitative analysis, perhaps

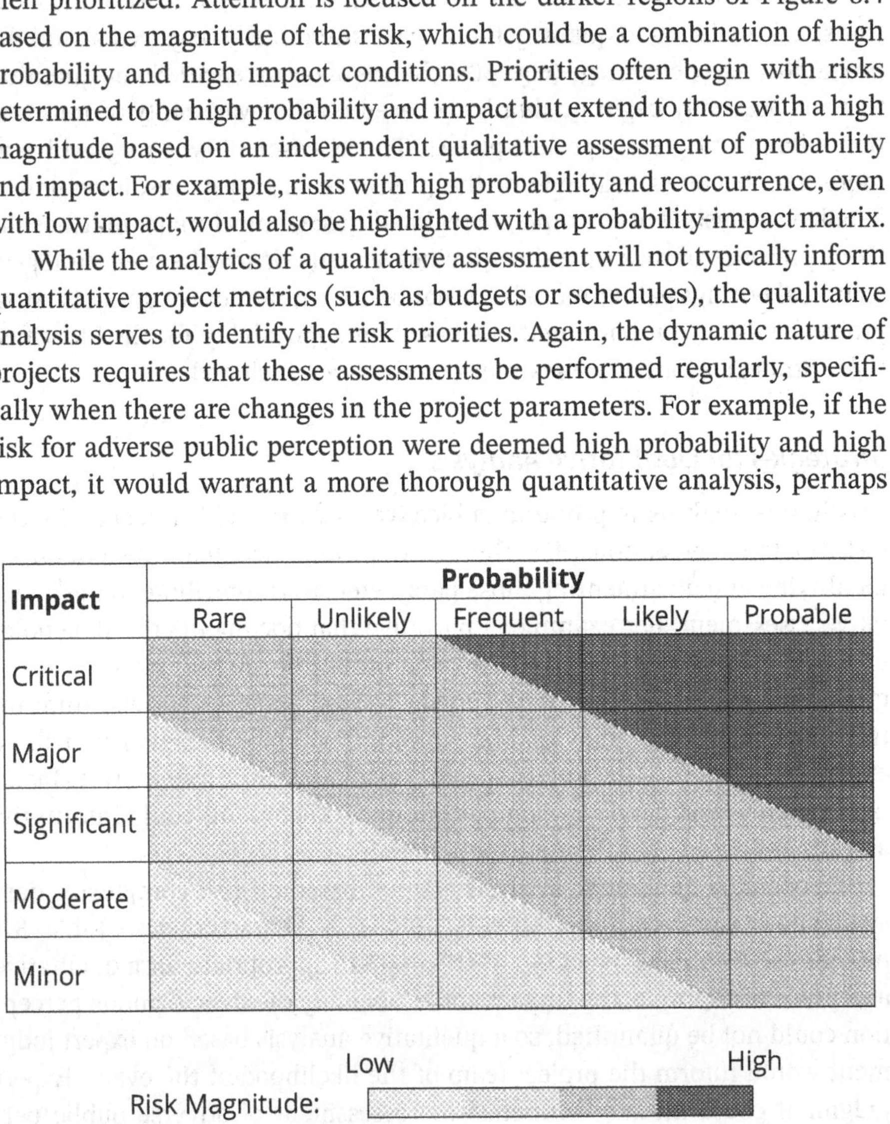

Impact	**Probability**				
	Rare	Unlikely	Frequent	Likely	Probable
Critical					
Major					
Significant					
Moderate					
Minor					

Low High
Risk Magnitude:

FIGURE 8.4 Probability-impact matrix.

with questionnaires or surveys to the community that contribute to a quantitative analysis of the risk. Additionally, this information could inform the project team with strategies, such as communicating with small community groups (e.g., homeowner associations) before a formal public hearing. While questionnaires and additional meetings require project time and resources, this approach can be deemed a worthy investment to mitigate high-priority risks.

Strategies for Quantitative Analysis

A quantitative analysis is more technically complex and relies on more data than a qualitative analysis but may provide a more trusting result (assuming there is confidence in the data and models). Not all risks can (or should) be evaluated with a quantities analysis, and the methods rely on detailed data to be effective. Referring to the sample risk events in the RBS shown in Table 8.1, the risk of design errors provides a good example of a potential candidate for quantitative analysis. However, forecasting the probability and impact of design errors requires the engineering firm to have accurate historical records of prior analogous projects (and project team personnel) that reference the number of errors typically reported and the associated schedule or cost impact.

This example of project errors is a common risk for an industry such as manufacturing, tracking the records of manufacturing defects to forecast future performance. In this way, a series of probability distributions can be evaluated through simulation (e.g., Monte Carlo) to establish confidence thresholds for expected results. Given enough reliable and historical data, these methods could be applied to civil engineering management practices; however, because civil engineering projects operate with unique conditions and are driven mostly by personnel work, there are significant challenges. Still, much civil engineering work relies on numeric analysis, and quantitative risk assessment can influence project decisions.

For example, consider a developer's site located on a beachfront property. This area holds a high value due to its desirability and accessibility to natural amenities, but the inherent environmental risk is not always quantifiable by existing policies or historical data. While some jurisdictions might impose development and setback regulations based on jurisdictional requirements, a stakeholder's risk appetite will influence the design and location of the development. A stakeholder willing to accept higher risk might choose to maximize the use of the property and build as close as possible to the shoreline (maybe while incorporating innovative coastal defense technologies). Conversely, a stakeholder with a low-risk appetite might decide to set back and elevate the development from the shoreline to mitigate potential future impacts, accepting the trade-off of reduced developable area and possibly lower immediate returns on investment.

As another example, suppose a developer needs to design and build a new roadway for a future development site. The developer has two options:

- Option A is a direct route from an existing highway to the development site but runs adjacent to several residential communities.
  - Base Cost Option A: $8 million.
- Option B is an indirect route with a longer distance but avoids the community adjacency; however, there is evidence of subsurface rock along the alternate route.
  - Base Cost Option B: $9 million.

While the base costs can influence the developer's decision, each option has risks that should be considered. With Option A, there's a risk of community opposition that could change the required scope (e.g., new requirements for sound mitigation walls) or could significantly delay the project as the developer negotiates with homeowners to alleviate their concerns about the new infrastructure. Schedule delays can cascade into other financial burdens, such as inflation risks, material shortages, and delayed rent from future tenants. Alternatively, Option B would avoid the community opposition with a slightly longer route but carries environmental risks, including costly excavation of subsurface rock. The rock excavation cost and required time are also potential risks, but the client has limited data on the quantities of rock.

In this case, two quantitative analyses can inform project strategy: a *decision tree analysis* and the *expected monetary value* (EMV) evaluation. A decision tree evaluates project options (Option A or Option B in the example) and assigns probabilities to potential outcomes. The probability values are informed by data and expert judgment, such as polls of community opposition to the project (for Option A) or geotechnical exploration to estimate rock quantities (for Option B). The impact of each risk is quantified – in this case, with a cost implication. The EMV is a calculation based on the probability of each event and the associated cost (in this example, monetary costs). A decision tree with the resultant EMV for the development road example is shown in Figure 8.5.

Based on these sample probabilities and monetary costs, the results indicate that Route A has a lower base construction cost ($8 million) but carries a high probability of a significant additional cost from community opposition, which results in an EMV of $11 million. Comparatively, the alternate Route B has a higher base cost ($9 million) and some environmental risks but a lower resultant EMV of $9.9 million.

While a decision tree and EMV are valuable tools in risk analysis, they should not be viewed as a definitive guide. Instead, it represents one of many approaches for evaluating the identified risks. Using decision trees, project teams can make informed decisions about the potential options, such as whether to allocate additional resources to validate assumed probabilities and cost impacts or implement strategies to mitigate and manage risks and enhance the likelihood of preferred outcomes. However, it's essential to recognize that this example method focuses on quantifying monetary costs and does not capture the full spectrum of project risks (schedule, public perception of the engineering firm, environmental impact, etc.).

Other critical factors and perspectives also warrant consideration. For example, the risk of each decision on the organization's broader goals and public interests extends beyond the immediate scope of the project.

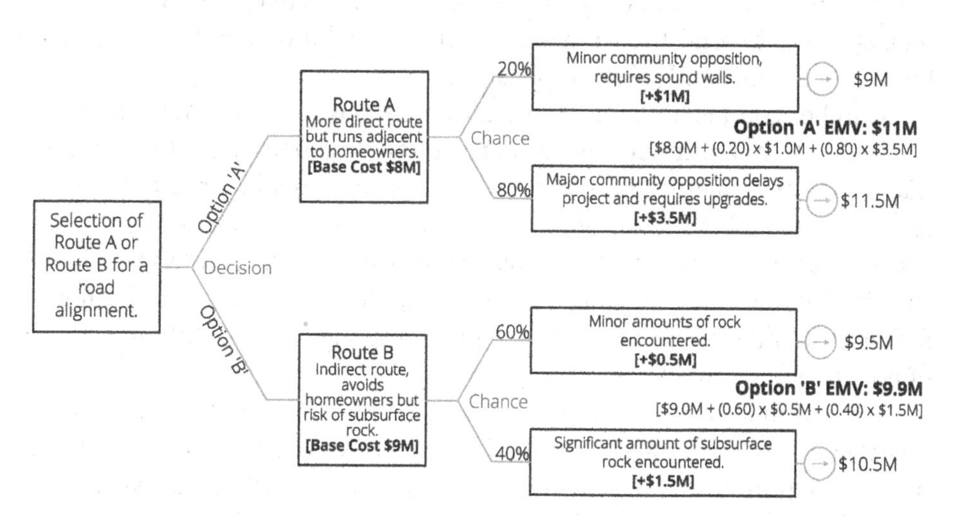

FIGURE 8.5 Decision tree with an EMV for an example road construction project.

Additionally, it's crucial to consider how each option might influence future decision-making processes within the project team and the community. Such considerations ensure a holistic approach to risk management, considering the financial implications and the strategic, social, and long-term impacts of each decision. Therefore, quantitative risk assessments, while helpful, should be integrated with multiple methodologies and perspectives to develop a comprehensive and multidimensional risk management strategy.

8.4.2 Risk Prioritization

Given limited resources and time constraints, project teams must strategically prioritize risks that are most disruptive to project objectives. This prioritization process allows for the effective allocation of resources by focusing on operations that address the most probable and significant risks.

Risk prioritization benefits from diverse perspectives and engaging in scenario-based planning. Different stakeholders will have varying perceptions of the project risks and the appropriate response strategies. For instance, while many transportation projects aim to improve vehicle highway mobility, this is often achieved by reducing site access points and intersections that could inadvertently impact local businesses. A retail outlet might experience decreased customer footfall due to restricted access, leading to reduced revenues (new risk, business perspective). Engineers must manage risks beyond the immediate technical solutions and consider the broader economic and social implications.

Another complexity in civil infrastructure and real estate development is the long-term nature of these systems, often designed to last for decades. Many project design choices and risk responses implemented today may result in trade-offs or limitations on future options, contingent upon unexpected scenarios. Scenario-based planning plays a crucial role in guiding stakeholder decisions, including engineering design solutions, as it allows for the evaluation of how risk priorities and responses might shift in the face of potential disruptions.

For example, in designing a new mixed-use development, engineers might consider the impact of future advancements in vehicle technology, parking requirements, or changes in regional population density. Scenario planning could reveal that specific design choices (like the inclusion of flexible-use building spaces) might either accommodate or impede the integration of these future conditions. Initiatives such as buildings with adaptable spaces could offer resilience against market changes that could be more costly upfront but allow for building spaces and site areas to be easily repurposed as residential, commercial, or retail uses in response to shifting demand patterns (subject to zoning regulations). These examples underscore the importance of incorporating flexibility and a forward-thinking perspective into design and risk management processes, ensuring that today's decisions deliver value and remain functional in the future.

Effective risk management involves continuous engagement with all stakeholders, regular re-evaluation of risks, and adapting strategies in response to changing circumstances and stakeholder feedback. By doing so, engineers can ensure that their risk management strategies effectively mitigate risks and align with the broader goals and needs of the communities they serve. This approach underscores the interconnected framework and the importance of adaptability in risk management, ensuring that solutions are sustainable and beneficial in the long term.

Effective risk management involves continuous engagement with all stakeholders, regular re-evaluation of risks, and adapting strategies in response to changing circumstances and stakeholder feedback.

Strategies for Risk Prioritization

One method of prioritizing risks is *multicriteria decision analysis* (MCDA). The MCDA process begins with identifying risks (Section 8.3.2) and assessing the probability and impact (Section 8.4.1). An initial list of risks can be prioritized through qualitative and quantitative risk analysis to determine the magnitude of the risk to project objectives (e.g., initially filtering any known low-impact and low-probability risks that don't meet an established risk threshold).

MCDA is used with a numeric scoring analysis to consider the diverse risk parameters (Section 8.2.5), which will inform risk prioritization. Similar methodologies are often used in project procurement (Chapter 2, Section 2.6) and go/no-go decisions (Chapter 3, Section 3.2). MCDA begins by establishing how various risk parameters could disrupt the project. These parameters will vary by project objectives but could include those shown in Table 8.2.

TABLE 8.2 Multicriteria decision analysis (MCDA) with risk parameters.

ID	Parameter	Score Range	Scale	
			Low	High
A	Magnitude	0–30	Low-probability and low-impact risk.	High-probability and high-impact risk.
B	Control	0–20	The project team can avoid the risk with proper initiatives.	The project team has no control and must accept the risk.
C	Detectability	0–20	Identifiers are available to indicate the occurrence and impact.	Not easily detected.
D	Connectivity	0–20	Risk is isolated.	Risk occurrence could create new risks.
E	Perception	0–10	Stakeholders are indifferent.	Stakeholders are concerned.

While not explicitly listed, the risk parameter related to time (temporal domain) is incorporated by conducting an MCDA at various stages throughout the project's life cycle. This approach assesses how present conditions affect both current and future project operations. As conditions change, this periodic reassessment of risks can lead to different prioritization scores, identifying new critical risks based on the current project status. Furthermore, the analysis includes additional criteria that shed light on how risks may impact various aspects of the project, such as schedule, cost, resource management, and quality, among others.

After determining the relevant risk parameters, the next step is to assign a weight to each, with larger values representing higher importance. This is typically achieved by assigning points that sum to a score of 100 across all parameters. Determining the weights for each criterion should be achieved through an inclusive and open discussion, which will inform stakeholder perceptions of relative importance based on the distribution of weights. For example, with the five sample parameters (A–E), the team could equally weigh each as shown (or stakeholders could determine that each parameter should have the same weight). Each parameter is scored, and the assigned weight is summed to establish rank. An example of this process is shown in Figure 8.6.

Risk ID	Risk Evaluation Parameter (weight, of 100)					Results	
	A (30)	B (20)	C (20)	D (20)	E (10)	Total Score	Risk Priority
1	25	15	20	15	5	80	#2
2	30	0	15	10	10	65	#3
3	20	15	0	20	5	60	#4
4	30	10	20	15	10	85	#1
5	25	10	10	5	5	55	#5

1. **Multi-Criteria Decision Analysis (MCDA) matrix established based on established Risk Evaluation criteria (A, B, C, D, E), assigned Weights (Σ=100) evaluated with various Risks (1, 2, 3, 4, 5) sample shown.**
2. **Sum the scores for each Risk and assign a rank based on the calculated score.**

FIGURE 8.6 MCDA for an evaluation of risks and their resultant scores for risk criteria (A–E) to inform prioritization

While the MCDA risk-ranking process provides a numerical score to help prioritize potential risks, it is important to note that these scores are more informative than definitive. The true value of the MCDA framework lies in the process itself, which stimulates in-depth discussions and investigations among project stakeholders, leading to the identification and better understanding of risks. As risks are evaluated and response strategies are considered, the initial scoring and prioritization may shift. For instance, the exploration of a risk response initiative might reveal prohibitively high resource requirements, impacting its feasibility and altering its controllability score. The primary benefit of this systems-based approach is that it fosters stakeholder engagement and documentation, thereby enhancing the overall risk management process.

8.5 Reserve Studies

Establishing contingency reserves is an essential component of risk analysis. Contingency reserves provide a buffer against the uncertainties inherent to project conditions, be it cost or schedule. Cost reserves are financial provisions set aside to cover unexpected expenses that arise due to identified risks. Schedule reserves include extra time added to the project schedule.

The project plan typically documents these reserves to inform internal project stakeholders because the contingency reserve is funded through the project budget. This transparency identifies how reserves are being used, and any changes to the reserve budget will be seen and help maintain trust and confidence among stakeholders.

Contingency reserves should be integrated into the broader risk management plan. Both cost and schedule reserves should be reviewed regularly as the project progresses. This review should consider any changes in the risk landscape and adjust the reserves accordingly. Their usage should align with the overall strategy for risk response and mitigation and should be discussed between the client and the project team. Unforeseen conditions typically warrant the use of project contingency funds, whereas engineering errors, omissions, and subsequent rework fall under the responsibility of the engineer. Such errors and omissions are often mitigated using an internal manager's reserve, as detailed in Section 8.5.3.

Cost and schedule reserves must be calculated based on a thorough risk analysis, regularly reviewed and adjusted as the project progresses, and closely monitored and controlled to ensure they are used effectively and for their intended purposes. This proactive approach to managing contingency reserves helps project teams navigate uncertainties and remain on track toward successful completion.

8.5.1 Cost Reserves

The importance of cost reserves lies in their ability to ensure that a project remains financially viable even when faced with unforeseen costs. Without these reserves, a project could quickly exhaust its budget, leading to financial strain or project termination. One simple method of establishing a cost reserve is by a percentage of the project fee (design or construction) based on historical data or expert judgment. Another method to establish the reserve amount is the use of the expected monetary value (EMV) approach, where the potential impact of each identified risk is multiplied by its probability of occurrence. This total gives an estimate of the required cost reserve (as shown in Figure 8.5).

The example from Figure 8.5 considers adverse risks, but similar methodologies can be used for opportunities that determine the potential gains. Whether EMV is used to estimate a gain or cost, the reserve is based on the statistical expectation that an event may financially affect the project due to specific risks. It's important to note that EMV provides an average expected value and does not guarantee that these exact costs will occur. As with many risk management strategies, these values are meant to inform the project team and not be used prescriptively. Additionally, regular review and adjustment of the reserve is required as the project progresses and as more information becomes available.

8.5.2 Schedule Reserves

Schedule reserves are crucial for absorbing the impact of delays due to risks from adverse weather, supply chain disruptions, unexpected design changes, or others. Adequate schedule reserves help maintain the overall timeline of a project, ensuring that deadlines are met despite unforeseen setbacks. A similar approach to EMV can be used for schedule reserve estimates, where the potential delay from each risk is evaluated in terms of time, factoring in the likelihood of each delay. The sum of delays estimates the additional time that should be allocated as a buffer to the tasks impacted by the risk.

Managing schedule reserves is particularly challenging due to time's irreversible nature. Unlike monetary budgets, time, once spent, cannot be regained or reallocated within the project. While accumulating a time reserve is possible in cases where tasks are sequential and independent, such as a team finishing a design phase ahead of schedule and moving on to subsequent tasks, an independent linear schedule is relatively rare. In many civil engineering projects, tasks are often interlinked, making it impractical to "shift" time from one area to another. For example, even if one team finishes their task early, this might not expedite other tasks that depend on different resources or personnel. The challenges of aligning schedules, critical paths, resource leveling, and task interdependencies, are explored in depth in Chapter 6 (Section 6.4).

> Time, once spent, cannot be regained or reallocated within the project.

8.5.3 Manager's Reserve

Sometimes, a project may elect to establish a "manager's reserve" for cost, which is not disclosed in the project plan or communicated to all stakeholders. Unlike the project's contingency reserve, the manager's reserve is typically not included in the project budget and instead an internal budgetary allocation. For example, from a client's perspective, a corporate-level reserve may be established to account for unforeseen engineering risks. In this way, the inherent variability of project performance (which could prompt change orders) can be accommodated. Conversely, it is not a standard practice to have a manager's schedule reserve – project due dates and float should be transparent to the project team, but the project may include policies that require a planned earlier delivery to consider potential schedule disruptions.

8.6 Risk Responses

Responding to risk in a project requires a proactive, strategic approach that aligns with the project and broader enterprise objectives. As risks are prioritized, stakeholders can evaluate the appropriate risk response. Organizations such as PMI document standard risk response strategies, as portrayed in Figure 8.7, for possible responses to adverse risks.

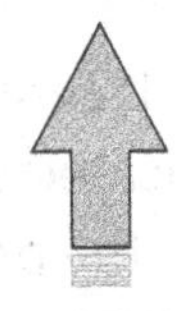

Escalate
The risk is no longer managed at the project-level and is instead managed at a higher level.

Avoid
Eliminate the threat or protect the project from the threat. This may require a change in the scope or objectives.

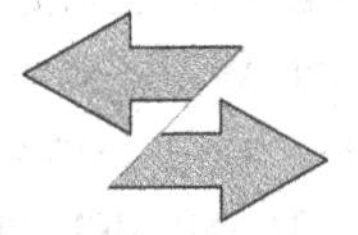

Transfer
Change of ownership of a threat, often with insurance, a third party, or performance bonds (likely at a cost).

Mitigate
Reduce the impact or probability of the threat, often with early action. This may require testing, reducing complexity, increasing redundancy, etc.

Accept
Document the risk, but no proactive action is taken. Often used for low-priority threats, or high impact but low-probability threats.

FIGURE 8.7 Standard risk response strategies.

The adage "the best decisions come from more than one good option" underpins the risk response process. This involves generating, assessing, and selecting from multiple response strategies to ensure the most effective approach is adopted. Though inherently solution-oriented, risk responses may not always constitute complete solutions themselves; instead, they are often steps toward mitigating, transferring, or managing risk. Suppose a site with a planned building is located adjacent to a flood zone. In that case, the building floor elevation could be raised (mitigate), or the owners could purchase flood insurance (transfer). These responses do not eliminate the risk of an extreme flooding event, but the responses serve to reduce the potential impact.

Risk responses must also be relevant and practical from stakeholders' perspectives. This means considering different stakeholders' unique interests, constraints, and objectives when responding. For example, a new roadway improvement project could face community opposition due to the potential traffic disruption during construction. In this case, an appropriate response could be meeting with community leaders to establish a traffic management plan (mitigate) and setting up open community channels to inform residents (acceptance). Both responses aim to align the stakeholders' perceived concerns.

As risk response initiatives are identified, the project team may encounter another prioritization responsibility. Implementing all risk responses is likely not feasible or necessary. However, some proactive responses could benefit the project objectives and operations. Also, risk response initiatives are not limited to addressing a single risk. Still, the project will work with finite resources and must consider how each risk response initiative will best manage project risk. The MDCA framework of Section 8.4.2 can serve to prioritize possible risk initiatives. The evaluation criteria could be designed to consider these conditions and include such parameters as an initiative's benefit in managing multiple risks, influence in controllability, ability to change probability or impact, cost-benefit, and stakeholder association (who benefits?).

A comprehensive inventory of all risk management strategies in the civil engineering setting is beyond the scope of this book; however, escalating, avoiding, transferring, mitigating, and accepting are risk responses commonly found in most scenarios. Some of these risk responses apply to both threats and opportunities (escalate or accept), while others use an inverse action to realize the opportunity (exploit instead of avoid, share instead of transfer, and enhance instead of mitigate). These responses are outlined in the following sections.

8.6.1 Escalate

Escalation is a risk response strategy that moves the risk from the project team to a higher level of power in the organization (program or portfolio). For civil infrastructure and real estate development projects, the concept of power relates to the ability to do something or act in a particular way to direct or influence the behavior of others or the course of events. For example, a project manager relies on organizational resources, such as availability of personnel. If there is a risk that key personnel will not be available, then the project manager could escalate the risk to a program manager who has control over allocating resources. At that level, program or portfolio managers would have the power to respond to the risk in various ways (e.g., allocate more personnel). With escalation, the risk is effectively moved outside the project team's accountability.

Often, the allocation of power can focus on one or more stakeholders' ability to control input or outputs affecting risk within the enterprise. This does not necessarily mean that a project manager's supervisor has control – in some cases, escalation may also be the first step when a risk is determined to be outside the project scope. For instance, if an unexpected archaeological site is discovered during the construction of any project, this could be a risk that needs escalation to legal and archaeological experts due to its potential impact on historical preservation and legal compliance. Escalation is also crucial when dealing with risks that have potential safety implications that exceed the project team's ability to manage, such as discovering unstable ground conditions in a bridge construction project.

Project managers must understand their ability to control inputs or outputs that facilitate risk management. In some cases, the perception of control over project operations and risk responses may be muted, limited merely to influence. Reconciling the relationship between that which can be controlled and that which a project manager does not control is critical to determining when a risk should be escalated.

8.6.2 Avoid

Avoidance of an adverse risk is a deliberate and proactive response strategy by the project team, which is most effective with early implementation. This response is appropriate for high-priority risks, likely those with high probability and high impact. Avoidance often requires changing a project attribute, such as modifying or removing some of the scope to eliminate the risk. Alternatively, some avoidance measures include new actions by the project team, such as clarifying ambiguous requirements that were the source of the risk. For example, design specifications may be authored to avoid concerns about construction methods, inspection, or testing.

Adverse risks can often be avoided by adjusting the project schedule, allowing stakeholders to control certain events' timing. This approach of "scheduling risk" can be practical in aligning project phases with optimal conditions or avoiding periods of risk. For example, in a development project involving a new retail center, scheduling risk can mean planning major construction activities outside the region's snowy season to avoid weather-related delays. This decision directly impacts both the probability (lowering the chance of weather-related interruptions) and the impact (reducing potential delays and associated costs). Another instance might be scheduling critical inspections or material deliveries to coincide with periods with fewer demands on resources, ensuring that these key project activities can be carried out without delay or complications. This strategic scheduling is a proactive way to manage risk by aligning project activities with periods that present the most favorable conditions or the least potential for disruption. Refer to Chapter 6, Section 6.4 for more strategies on schedule management.

Exploit

When responding to an opportunity (positive risk), the project team will seek to exploit rather than avoid the risk. The intent is to maximize the benefits with an intentional action or investment. For example, if there is a temporary decrease in construction material prices, the team can expedite purchases to exploit the favorable market condition. The team will proactively modify project attributes to realize the opportunity.

8.6.3 Transfer

Stakeholders can transfer, allocate, distribute, or shift adverse risks (threats) in various ways. Transferring a threat entails allocating it to another party better equipped to manage it, often through contractual agreements or insurance. For example, in a large-scale urban development project, the risk of construction delays due to unforeseen ground conditions might be transferred to the construction contractor, who agrees to bear the cost of delays through liquidated damages (a set monetary price for unit time of delay). Contract indemnification provisions, perhaps one of the most common tools of transferring risk, reflect the tension of allocation in terms of scope of liability (i.e., negligence and willful misconduct), insurance status, and mechanisms relating to defenses of claims.

Another common practice is purchasing insurance policies to cover specific risks like property damage or workplace accidents. Project team members can adopt or impose insurance requirements intended to ensure better funds will be able to support the contract indemnification obligations. The effort to transfer risk among stakeholders must account for the interplay of the various obligations and resources of the project.

For example, suppose a construction contractor hires a subcontractor for specialized blasting of rock. The contractor is concerned about the risk of this work and has negotiated a strong indemnification clause on the contract. According to the clause, the subcontractor would be responsible for damages, claims, and losses arising from their work on the project. However, assume the contractors overlooked a crucial aspect during negotiations: they did not require the subcontractor to obtain (or maintain) adequate insurance coverage. If, during the project, the subcontractor was operating with thin financial margins and encountered issues with blasting that led to structural damage to adjacent properties, a risk exposure has been created.

As a result of these damages, the adjacent property owner can file a substantial claim against the contractor, who in turn will seek indemnification from the subcontractor (as per the contract, through an attempt to transfer risk). However, the subcontractor, lacking sufficient insurance coverage and facing financial challenges, can not fulfill the indemnification obligations and this leaves the contractor facing substantial claims without any financial support from the subcontractor or an insurer.

This example highlights that project managers must understand the relationship between contractual indemnification and insurance requirements to effectively allocate risk among the enterprise constituents. This approach does not eliminate the risk but rather assigns the financial burden and management responsibility to a third party. Transferring risk is particularly useful in complex projects where specific risks fall outside the core expertise of the project team.

Share

The inverse of transferring a threat is to share an opportunity. In some cases, an engineering firm or other stakeholder may seek to share the benefits of an opportunity. This could apply to strategic teaming arrangements, such as establishing a joint venture for a project. Finding the right team could improve the chances of project success. For example, a prominent engineering firm seeking to work on a new sports stadium with sustainability objectives may share the opportunity by teaming with a small specialized green engineering firm known for innovative solutions. In this case, the two firms agree to share the potential reward, which benefits the project's success by adding specialized expertise, improving the project's marketability, and benefits the smaller firm's portfolio and business.

8.6.4 Mitigate

Mitigation is a typical response to a threat that reduces a risk's likelihood or impact. This might involve implementing additional safety measures, enhancing quality controls, or adopting more robust design standards. This action is deliberate, often requiring an investment of resources and time to protect the project from the identified threat. For example, an engineering firm may allocate additional funds for quality management when the project design is determined to be complex and at risk of a larger number of potential errors. Mitigation also involves contingency planning, where alternate plans are prepared in case of a risk occurrence.

Mitigation is a typical response to a threat that reduces a risk's likelihood or impact.

For example, having a backup supplier for critical construction materials reduces the risk of project delays. Effective mitigation requires a thorough understanding of the potential risks and proactive planning to address them before they materialize.

Enhance

The inverse action of mitigating a threat is to enhance an opportunity. In this way, the project team would intentionally invest and act to increase the probability and likelihood of an identified opportunity. For example, an engineering firm may see a benefit in implementing a new design technology and choose to enhance the opportunity by investing in training and hardware. In this way, the project team is better prepared to realize the benefits of the opportunity.

8.6.5 Acceptance

Accepting a risk is an appropriate response when the cost or effort of other risk responses (mitigate, avoid) is greater than the impact of the risk itself. This approach is typically applied to high-impact but low-probability risk, a low-impact risk, or an otherwise uncontrollable risk. For instance, the risk of minor weather-related delays might be accepted during the construction phase of a project, with the understanding that such delays are manageable within the overall project schedule. Acceptance does not mean ignoring the risk; it involves recognizing the risk, understanding its potential impact, and planning to accept it if it occurs. This strategy often requires continuous monitoring to ensure the risk remains within acceptable limits (probability and impact) throughout the project life cycle.

In Appendix A, Part 8 of the Millbrook Logistics Park, the client outlines the project's challenges and the decision-making methods used. This section is presented from the client's perspective, framed as a discussion with their partners. It details the rationale behind certain decisions made in response to uncertain market conditions and financial considerations. Specifically, it discusses the implications of accelerating project operations to achieve earlier completion. This acceleration is acknowledged to increase design costs and add complexities to design execution due to the need to manage concurrent operations.

CHAPTER 9

Principles of Effective Communication

Verbose and complex discourse is indicative of superficial comprehension.

CHAPTER OUTLINE

Management Essentials for Civil Engineers: A Practical Guide to Business, Communication, Ethics, and Risk, First Edition. Cody A. Pennetti, C. Kat Grimsley, and Brian M. Grindall.

This chapter will build on the communication concepts introduced in Chapter 1, including a review of different audiences and forms of communication. It will provide further discussion on the importance of word choice and tone as well as offer suggestions for best practices in different forms of communication. The overarching chapter focus is on the style, context, purpose, and best practices of communication.

9.1 Introduction

Engineers develop the skills necessary to solve complex technical problems through their formal education and project work. These skills require communication through technical narratives, calculations, and graphic design. However, as engineers advance in their careers, they often find themselves in various business situations requiring extensive use of many different forms of communication.

Engineers who take on the role of project managers need to be well-versed in communication principles due to the significant range and volume of communication involved in more senior positions. Examples of project communications include team or client meetings, team coordination, emails, approval hearings, and community discussions. Professional communication encompasses all formal interactions, whether written, verbal, or visual. This can include emails, reports, presentations, client calls, community charettes, approvals hearings, team meetings, and many other scenarios requiring communications. Any interaction during the workday, including work events held after-hours, should be treated as professional communication.

The results of any of these interactions are especially important because they can materially impact the success of a project and the reputation of the engineering firm (and the individuals). For example, when project managers handle external communication, their style and tone should reflect the identity of the engineering firm. While individuality and personality remain important for many forms of communication, formal communications are influenced by the firm's preferences for style, voice, and presentation. Professional, concise, and clear communication is a learned skill that can be improved with training. Communication is a nontechnical professional skill that can help engineers succeed in these encounters and advance project results for their clients.

9.2 Dimensions of Communication

Communication takes many forms and can occur through written, verbal, nonverbal, or visual expression. Often, multiple forms of communication are used simultaneously. In any form, communication serves several purposes within a particular task or project stage. Table 9.1 provides examples of some of the different professional communication goals an engineering project manager may be asked to participate in at different times during the project life cycle. Note that there can be significant overlap between many of these examples, depending on the specific goals of a particular communication. For example, successfully *justifying* the rationale behind a particular design choice may also be critical in *negotiating* project approvals.

Because of their technical training, engineers may be more comfortable with fact-based communication, such as specification lists, plan notes, and narratives, and less familiar with other forms, such as negotiations, describing value, and persuasive communication. However, as they are promoted throughout their careers, engineers need to be aware of – and develop competency in – a range of different forms of communication. The audience and goal of a particular communication will typically dictate the appropriate approach to be used during the exchange.

Engineers need to be aware of – and develop competency in – a range of different forms of communication.

For example, if the engineer needs to communicate basic technical information, then providing a list of factual information in a straightforward manner (*informing*) may be most appropriate. In contrast, if the goal

TABLE 9.1 Examples of different communication goals.

Informing	Leading	Negotiating
Justifying	Motivating	Persuading
Demonstrating	Collaborating	Promoting
Confirming	Brainstorming	Marketing
Delegating	Consensus building	Managing expectations
Warning	Relationship building	Resolving conflict

of a communication effort is to gain community support for a project, then simply listing technical facts is unlikely to achieve a successful result. Rather, the engineer will need to adopt a more appropriate style to help concerned citizens understand why certain project elements are designed in a particular way (*justifying*), how certain technical choices improve the project and result in a benefit to the community (*promoting*), or otherwise demonstrate the value of alternative solutions in a compelling way (*persuading*).

As another example, suppose a community member insists that more vegetation and green space should be included in a site design. In that case, it may be helpful to explain that the site infrastructure requires dedicated utility corridors that must be accessible and clear of vegetation. Similarly, trade-offs may be established that prioritize safety with high visibility at intersections at the expense of reduced streetscape plantings. While there is a technical basis for these decisions, effective communication requires more than just presenting the facts. It involves a critical assessment of the audience's perspective, coupled with attentive listening to their concerns, to enable engineers to respond effectively. This approach ensures that the communication is factually accurate and addresses the specific needs and queries of the audience in a way that is respectful and productive.

Communication has a powerful impact on team members, clients, the public, and other stakeholders. Engineering projects will often have conditions where stakeholders disagree on an approach or a solution and the manner in which these disagreements are resolved can have a real and significant impact on current operations and future projects if either party perceives the exchange as abrasive or dismissive. A pointed negative comment, especially as an emotional response or legitimate frustration, may be remembered well into the future and can result in loss of professional opportunities for the engineer or future work for the firm. For these and many other reasons, it is important always to consider word choice, tone, and the appropriateness of emotion in an exchange. In practice, this requires a high degree of self-awareness on the engineer's part and may mean taking slightly longer to respond in certain situations; however, making best efforts to ensure positive communication is well worth the extra time needed to edit or self-censor when appropriate.

9.2.1 Word Choice

Colloquial language includes a range of expressions, including slang, esoteric references, clichés, profanity, poor grammar, poor spelling, or vague expression. While these *may* be acceptable outside the workplace, it is not considered a best practice to use colloquial language in professional communication. Even if a personal rapport develops between engineers and stakeholders, it is important to recognize that most project discussions remain subject to scrutiny in case of legal disputes involving an engineering firm and should, therefore, remain formal and professional. In particular, colloquial language should be avoided by engineers in outward-facing positions that interface with diverse stakeholders. If certain terms are perceived as offensive by a key stakeholder, it could lead to the firm's termination from a project or even the termination of the engineer's employment. Additionally, some forms of communication can be shared beyond the initial recipients. What may have seemed like an informal and casual discussion during the initial conversation could potentially disadvantage the engineering firm when viewed outside its original personal context.

While the use of colloquial language in professional engineering communication can be problematic due to its potential to be offensive or noninclusive, another critical concern is its lack of precision and clarity. Casual phrases often fail to convey the specific, technical information necessary for clear understanding in a professional context. In engineering, where accuracy and detail are paramount, it's essential to use language that is explicitly informative. Operationally, using informal or colloquial language in professional communication can create confusion over who is assigned a particular task, what a task may entail, what time constraints may exist, or what implications may result for a client, all of which can jeopardize project results. Table 9.2 provides several examples of phrases using colloquial language and suggests how the same sentiment can be better expressed using professional language. This demonstrates the importance of choosing words that not only avoid misunderstandings and clearly communicate the intended technical information.

TABLE 9.2 Examples of colloquial versus professional language.

Colloquial Language	Professional Language
We really need to ramp up our efforts.	We must dedicate more time to resolving the drive lane conflict.
The project has many speed bumps.	We've identified the following benefits and challenges . . . [specify]
This is right up our alley.	Our firm has the demonstrated expertise to address this challenge.
I threw myself into the calculations.	I dedicated substantial time to this analysis.
The hearing will be a slam dunk.	We have a high level of confidence that we can obtain approvals.
Looks good – let's run with it ASAP.	I recommend we submit version B of the site plan to the client no later than March 1.
We've got the right man for the job.	We have the right person for the job.
It's on our radar.	The team is aware of the June 15 submission deadline.
We need to nip this in the bud.	We should proactively consider other solutions.
Let's kick the can down the road.	We should reevaluate at [specific date].
One of our guys messed up the math on that.	There were errors in our calculations, which we are addressing.
We need to dumb this down.	We should take the time to clarify the content so it is more accessible.

9.2.2 Tone

Considering the tone of a message is equally important to word choice in professional exchanges. The tone of communication can either motivate an audience or foster disenfranchisement. Similarly, tone can either escalate conflict or help resolve it. Tone can be difficult to decipher because of its nuance, and a message's meaning can be disconnected from the underlying word choice. For example, navigating conflict when a party is passive-aggressive can be challenging. When

The tone of communication can either motivate an audience or foster disenfranchisement. Similarly, tone can either escalate conflict or help resolve it.

communicating in person, tone can be expressed through word choice, volume, inflection, level of eye contact, and other body language. A benefit of in-person communication is that it is usually possible to clarify misunderstandings related to tone during the exchange. A drawback, however, is that the exchange often happens quickly, so misunderstandings or misgivings caused by tone can rapidly escalate. Hasty responses can become emotional and create conflict, particularly if at least one party to the discussion does not seek clarification before responding.

Team members may interpret a manager's tone in a variety of ways. A negative tone can manifest in reactions such as defensiveness, reciprocally negative attitudes, passive silence, nervousness, or heightened anxiety about the project or individual work. A negative tone and its impact on team dynamics can make any meeting far less productive than it could have otherwise been. These conditions can create a risk for the project if the meeting includes clients or other key stakeholders. Engineers at all project levels should practice self-awareness and focus on expressing a positive tone and matching nonverbal communication in live exchanges. This skill is often learned and requires intentional cognitive effort to monitor tone and evaluate how the audience receives the messages.

Nonverbal communication is especially important to monitor during live exchanges. For example, a manager who is experiencing high levels of stress from other sources may inadvertently set a harsh tone for a meeting through body language (e.g., sitting with arms crossed), facial expressions (e.g., frowning), appearing intentionally distracted or disinterested, or speaking in a terse manner. People have lives and responsibilities beyond those visible to colleagues, which can influence actions, attitudes, and energy in the workplace. The project manager must recognize when external stressors will influence operations and, most importantly, when the stress will affect the team. Similarly, a project manager should realize when a team member may face external (or work-related) stress and need support. In these cases, most organizations provide policies established by human resources on how managers can best support their team members.

The frequency of communication can also be a factor in setting the tone. Numerous scenarios exist in which the tone of cumulative, ongoing exchanges can affect the work environment or project outcomes. For example, repeated calls or emails in a short period can be perceived as demanding, even if the intent behind them is driven by legitimate urgency.

In written communication, tone is conveyed by word choice, punctuation, font, and style. It is arguably harder to convey tone accurately in written formats where word choice may be misunderstood, and lack of live exchange makes clarification impossible. For this reason, it is best to avoid sarcasm in emails and other written forms because doing so may create conflict if the recipient misunderstands the intention. It is also best to use courteous phrasing (e.g., "Can you please . . ." or "I would appreciate it if . . .") to avoid the appearance of inappropriately commanding subordinates or delegating tasks to a peer or client outside of the engineer's authority.

To highlight an example of tone in email, Figure 9.1 shows two emails communicating urgency to the recipient, Shawn. Example A has a demanding tone and does not explain the basis of the need. If a peer sends this email, it may be interpreted by the recipient as overbearing, which can damage team dynamics; this may be worsened if the sender copies Shawn's manager

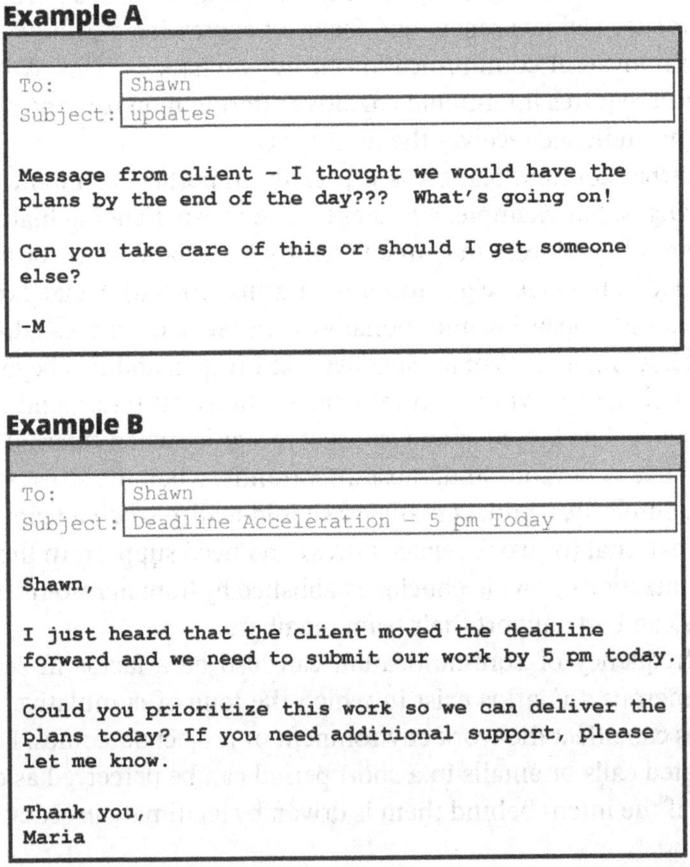

Example A

To: Shawn
Subject: updates

Message from client - I thought we would have the plans by the end of the day??? What's going on!

Can you take care of this or should I get someone else?

-M

Example B

To: Shawn
Subject: Deadline Acceleration – 5 pm Today

Shawn,

I just heard that the client moved the deadline forward and we need to submit our work by 5 pm today.

Could you prioritize this work so we can deliver the plans today? If you need additional support, please let me know.

Thank you,
Maria

FIGURE 9.1 Examples of tone in email.

in the email, making it appear that Shawn is somehow behind in her work. This format can also be interpreted as a sign of a hostile or unreasonable work environment, which can harm morale. In contrast, Example B follows best practices by briefly explaining the urgency, using respectful language, and including a mechanism for addressing the possibility that the work won't be ready on an accelerated timeline, regardless of the client's request. Although the message in example B is longer, it is not verbose, and the extra language adds valuable context to the communication. In contrast, the message in Example A could quickly spiral into a series of rapid responses seeking clarification or defending against a perceived attack.

Note that the same message provided in Figure 9.1 may need to be adapted if the team uses communication platforms that promote short messages rather than full emails; however, the principle behind the need to provide relevant information and communicate respectfully remains important regardless of the communication vehicle. Indeed, when communicating with a larger group through a messaging platform, it becomes even more important to take the time to write clearly to avoid confusion or potentially alienating team members.

9.2.3 Emotional Intelligence

Expressing positive emotions in the workplace can be encouraging and helpful, while negative emotional responses tend to damage the work environment and relationships. Unfortunately, stressful project deadlines, difficult interpersonal dynamics, and other challenges often lead to frustration and conflict in professional environments, which increases the risk of negative communication. Engineers can significantly enhance the effectiveness of their communication by cultivating emotional self-awareness. By managing emotions and responding thoughtfully, engineers can defuse tense situations, leading to better outcomes and stronger professional relationships. In a supervisory role, engineers must manage their own emotional responses and help manage inappropriate behavior from their team. The ability to recognize, understand, and manage emotional communication is called

By managing emotions and responding thoughtfully, engineers can defuse tense situations, leading to better outcomes and stronger professional relationships.

emotional intelligence (EI) or an *emotional quotient* (EQ). This is an important leadership trait and critical to effective communication.

Engineers can use numerous de-escalation techniques during stressful or contentious exchanges. The examples in this section create a time buffer and, in some cases, even a physical separation so that the parties can briefly disengage from the discussion to manage emotions and responses.

During a verbal exchange, such as a call or meeting:

- Ask for a "sidebar" to confer with colleagues; this is a pause in the larger discussion that allows the engineer to gather their thoughts and confer with colleagues privately. Leaving the room during a sidebar discussion is acceptable, although not necessary.
- Ask to put a "hold" on the topic and continue to the next point, understanding that the group will return to the contentious point toward the end of the discussion – or even on a different day. This helps to ensure that an individual contentious point does not derail the larger discussion. Writing down such deferred topics is important so they are not forgotten, which could otherwise be perceived as dismissive.
- Ask for a 5-minute break (to get water or coffee, stretch, or use the restroom, etc.) and then briefly leave the room or call.
- Ask for additional information or context to shift the focus briefly; this can help deescalate while confirming that no key points have been missed, resulting in a misunderstanding.
- If pausing or deferring the discussion is impossible, ask for a moment to think or reflect on what has been said. This may feel awkward, but it usually produces far better results than responding immediately under pressure.
- If the situation is untenably aggressive, politely acknowledge that the discussion is not currently productive and inform the other parties that the meeting or call will need to be continued later.

In response to an email or other written communication:

- Wait to respond; this should be at least several minutes or even up to a full day, depending on the issue's urgency and the time needed to process emotions. Note that this can feel uncomfortable when using instant messaging communication platforms, as team members are often conditioned to expect an immediate response. Taking a break and physically moving away from the computer can be a helpful strategy to create some time before sending a response.

- Send a brief response acknowledging the email and informing the recipient that a more complete response will be forthcoming.
- Draft an emotionally driven response as a method of venting frustration with a separate word processing tool (so as not to send it accidentally). Revise and send the message after managing emotions.
- Ask a colleague or manager for input on the situation and to proofread the response before sending it to avoid escalating.
- Ask for additional information or context to confirm that nothing creates a misunderstanding.

Difficult and contentious discussions do sometimes need to take place. Conflict isn't always bad – often, it can lead to creative solutions that respect multiple perspectives. The expectation is that emotional responses will be managed and communication will remain professional and productive, even in tense situations. Engineers should address the underlying issue during difficult conversations rather than responding personally or defensively. If something difficult needs to be communicated, it is a best practice to present the issue factually and explain its impact on the individual, team, or project. Focus can be placed on *why* a resolution is needed, and the conversation can be framed as an effort to find a solution in partnership with the other party rather than an opportunity to exchange accusations.

Conflict isn't always bad – often, it can lead to creative solutions that respect multiple perspectives.

Engineers should be sympathetic and supportive of stakeholder issues. In some cases, a sincere apology is warranted, and stakeholders may appreciate an engineer who is honest and responsible for mistakes. However, constantly apologizing to placate a client or out of nervousness to soften a situation can cause the engineer to appear insincere. Conversely, an engineer who never appears contrite may seem uncaring or irresponsible, jeopardizing professional relationships. Further, engineers need to be aware that there are instances when an apology could imply accountability and liability for any resulting damages. A solution-focused response to a difficult situation is best, which includes acknowledging and owning a mistake while keeping the project moving forward.

Project managers are the voice of a team. They should be prepared to maintain accountability during difficult conversations – in no case should the project manager blame other team members for issues when communicating with stakeholders. For example, if a team member omitted critical information from a plan that the client later identified, the manager should accept accountability for the omission and commit to resolving it with the team. Attempting to blame other team members will make the manager appear disorganized and disconnected, which harms the team and the engineering firm.

Examples of difficult messages engineers may have to deliver in their careers, include:

- A problem or cost overrun arises that is related to the work of the engineer or the team and must be explained to the client and senior management.
- An engineer is asked to expedite delivery and sign plans without time for formal quality control.
- A project deadline will be missed (regardless of whether the cause was within the engineer's control).
- Work must be redone (regardless of which party caused the need for rework, this message is not likely to be well-received by the client).

The challenges of tone and word choice highlighted in this section identify why difficult conversations are best had in person. Emotional intelligence is critical when discussing conflict or conditions that could affect project cost and schedule. Note that not all stakeholders will have the same level of emotional intelligence to manage an emotionally driven reaction, and this creates a potentially even more challenging situation that the engineer must manage. Active listening and empathy will be appreciated if practiced effectively in these cases. Engineers should prepare for critical feedback during these conversations. To help offset the difficulty in managing these situations, identifying solutions before a difficult conversation can show a proactive approach to addressing the situation and demonstrate that the engineer recognizes accountability, or if not directly accountable, acknowledges how the situation could affect the client. In all cases, prompt communication is appreciated.

9.3 Branding

An important purpose of communication from a client and engineering firm's perspective is to support corporate branding, which can influence a firm's recognition and reputation with stakeholders, help with recruiting, and improve business development, among other goals. Branding typically falls under the purview of a company's marketing, public relations, or communication departments, but an engineering firm's brand standards will affect engineers' communication. This may include visual standards for consistency in communication, such as colors or logos that must be used on reports, email signatures, templates, or standard design drawing files. It may also include formatting standards or requiring language consistent with an engineering firm's social or ethical policies. Indeed, adherence to a corporate vision statement or code of ethics may impact how an engineer is required to communicate certain corporate operations and cultural norms to teammates, clients, or other stakeholders.

9.3.1 Logos

A logo is one of the most important aspects of its corporate identity. Some logos are registered with the US Patent and Trademark Office, which grants legal rights over how and where the logo is used. Logos are intended to be used from source material, not modified or redrawn. Most engineering firms and clients will have images file for several logo versions based on use, such as color, black and white, vertical formats, and others. Care should be taken with logos to avoid distorting the image and scale. Other guidelines may govern best practices, such as leaving space around a logo, establishing the proposer scale, avoiding logo placement over a busy or patterned background image, and others. Figure 9.2 provides an example of unauthorized alterations of a sample logo.

The logo from a teaming partner and client should also be respected, and in many cases, using logos from an external source will require permission. Formal requests to use a logo can help offset the risk of using an outdated logo, a logo from the wrong corporate entity, or an altered version. Many organizations have a published policy on how their logo can be used and an available source graphic. For example, federal agencies may

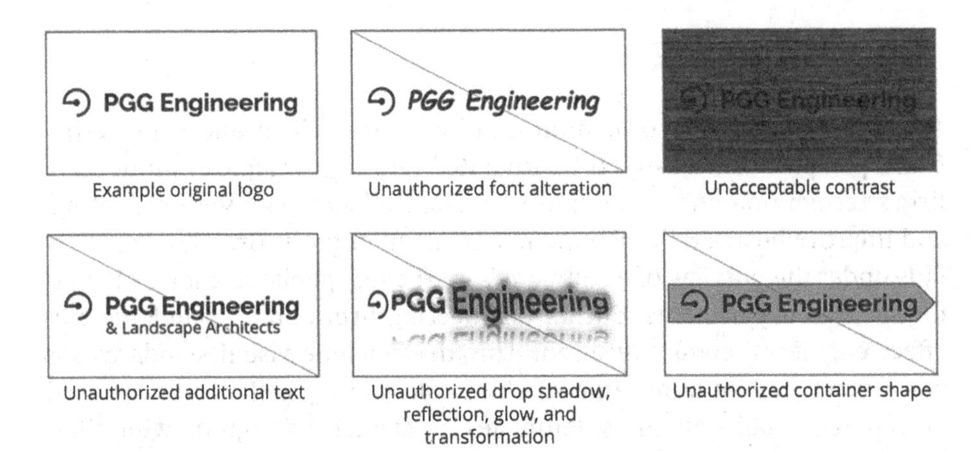

FIGURE 9.2 Unauthorized alterations of a logo.

authorize the use under certain conditions (no alterations, no use of social media, no recoloring, etc.). In most cases, it's best to avoid using an external logo without written permission.

9.3.2 Styles

The style of a firm is often established with published communication guides available to personnel. These guidelines will stem from the engineering firm's vision. A writing style guide can include references to tone, messaging, nomenclature, and grammar for external communication. Has the firm determined their persona is honorable, inspired, and proven, or are they considered curious, optimistic, and intense? Does the firm prefer a corporate and professional style or a more personal voice? These guides may also suggest terminology, such as active voice, first person (using "we" to refer to the organization), and others. This content often includes reminders on common writing errors, formatting tips, standard abbreviations, etc.

An engineering firm may also have a library of photos, graphics, and icons. These resources will save time for authored documents such as proposals and reports. The content libraries provide a consistent depiction of the organization, sometimes focused on personnel, projects, actions, or illustrations consistent with branding colors and fonts. Does the organization prefer to show personnel in professional business attire and a serious demeanor or casual attire and energetic expressions? Styles will vary, but most engineering firms will recognize that written content should be factual and client-centric (or, in the case of a public agency, it may be community-centric). For many individuals, learning the corporate writing style will require practice and the recognition that communication is not the voice of one individual but of the firm's personality.

Similarly, when an engineering firm manages communication tasks on behalf of a client, such as producing public notices or job site posters, it's essential to align with the client's preferred style. This requires coordination to ensure that the messaging accurately reflects the client's character and objectives, blending the engineering firm's expertise with the client's unique voice and vision.

9.3.3 Templates

Templates are an important organizational process asset. Engineers will be most familiar with standard templates for design production and spreadsheets for common calculations. Most organizations will publish templates that employees must use to maintain consistent branding. The templates also reduce the time and effort required by team members who would otherwise feel obligated to create new content. These templates will embed logos and brand colors, and will be designed to follow corporate styles. Styles extend from written content into templates for presentations, standard email signatures, letterhead templates, social media templates, and others. Similar to restrictions on logo reformatting (Section 9.3.1), these templates are not intended to be altered by personnel. The communication style should be recognizable as belonging to the engineering firm regardless of the individual author.

Engineers should expect that most authored content for a project has likely been developed for other projects, providing an organizational process asset to make similar other tasks more efficient. While the content should not be directly copied from other projects, these references serve as valuable resources and examples for content development. For example, a project manager could look to previously awarded proposals that include strong narratives and supporting graphics. Many organizations will have project plan templates and record copies that could prompt ideas about risks, quality control processes, resource management, and other content. Before starting a new document, project managers should look to templates for formatting and investigate other organizational process assets.

9.4 Considerations in Team Communication

An engineer's career often begins with a technical focus in a production role with limited requirements for communicating externally to clients or teaming partners. In advanced roles of project managers, an engineer's role includes responsibility for project performance, team leadership, and overall project success. These responsibilities require substantial nontechnical knowledge and specialized communication strategies to motivate and support team members.

9.4.1 Communicating as a Manager

It is important to understand how to operate and communicate in a supervisory role to best serve the project team's productivity and deliverables. A common hazard for many newly promoted professionals is the tendency to confuse their new role as one with unimpeachable autocratic authority rather than as a role of service to empower the team and help facilitate outcomes. Other new project managers may be insecure about their ability to perform in the new role or may legitimately be nervous about project results and the professional consequences of poor outcomes. As a result of any of these underlying causes (and others), newly promoted engineers may be tempted to micromanage by overly controlling their teams, processes, and project operations. This can lead to unproductive communications, harming team morale and project results. The following lists provide examples of traits to build and avoid to enhance management effectiveness.

A common hazard for many newly promoted professionals is the tendency to confuse their new role as one with unimpeachable autocratic authority rather than as a role of service to empower the team and help facilitate outcomes.

Traits to build:

- *Clarity*
  - Project managers should strive for clear and concise communication to ensure that information is easily understood by all stakeholders.
- *Listening skills*
  - Effective project managers practice active listening, showing a genuine interest in what others have to say.

- *Empathy*
  - Empathetic communication involves understanding and acknowledging the emotions and perspectives of team members and stakeholders.
- *Adaptability*
  - Being adaptable in communication means adjusting your style and approach to suit the needs of different individuals and situations.
- *Respect*
  - Respectful communication involves treating all team members and stakeholders with dignity and valuing their contributions and opinions.

Things to avoid:

- *Blame*
  - Avoid assigning blame or finger-pointing when issues arise, especially in public settings, like team meetings. All team members are responsible for success, but the project manager is ultimately accountable.
- *Lack of documentation*
  - Do not neglect to document change, decisions, and important communications to prevent misunderstandings or disputes.
- *Overcomplication*
  - Avoid overloading communication with unnecessary jargon or complexity. Keep messages simple and straightforward.
- *Second-guessing*
  - Do not attempt to control all decisions but discuss the work with input from the team. Learn to build trust and allow the team to feel comfortable to ask for review or support.
- *Hasty reactions*
  - Refrain from reacting impulsively to challenges or criticisms that lead to a curt, rude, or dismissive tone. Take time to assess situations and respond thoughtfully.

Some situations do require firm guidance or heightened scrutiny from project managers. For example, when onboarding a recent college graduate who does not have substantial professional experience, the new hire may appreciate additional guidance. However, normal team operations typically do not require constant and direct supervision from engineering leads. The expectation of personnel and the best coordination style will vary by individual.

While micromanaging is an unproductive leadership approach, the opposite, whether a laissez-faire approach or a people-pleasing attitude, can be equally unproductive. Project managers must find balance between empowering, supporting, and directing their teams. One way to achieve this is for the project manager to begin by practicing self-awareness and honestly and objectively evaluating their communication style, how their messages are being received, how the team is performing, and the reasonableness of their expectations. If, after reflection, it seems that a leadership trait would benefit from self-improvement, a project manager can ask their team for input and start by making incremental adjustments to improve clarity about expectations, deadlines, or other project needs without creating demands. This can be extremely challenging in practice because leading a highly functional team requires leaders to work within complex team dynamics and navigate different team member personalities to harness the team's professional strengths.

9.4.2 Formal and Informal Communication

Project managers must be prepared for both formal and informal team communication. *Formal communication* is project-related and involves communication about requirements, budgets, deadlines, deliverables, or other issues. Many examples of formal communications are familiar: emails, calls, meetings, and presentations. These communication scenarios extend beyond project-related activities and include responsibilities associated with a senior role with an engineering firm.

Performance reviews and human resource (HR) complaints are other formal communications with which an engineering manager may be involved. When the situation arises, engineers in management roles must be fair but impartial and document behavior (positive or negative), complaints, and other HR issues factually and objectively. This can be challenging if, for example, team members have different personalities or different beliefs than the engineer's own; however, individuals must be evaluated on professional contributions

Individuals must be evaluated on professional contributions and not penalized or rewarded based on whether they have a strong personal connection with a project manager or a shared belief system.

and not penalized or rewarded based on whether they have a strong personal connection with a project manager or a shared belief system.

Project managers are not only responsible for project solutions and formal HR documentation; they are also responsible (explicitly or otherwise) for their team's morale, mental health, inclusivity, and overall productivity. It is important to foster an environment and team culture that supports these goals, often achieved through informal communication and team building.

Informal communication may include messaging to set a positive team culture and sense of belonging, recognize individual professional achievements, or host non–work-focused activities where team members can establish more meaningful connections. Informal communication may also enable team members to share ideas or express concerns about project requirements, deadlines, client expectations, or professional disappointments without fear of judgment or censorship.

In well-aligned teams, conflict may be resolved early with these informal discussions. This approach could avoid undue friction with a client or between team members. An engineer who can lead a team by establishing healthy and productive informal communications can improve team dynamics and productivity while ensuring that individuals feel appreciated. Team members are more likely to be dedicated to their work and employer if they realize a benefit from workplace well-being, are professionally satisfied, and share mutual professional respect with their supervisors.

An engineer who can lead a team by establishing healthy and productive informal communications can improve team dynamics and productivity while ensuring that individuals feel appreciated.

There are three key points for engineering project managers to note regarding informal communications. First, "informal" does not mean the communication is necessarily colloquial or that unprofessional conduct is acceptable; rather, it means that the communication takes place outside of a formally structured context (like a client meeting) or does not involve required corporate forms or protocols.

The second important point about informal communication is that, while it fosters a collaborative and friendly environment, professional discipline still needs to be maintained. It can be challenging for any leader to create a team culture that balances the need to create an encouraging and positive environment while maintaining professional expectations.

For example, suppose the tone becomes too friendly or collaboration devolves into socializing too often. In that case, the team risks becoming too casual in their interactions and may start to spend too much work time in debates or social conversations rather than being productive and respecting personal-professional boundaries.

In addition to productivity penalties, this environment may be perceived as undesirable and distracting to team members with introverted personalities who do not want to be forced into social interactions. Conversely, if the tone is continually too rigid, team members may find the environment hostile and leave to join another team or firm. A similar challenge comes from maintaining a professional tone with clients over a long period of time when it is important to keep communication positive but firmly realistic.

> A consideration of managing informal communications is the need to be aware and inclusive of all team member preferences – not just the majority.

The third critical consideration of managing informal communications is the need to be aware and inclusive of all team member preferences – not just the majority. If care is not taken in this regard, a communication that is interpreted positively by some team members might simultaneously be considered exclusionary and unfair by others. This can damage team dynamics and productivity by creating mistrust, or it could lead to an HR complaint being filed by a team member.

For example, a project manager might propose a monthly after-hours happy hour as a team-building event and an opportunity to recognize individual team members who have made outstanding accomplishments over the preceding month. This idea may be incredibly well received by team members with extroverted personalities, no family obligations, no reservations about alcohol consumption, and by those who are motivated by receiving public recognition. However, the same event may be perceived as an inappropriate forced socialization by team members who have introverted personalities, need to prioritize family obligations in the evenings or would rather recognition be handled privately between each employee and the team lead.

Although informal events should always be optional, simply labeling them as such is not an adequate solution. Doing so creates a scenario in which certain team members will always be excluded, and others will always receive more time to strengthen their relationships; this does nothing to resolve the potential for damaged team dynamics. Rather, if team-building events are part of an engineer's leadership style, communication must reflect an inclusive, rotating program of team-building opportunities throughout the year. This approach could include a happy hour, a lunchtime learning event during work hours, a coffee event, a weekend picnic that welcomes employee family members, an office bulletin or a gift certificate program to recognize accomplishments, or a poll allowing team members to share other ideas.

9.5 Communicating in Meetings

Meetings, whether in-person or virtual, *can* be the most effective format for discussing project issues and communicating with multiple team members; however, meetings can just as easily devolve into an unproductive use of the team's time or irresponsible use of client funds (in a time and materials or hourly contract). While scheduling a meeting and discussing a general topic is easy, planning and managing an effective meeting is deceptively challenging. The ease of setting up meetings can inadvertently encourage too many meetings. The project manager is responsible for ensuring that each meeting is necessary, organized, efficient, and adds value to the project. This section outlines best practices for keeping meeting time productive, managing the exchange of information, and establishing expectations about how ideas will be shared.

The first step for a successful experience is inviting the right people to the meeting. While this may seem obvious, it can be tempting to schedule meetings quickly and invite many personnel without considering what issues might arise and who might be best suited to add meaningful value to the conversation. Inviting people who do not need to attend will waste their time; conversely, forgetting to invite a team member with critical insight can lead to poor outcomes or the need to redo work later. With civil infrastructure and real estate development projects, meetings are often used to coordinate design and construction operations. These meetings benefit from contributions from the entire team for coordinated efforts, but this does not mean that all team members need to attend reoccurring meetings, nor do they need to attend for the entire duration.

For comprehensive meetings with many disciplines and stakeholders, the team will appreciate *time-boxing* meeting topics, which involves strict adherence to scheduled discussion topics. This approach can facilitate team members who need to participate in some conversations but not all. This approach is especially valuable for reoccurring coordination meetings. For instance, if the entire team is involved in a weekly coordinating meeting, the team can have a placeholder for a continuation meeting (also set weekly, perhaps on a different day and time). This way, if the discussion on one topic goes too long, the continuation meeting placeholder can be used with a smaller team and a focused agenda.

Topics can always be revisited as attendees join or items are dropped from the meeting agenda. For example, civil engineers will benefit from coordinating with the building team about ingress points, building utility connections, and site accessibility; however, civil engineers likely do not need to participate in conversations about interior finishes, elevators, and bathrooms.

9.5.1 Meeting Agendas

To ensure a productive meeting, it is essential to provide a well-defined meeting agenda ahead of the scheduled date. This additional effort on the part of the meeting initiator allows the meeting participants (potentially dozens of individuals) to understand how the meeting's time will be allocated, promoting better preparation and smoother proceedings. Conversely, without a clear agenda, meetings tend to become disorganized, leaving participants unprepared or unsure of the meeting's relevance, ultimately resulting in inefficient use of client and personnel time. Ideally, the agenda should be circulated along with the meeting invitation.

Without a clear agenda, meetings tend to become disorganized, leaving participants unprepared or unsure of the meeting's relevance, ultimately resulting in inefficient use of client and personnel time.

A meeting agenda does not need to be formal or exhaustive. However, the meeting initiator is responsible for justifying the use of everyone's time by preparing an agenda that ensures the meeting is efficient and productive. This approach places the burden on the meeting initiator rather than the participants and helps determine if a meeting is truly necessary.

Among the different agenda elements, it is a best practice to include information describing the objective at the beginning (and in the subject line) to clarify why the team is meeting. Similarly, the agenda can describe expected next steps and action items, which can be refined at the end of the meeting. Noting action items with a responsible party and a deadline ensures that the participants acknowledge expectations. Verbal action plans and agreements are easily forgotten. It is also important to allow time for open comments so team members can raise important ideas or concerns not specifically covered by the agenda. While a formal agenda may not be possible or necessary for all meetings, listing the objective is essential. An example of a well-defined objective is identified here:

> *"We are meeting to discuss the cost consequences of accelerating the schedule by two weeks, with an expectation that we will establish how to calculate the impact on budget and resource availability to provide the client with a change order."*

At the beginning of each meeting, the project manager should thank participants for joining, restate the purpose of the meeting, and make clear who will be running the meeting. With a small number of participants, especially with unfamiliar parties, it is customary to begin with introductions. For large meetings with time constraints, it may be better to provide a roster of attendees and their affiliations with the meeting invitation. The names and roles of participants should be gathered to create a record of who attended the meeting, and the meeting notes should be circulated after the meeting via email or by posting to a team working page.

Assigning a time limit and a responsible party to each agenda item is crucial. This helps limit confusion about which team member is responsible for different topics and gives them time in advance to prepare. This strategy can also help the project manager with time management to determine the appropriate amount of meeting time and manage the progression of the meeting.

During the meeting, the project manager must keep the discussion focused on the agenda while providing enough flexibility for brainstorming and conversation on potentially relevant topics. This can be more difficult to achieve than it might seem. Documenting peripheral ideas in a separate list is one strategy that can help keep the discussion focused while still signaling to team members that their ideas are valued; these ideas can be reviewed at the end of the meeting or captured as part of the meeting notes so they aren't forgotten. If it becomes apparent that a topic, whether directly related to the agenda or tangential, has prompted substantial discussion, scheduling a continuation meeting dedicated to that topic is better than attempting a quick resolution.

In all cases, documenting key discussion topics, results, and action items is a crucial aspect of all project meetings. In situations where a dedicated scribe isn't assigned, the project manager should take responsibility for summarizing these key points. Meeting notes play a vital role in ensuring transparency and creating a record of the project's progress, which can be invaluable for future reference, particularly in scenarios involving disagreements, confusion, or legal disputes over project directives and actions.

For some projects, there may be a formal requirement to document detailed meeting minutes, which stakeholders then share, review, and approve. Often, these minutes are distributed assuming they will be considered accurate unless disputed within a specified timeframe. In other scenarios, only a brief and informal meeting summary is warranted and can be beneficial, serving as a record of what was discussed and agreed upon; this can often be captured in an email following the meeting.

However, it is important to understand that meeting notes are not authoritative documents on their own. While they are helpful for record-keeping and transparency, they do not replace formal change management processes. Any changes to the project's scope, schedule, budget, or actions assigned to meeting participants must be officially escalated and documented through formal modifications to the project's governing documents, such as the contract, project plan, or scope of work. This formalization ensures that all changes are officially recognized, agreed upon, and integrated into the project's structure and strategy.

9.5.2 Managing Team Dynamics in Meetings

The project manager is responsible for keeping meetings focused on agenda topics and using appropriate communication strategies to manage the dynamics of personnel while ensuring the voices of each team member are heard. Specifically, project managers need to recognize that a group meeting format tends to favor extroverted participants. For example, it may be easy for extroverted team members to share their thoughts and receive input in a group setting, even on contentious or personal subjects; however, introspective team members may be uncomfortable in such settings and prefer to communicate in one-on-one sessions with their team members. For communication to be inclusive and effective, managers should be aware of the different communication styles of their team members and attempt to accommodate them.

For communication to be inclusive and effective, managers should be aware of the different communication styles of their team members and attempt to accommodate them.

A best practice for managing team dynamics is to inform the team that everyone will be asked to share their thoughts. This can be facilitated by including a list of discussion topics with the agenda before the meeting. This approach gives team members time to prepare their thoughts before the discussion, which may help them when asked to speak publicly. In the meeting itself, rather than only inviting open comments, it can be helpful to deliberately invite comments from each person, in turn, to make sure that each team member has been given the opportunity to speak. This tactic has the advantage of communicating to the team that everyone's input is valued and one person's input isn't more important than anyone else's. This tactic can be modified for cases where, for example, one team member's work is the meeting's primary focus, and that team member needs to take a more active role in the discussion.

In a meeting environment, some team members will volunteer solutions or answer questions more quickly than others because of their personalities, but a quick answer should not be confused with a competent one. While some team members answer quickly because they have an appropriate answer to a particular question, there are many other reasons that people may provide a quick response. For example, some might give a quick answer to seem like they are contributing or because they perceive themselves as competing with other team members. Other people may be apprehensive about silence and respond to alleviate their stress over an uncomfortable silence, even if it means providing incorrect information or a suboptimal suggestion. Project managers must be able to recognize these scenarios to improve communication outcomes and determine whether to rely on the answers they are receiving. In contrast to those who answer quickly, some team members are more reflective and will need time to process before responding. Project managers should not assume that a team member who takes longer to respond is less competent.

The project manager is also responsible for allocating an appropriate amount of time for feedback. For example, taking an intentionally timed pause after a question or between comments can encourage feedback from all participants. Communication through body language and facial expressions can indicate that a participant needs time to gather their thoughts before speaking up. Project managers must be able to recognize these scenarios to improve communication outcomes and determine whether they have successfully created a functional meeting environment. If a team member never contributes during meetings (or even after meeting events), it may indicate their attendance is unnecessary or that there is a personal or performance problem that the manager needs to address privately.

The greater context and audience for a meeting are important to consider. Project managers generally have the authority and responsibility to administer group dynamics during internal meetings. However, in external meetings with clients or other stakeholders, the individual behavior of each team member becomes more critical. For example, during an internal team meeting, if an overeager team member provides a hasty or impulsive response, the team can calmly and politely address the input. However, a team member making a mistake during a client meeting may create confusion or raise doubts about the engineering team's comprehension of the topic. Omitting key information, potentially because a participant is not confident in discussing the topic, could have similar consequences. For these reasons, project managers should work with their teams to prepare for external meetings and generate agreement about the approach, appropriate terminology, effective tone, and responsibilities.

In external meetings with clients or other stakeholders, the individual behavior of each team member becomes more critical.

9.5.3 Public Meetings

Public meetings with the community or as part of the formal government approval process warrant special consideration for two main reasons. First, many public meetings may be recorded, creating lasting evidence of what and how the team presented. Further, anything the team says or provides, including work that may be presented by other team members (e.g., land use attorneys or clients), becomes part of the official record. For these reasons, any work being presented should undergo a rigorous review. The review process should check content for accuracy and formatting (per branding guidelines) while considering how information might be reused beyond the meeting event. For example, could a draft concept layout be published by a third party and misconstrued as a final design? In that case, the project team could proactively add watermarks ("draft"), notes, and other disclaimers to mitigate against unexpected use by others.

A second reason public meetings require special attention to communication is the type of audience. As originally outlined in Chapter 1 (Section 1.5.1), engineers may need to communicate with different types of audiences during a project: external, internal, technical, and nontechnical. Recall that a technical audience includes any stakeholder with sufficient technical background to understand the engineering issues, constraints, and concepts fully. In contrast, a nontechnical audience includes those stakeholders that do not have or need expert knowledge. Public meetings generally involve external and nontechnical audiences (although some audience members could be experts). While the occasional citizen participant may be a professional engineer, it is best to assume that the audience in a public meeting does not possess a deep understanding of engineering principles. Communicating the technical rationale behind civil engineering design features and site layouts to nontechnical stakeholders can be challenging for a technical professional. Avoiding jargon, using familiar terms, and thoroughly explaining design concepts will best serve the audience. Team members should always exercise professionalism, present and respond with formal language, maintain a courteous tone, and exercise emotional intelligence even when challenged or faced with opposition.

9.5.4 Responding to Questions

Responding to questions is a key element of meeting communications. Project managers can expect to field a constant stream of questions from a wide range of stakeholders, including junior members of their own team, clients, partner consultants, senior management, the public, and others. A project manager may feel pressured to appear omniscient, particularly when responding verbally in meetings. However, engineers should be ethical in their responses and qualify limitations of knowledge. Inventing or guessing

When faced with a challenging question, an engineer should acknowledge if the answer is not immediately available but commit to providing one as soon as reasonably possible.

an answer will create problems later in the project that can damage the engineer's reputation or the client's trust. Rather, when faced with a challenging question, an engineer should acknowledge if the answer is not immediately available but commit to providing one as soon as reasonably possible. Similarly, if there is insufficient time to respond to a request, it is a best practice to provide a reasonable date by which the information can be provided. Examples of strategies that can be used in these scenarios include:

> *"I don't have that information, but I can get it for you by the end of the week."*
>
> *"I need to confirm a few things before I can give you a complete answer, but I will look into it as soon as I get back to the office."*
>
> *"That's a great question, but it might be outside of the scope we're working under. Let me check and get back to you with some ideas about how we can address this."*

These techniques become especially important in unforeseen or contentious circumstances. Should such circumstances arise, the project manager should not attempt to solve a problem on short notice or offer immediate assurances; doing so could put the team in a compromised position where the promised result is impossible to meet, lead to other unforeseen consequences, or fall outside the project services contract. The sheer volume of project documents, codes, and guidelines and the complexity of multidisciplinary teams will make it extremely difficult to have confidence in a hasty answer. For example, a quick reference to an important code or detail could inadvertently cite outdated (but still publicly available) documents. Similarly, overlooking an interdependency of design work could create challenges for some team members when attempting to resolve an issue quickly. Rather, it is a best practice for a project manager to acknowledge that there is reason to review the information and let the parties know that it will be addressed quickly by the team (or to the appropriate level of urgency the situation warrants).

9.6 Project Communication Documents

Many different project documents are used to communicate information throughout the project life cycle. These range from emails, memorandums, narratives, calculations, details, requests for information (RFIs), plan notes, and others. The effectiveness of the communication across all documents can contribute to (or detract from) project success. Thus, engineers need to be cognizant of the technical elements of these products and their communication functions as well. In civil infrastructure and real estate development projects, many project documents are publicly available. This condition should always be considered with any document produced for an external audience, such as plan review comments, comment responses, waivers, and design plans. The principles of effective communication identified throughout this chapter are applicable to project documents, and additional considerations are included in this section.

9.6.1 Emails and Other Written Documents

A project manager will participate in many internal and external project-related email communications. While the project team may use several messaging communication platforms, emails are most common for communicating with clients, senior management, external project partners, and other stakeholders. Email responses should be sent promptly because this helps maintain project momentum. A quick response may also be warranted if the sender reasonably expects immediate action. A response should indicate if immediate action is or is not possible. Sometimes, clarifying the urgency or need may uncover that additional time can be accommodated or inform the sender that the request will take longer than anticipated. When an immediate answer to an email inquiry is impossible, the best practice is to include a reply that the communication has been received and will be addressed at a scheduled date.

Generally, all emails should receive a response within a business day, although the exact amount of time permissible for a response will depend on the level of urgency of the topic (sometimes within minutes) as well as the corporate policy or culture of responsiveness. However, engineers need to distinguish when a response is legitimately required as a matter of urgency. Corporate culture and client expectations can occasionally put undue pressure on engineers to provide immediate results. These operating conditions can lead to stress, premature responses, poor outcomes, and team burnout, ultimately detracting from time dedicated to completing project work. A project manager needs to understand the context of inquiries and can inform the team of the priorities and urgencies in the greater context of the project schedule and the awareness of the client's expectations.

Another important element to consider in written communications is conciseness. Many professionals receive hundreds of emails each day, so effective email communication should highlight the main points quickly and clearly. Recipients should not be required to read wordy paragraphs to understand the material points. Additional or supporting information can be provided as an attachment, link, or further down in the body of the email for the recipient to read if they choose. The subject line of an email should also be used to convey important information. For example, if planning the materials for an upcoming meeting, instead of using the subject of "Meeting," it could be "Seven Acres Client Meeting Prep – June 29."

Effective email communication should highlight the main points quickly and clearly.

When requesting a response, action, or information, emails should list a specific recipient. An email addressed to "all" will confuse the recipients regarding who is accountable for the action. Many recipients may deprioritize emails not directly addressed to them and may expect they are included for reference only. An email with multiple recipients should include specific directions to each to clarify which individual is expected to provide the response, action, or information. Without assignment, there is no accountability to the recipients.

The list provided in Table 9.3 can help engineers at all levels craft successful email communications.

Email messages that are rude, angry, accusatory, or otherwise unprofessional may make a strong impression on the recipient and can be saved, recalled, or even forwarded well into the future; therefore, it is critical to keep email communication professional and respectful.

When communicating via email (and many other messaging platforms), it is important to remember that emails are saved on electronic storage devices essentially in perpetuity. Different parties can view these messages, and many organizations will have a policy to share email messages for project records. Email messages that are rude, angry, accusatory, or otherwise unprofessional may make a strong impression on the recipient and can be saved, recalled, or even forwarded well into the future; therefore, it is critical to keep email communication professional and respectful. Internally, such emails can form the basis for an HR inquiry as evidence of employee actions. A hasty and emotional response can have a lasting

TABLE 9.3 Effective email communication.

Make sure each email is:	In order to:
Addressed to the correct recipients with names spelled per address or signature. Address to one individual when possible to assign accountability. Do not presume pronouns.	Maintain basic professionalism.
Contains a greeting and sign-off, even in extended or rapid exchanges.	
Professionally written using appropriate language, complete sentences, and correct spelling and grammar.	
Clearly articulates the purpose of the communication and provides next steps as necessary, including dates for expected replies or results.	Ensure successful results.
Provides the necessary, but not excessive, background or context.	
Highlights any relevant deadlines and makes any requests of recipients respectfully with sufficient lead time. Provide a deadline if requesting information.	
Uses a format that can be quickly read, such as numbered lists, when appropriate.	
Includes the correct attachments, if any, and a reference to how the recipient should use or review them.	
Provides a record that may be relied upon several years into the future, even after the engineer may have left the firm.	Facilitate higher-level goals.
Written a way that is defensible if it were to be forwarded (with or without the engineer's knowledge).	
Positively supports the current engineer-client or other engineer-stakeholder relationship with a responsive and professional tone.	
Is reviewed for and acknowledges any existing or potential ethical implications.	
Advances the firm's/team's reputation and contributes to the future likelihood of being awarded additional assignments.	

impact on an engineer in performance reviews or long-term effects on team and client relationships. External email exposure can also be a significant cost to the engineering firm. For example, if an inappropriate email is sent to an external audience, it can affect how clients, the public, and municipal partners view the firm. The resulting damage to the company's reputation may be unknown, depending on whether or not the email is brought to management's attention. If a project issue warrants legal action, emails can be recalled showing what information was exchanged and whether an engineer's actions were cooperative or obstructionist.

Engineering professionals at all levels should assume that every email (and similar electronic communication) they send is neither private nor a one-time event. Emails (and other electronic media) are saved and circulated, and they can be recalled at any time. Project managers should regulate their own emails and educate their team members on the importance of using strictly professional communication in emails.

Engineering professionals at all levels should assume that every email they send is neither private nor a one-time event. Emails are saved and circulated, and they can be recalled at any time.

Serial Questioning and Deflection

Writing emails with clear actions and deadlines will establish an efficient communication channel. While email messaging is intended to be concise, it's important to include the relevant information to avoid a pattern of serial questioning. This occurs when the message content is disorganized, and the recipient is forced to ask clarifying questions to understand the author. This pattern can begin immediately after a message is delivered and requires the original sender to find more time to draft a new message with the missing information. In worse-case examples, some parties may even use this disorganization to deflect or delay responsibilities.

For example, suppose a project manager asks a subconsultant for a report but omits some information on format or deadlines. In this case, the recipient may immediately respond that more information is needed and will likely not react to the request until that information is provided. This is more problematic when the recipient provides a delayed response seeking additional information, which leaves the sender little time to clarify and respond. Suppose a request for a report is sent on Monday and the document is expected by Friday, but not all the relevant information is included in the email. A possible outcome is that the recipient might reply late Thursday stating that more information is needed and claiming they can't proceed until that information is provided. Because of the original lack of critical information, the likely outcome of this communication exchange is missing the Friday deadline.

These serial questions and deflections can be mitigated in a few ways and addressing them is particularly beneficial when working with a new contact or with someone known to use these tactics. First, interim deadlines can be established to prompt a response and identify the urgency. For example, sending a request for action or information by a deadline and indicating any clarifying questions should be noted earlier. This establishes implied consent, requiring the recipient to identify any concern or challenge. Second, a request with a long lead time (e.g., several weeks) should be periodically followed up with a polite and supportive note. This maintains focus on the request without pestering the recipient. As a final mitigation technique, it is beneficial to prioritize a list of requests or inquiries. This technique can inform the recipient about dependencies, which may help manage resources to provide the requested information or action. Similarly, if the recipient has a series of questions, prioritizing the inquiries can facilitate a focused effort to share the requested information. For especially urgent or complex requests, it may be best to follow up on written correspondence with verbal communication by calling or meeting with key parties.

9.6.2 Plan Review Comments and Response Letters

Comment response letters are a crucial yet often contentious communication for engineers. Reviewers provide written feedback on the design when engineering plans are submitted to supervisors, quality control (QC) teams, clients, or public agencies. These comment letters are meant to offer constructive guidance based on the project's requirements and current work state. However, the nature of these comments – their curtness, tone, format, and content – can sometimes lead to contention. Negative tones in comments, in particular, may elicit defensive responses from engineering teams who take pride in their work. Despite the inevitability of errors and omissions, how feedback is presented can significantly impact subsequent discussions and actions in addressing the comments.

However, it is also expected that engineering teams and project managers will defend their design decisions and not always concur with changes directed through comments, whether authored by the client, QC reviewers, or public agencies. The response to the comments could initiate some challenging conversations with clients over disagreements on proposed solutions, confusion about the project scope, or misunderstandings related to technical criteria. In other instances, jurisdictional reviewers may interpret regulatory requirements differently and request (potentially) unnecessary changes. There are also situations where the reviewers might be mistaken, potentially leading to legal implications if their suggestions result in noncompliance with local codes or introduce new risks to the project.

For instance, suppose a transportation reviewer mistakenly refers to an outdated version of a traffic safety manual when evaluating a roadway design project. Based on this outdated information, the reviewer requests significant modifications to the design team's plans, insisting on adherence to standards and threatening to withhold permit issuance. If the design engineers accept these changes without question and proceed to alter their plans accordingly, several issues could arise. First, these unnecessary modifications might result in increased project costs and potential delays as the team works to accommodate the reviewer's incorrect guidance. Additionally, implementing outdated safety standards could decrease the roadway's overall safety, contradicting one of the project's primary goals.

Note that, in this scenario, the engineers would fail to meet their professional standard of care by not verifying the relevance and accuracy of the reviewer's comments. It's important to remember that all stakeholders involved in the comment process aim to enhance project quality and ensure compliance. However, the project manager bears ultimate responsibility for the final deliverables.

The engineering team should ensure that any comments and requested changes will maintain the project's integrity, including safety, quality, and compliance with the contracted scope of work. While these plan review comment exchanges can be confrontational, the best solutions are resolved through respectful and patient communication, seeking to benefit from any conflict to find effective resolutions.

While these plan review comment exchanges can be confrontational, the best solutions are resolved through respectful and patient communication, seeking to benefit from any conflict to find effective resolutions.

Authoring Comments

When writing comments, a best practice is to refer to a requirement and how the design element is deficient. This means the reviewer who authors the comment should ensure they are familiar with the requirements. The design engineer will quickly dismiss invalid comments and could establish the perception that the reviewer is unfamiliar with the project and therefore will not provide valid or constructive comments. The comment structure should balance between being too curt and too verbose. A short, direct comment, such as "wrong. fix this." may sound autocratic; however, a wordy comment (or a comment with multiple parts) will only confuse the reader.

As noted with quality management practices (Chapter 5, Section 5.7), the corrective actions from comments should seek continuous improvement for the team and the project work. The source requirement should be identified for technical comments, and an effort should be made to address how the error was established. Finding the root cause helps teach and support the design team.

Corrective actions from comments should seek continuous improvement for the team and the project work.

It is best to maintain a consistent structure that follows: "Per <requirement>, <design deficiency>, <corrective action>."

A comment with an inquiry could help find a creative solution but should include an identifier to clarify that it's only suggestive in nature. For example, "Could this pipe be realigned to reduce materials?" may not warrant a formal response, but instead, it provides some constructive ideas. A comment should not have an uninformed question, such as, "Do you need to change this detail – what does the contract say?" These comments are not constructive for the project and only introduce additional work to the team that is likely required to respond to all comments.

Responding to Comments

As comments are received, the project manager and design team should respond in the same professional tone. Many quality management and plan review processes require all comments to have a response. The response provides acknowledgment by the design team that the information was received, and action will be taken, or that there is a legitimate reason not to act. Some comments may only identify a graphical error that can be noted without additional information. In that case, a simple "will comply" will satisfy a prescriptive comment. With a more technical comment, the response may need to include information on how the comment will be addressed to satisfy the requirement. A standard categorization of comments will also facilitate discussion, such as coding each response with:

- *Concur.* Action will be taken to address the comment.
- *Nonconcur.* The engineer believes that no action is required for a noted reason.
- *To be discussed.* The comment is unclear or warrants additional coordination.
- *For information only.* The comment is already satisfied, and information is provided to indicate how it is addressed, often referring to where in the documents it is addressed.

Given the challenges with written communication, the review process is best supported by verbal discussion. This could occur before the review begins when the project manager can identify the project status, known gaps, and expectations (especially for interim submissions). Similarly, meeting after the comments and responses are exchanged will resolve "nonconcur" issues or clarify any comments that require clarification. Communication during the review process should use factual references to requirements and eschew personal criticisms.

9.7 Graphics in Communication

Engineers are likely familiar with communication methods that use graphics. This includes engineering site plans, details, sketches, data charts, maps, model renderings, and other presentation content. This section provides some best practices for communication with graphics. Graphic design is an art associated with a wide range of creative disciplines. Knowledge of art theory and awareness of best practices with visual media will benefit an engineer's ability to communicate with various project stakeholders. These skills may require additional training but should be considered an investment in effective communication practices. These practices are important because of the challenges of communicating complex technical information to nontechnical audiences while retaining honest and ethical communication through visual media.

9.7.1 Standard Designs and Intuitive Visuals

The engineering industry has an array of standard design symbols, line types, and other plan design content. Such resources include the United States National CAD Standard® (NCS) and National Building Information Modeling Standards – United States (NBIMS–US™). An engineering firm may refine these industry standards; however, the best practice is to use a common visual language that can be recognized and understood without translating unique styles created by personnel. Even if the plans include legends and annotation, using industry standards will reduce the cognitive effort required for an audience to interpret information. Investigating and incorporating visual standards is beneficial across many communication tactics. For example, Figure 9.3 depicts two options for a directional north arrow, where the left is intuitive (one arrow and "N" for north), the one on the right is decorative but makes it difficult to discern the direction (identified with a small circular marker on the border).

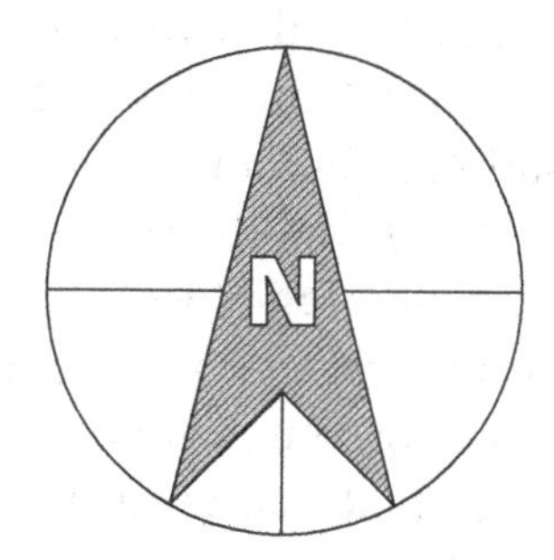

FIGURE 9.3 Sample directional north arrows, comparing an intuitive style (left) with one that requires additional cognitive effort to decipher.

This practice extends to other industry standards, such as selecting colors that are intuitive to the audience. For example, red is often associated with actions to "stop" or to represent negative conditions (such as poor financials, warnings, or errors). Other examples include the colors selected to represent utilities. Different color associations are used with utility designations as published by various agencies, which is a helpful reference when depicting those utilities in colors on plan sheets (orange as communication, blue as potable water, yellow as gas, etc.). Similarly, there are published and accepted standards for date conventions, such as those published by the International Organization for Standardization (ISO 8601):

YYYY-MM-DD

Where "Y" is the year in a four-digit format, "M" is the month in a two-digit format, and "D" is the day in a two-digit format. This uniformity, including date convention, mitigates confusion or complexity when reviewing documents.

9.7.2 Graphics for Legibility and Accessibility

When selecting content styles, engineers must consider legibility and accessibility for the intended audience. Many jurisdictional agencies have established minimum text heights for plans (often 0.1 inches) and suggestions on consistent text styles to improve plan legibility. The choice of font, text orientation, and page formatting may seem inconsequential compared to the technical content. Still, these procedural methods create a unified format that is easy for readers to interpret. Sources from the United States National CAD Standard® (NCS), National Building Information Modeling Standards – United States (NBIMS–US™), and the US Army Corps of Engineers (USACE) provide publicly available graphics design standards with best practices. Small, skewed, or mechanical style fonts (simplex) can be challenging to read and complicate visual communications.

When selecting content styles, engineers must consider legibility and accessibility for the intended audience.

Many engineering graphics follow a grayscale format, but color plans and charts should consider accessibility. For example, using patterns and colors in a chart (e.g., a pie chart) can benefit cases where some content is distributed in black and white or when readers have a color vision deficiency. With the examples referenced in Section 9.7.1 (such as color-coded utilities), the graphics should still include a line type with an indicator (such as "W" for water). Similarly, while red and green often serve as common representations for traffic operations, the colors do not translate well for some audiences with color vision deficiency. Many design tools provide alternative color palettes that can be used in addition to or instead of traditional color palettes.

9.7.3 Ethics with Visual Representations

Many of the published guidelines for ethical data visualization apply to engineering plan graphics and content. Even if unintentional, many graphics can be deceptive and misleading to the audience because the author of the content will have the context that the audience does not have. This can cause confusion or misinterpretation, so created content benefits from testing the visuals with an unfamiliar audience. For example, engineers from another team can review the graphics without any background context to test how it will be interpreted to a larger audience.

Many visuals created in civil engineering are intended to depict an attractive future condition of a project location. This includes showing future plantings, terrain, and building visibility. While these graphics are often designed to persuade the audience to support an initiative, the content should be honest and transparent. For example, showing a "tree save" area on a site plan should represent reasonable expectations for healthy and mature vegetation that can be protected from construction operations and should not arbitrarily outline surveyed tree canopy. Similarly, renderings that show proposed trees should include multiple phases that accurately represent the planted conditions and not just the anticipated 10- or 15-year tree size and canopy.

The project team should investigate other unintentional misrepresentations on plan graphics. This might include how site features are represented with colors or how content is annotated. For example, the project team should aim to balance marketing and realism in graphics, recognizing that not all stormwater features will have clear blue water. Similarly, omitting critical elements like utility poles, roof-mounted building equipment, or signage could also be interpreted as misleading to an audience evaluating the site's aesthetics.

The narrative in Appendix A, Part 9 of Millbrook Logistics Park, delves into a series of challenging discussions between the client (TerraHaven) and the engineering team (ApexTech). Upon uncovering troubling developments regarding the project's progress, the engineers must navigate these difficult conversations to forge a constructive path forward. This section highlights how emotional intelligence significantly shapes communication, demonstrating how the engineering team can effectively exchange information and enable better decision-making under pressure.

CHAPTER 10

Leadership, Power, and Stakeholder Relationships

Lead with integrity, adapt without compromise, and continuously improve.

CHAPTER OUTLINE

Management Essentials for Civil Engineers: A Practical Guide to Business, Communication, Ethics, and Risk, First Edition. Cody A. Pennetti, C. Kat Grimsley, and Brian M. Grindall.

This chapter begins by examining various leadership styles and forms of power, essential for understanding the dynamics of effective team and stakeholder leadership. It then delves into stakeholder management, a critical aspect of successful project leadership, with a focus on the importance of strategic planning in this context. Best practices in stakeholder relationship management are covered, offering insights and tools applicable to all facets of project operations. Engaging with diverse groups and navigating complex interpersonal dynamics are key to ensuring project success and fostering positive stakeholder relationships.

10.1 Introduction

Good project managers use both management and leadership skills while also maintaining an awareness of the broader context of project operations. Project managers are often the supervisors of other personnel in a department of an engineering firm, with responsibilities for the project's success and for serving as a leader to their colleagues.

Management skills are applied to project planning and operations, such as managing resources, schedules, and costs. In contrast, leadership skills are used to coach, support, and promote the project team members. This distinction between the application of management and leadership is essential. In civil engineering projects, almost all project work is developed by *human resources*, the expert professionals who use tools and techniques to imagine, design, and create the project deliverables. While team personnel fall under the category of "resources," it's essential to understand that leading people fundamentally differs from managing mechanical resources. Influential leaders in engineering inspire, mentor, and empower their teams while managing projects. Balancing management and leadership skills in engineering projects determines success and organizational growth.

Leading people fundamentally differs from managing mechanical resources.

10.2 Projects in Programs

Projects do not exist in isolation; they function within a shared environment of resources and stakeholders, often overlapping with other organizational operations. This interconnectedness arises because each project is not just about its specific requirements but is also tied to the broader objectives of both the engineering firm and the client. As discussed in Chapter 2 (Section 2.1), much of this realization comes from understanding a project's origin and purpose. This involves looking beyond its project-specific scope to understand *why* it is necessary in the first place. For an engineering firm, the project may align well with the portfolio of services, personnel skill sets, and goal of strengthening a portfolio of work. From the client's perspective, one project could be the first step in a growing geographic market, improving regional transportation operations, or increasing infrastructure capacity to accommodate new growth. This broader view aligns projects with an organization's strategic direction, where success is measured at a larger scale.

It's crucial to recognize that prioritizing the success of one project over others can be counterproductive to an organization's overarching strategic objectives. In this context, when multiple projects collectively contribute to a strategic goal, they are grouped into a *program*. Further, a *portfolio* encompasses a collection of related projects and programs, sharing common objectives and resources.

It's crucial to recognize that prioritizing the success of one project over others can be counterproductive to an organization's overarching strategic objectives.

This structured hierarchy of projects, programs, and portfolios is defined differently from the perspectives of the client and the engineering firm. Each entity views and organizes these elements in a manner that aligns with its unique strategic and operational needs.

This association of projects and programs informs the coordination of operations, resources, and delivery that is best realized through a programmatic effort. The geospatial associations of projects (i.e., where the projects are located) are also significant in civil engineering work because a project's location may influence the program and portfolio hierarchies (perhaps regardless of client or scope). Figure 10.1 expands from the initial portfolio, program, and project graphic in Chapter 1 and depicts a possible cross-section that considers how projects are aligned with disparate clients and engineering firms.

Engineering firms and clients will have disparate organizational strategies, which means the structure of portfolios, programs, and projects will also differ. A civil engineering firm may structure its portfolios by market segment (e.g., residential, education, transposition, etc.), with each program representing a different client in that market segment. The engineering firm could have various projects with each client, as shown in the vertical structure of Figure 10.1. Conversely, a client may organize portfolios based on

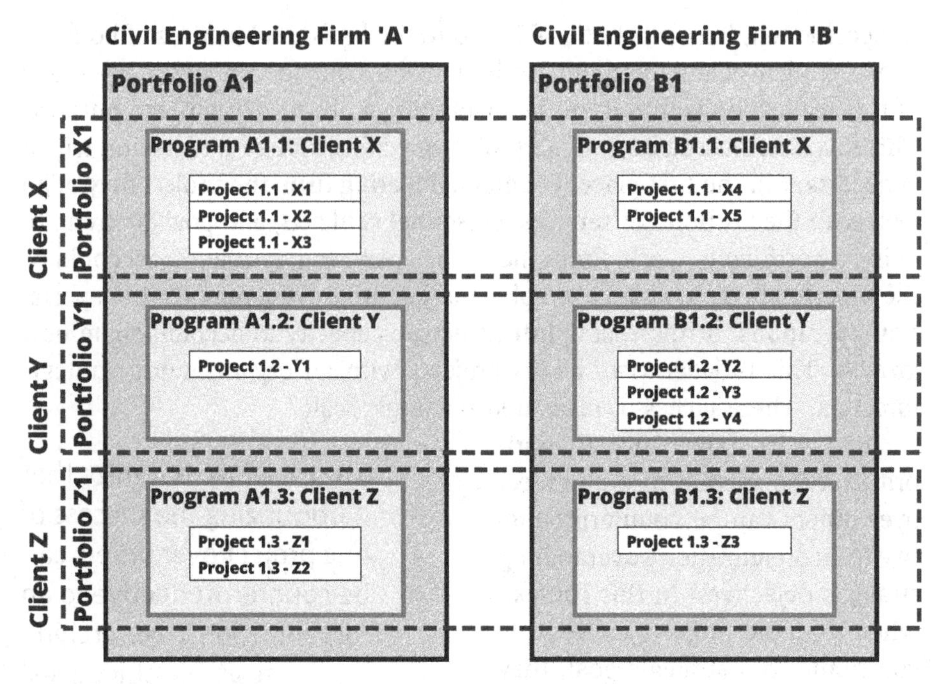

FIGURE 10.1 Aligning projects between clients and engineering firms.

geographic regions, perhaps with different programs based on which engineering firm is working on which projects or project types, as shown in the horizontal structure of Figure 10.1.

For example, suppose a transportation department (client) has a program to improve intersection safety along a highway corridor. Each intersection would be a project with a unique team, project manager, requirements, and design (perhaps with different engineering firms). However, because these multiple projects are aligned with shared resources and stakeholders, they benefit from a higher level of oversight at the program level. Additionally, suppose the transportation department has a different program for bicycle route planning with a few projects that overlap (spatially) with the intersection safety projects. In this case, both programs would benefit from portfolio management aligned to the geographic area. Projects in a portfolio are more than a categorization; they require governance

Projects in a portfolio are more than a categorization; they require governance that evaluates and balances the various projects' resources, schedules, objectives, and costs.

that evaluates and balances the various projects' resources, schedules, objectives, and costs.

From an engineering firm's perspective, strategic operations may revolve around one core client representing a substantial share of the firm's revenue or having other perceived value (e.g., growth market or high visibility). With finite resources, an organization must prioritize programs and projects to allocate resources accordingly. For example, one engineering firm cannot complete all traffic and transportation projects in a geographic region. Instead, the firm will evaluate resources and qualifications to determine which projects they should pursue (based on other criteria in a go/no-go a pursuit, as described in Chapter 3, Section 3.2).

The triple constraints (scope, schedule, and cost, introduced in Chapter 1) will scale to the level of programs and portfolios because resources are shared across the origination. For example, reassigning personnel from Project A to Project B could improve the schedule of Project B while causing delays with Project A. These portfolio and program-level decisions continue to emphasize delivering value and achieving success at the organizational level. To support these and other coordinated outcomes, project managers must develop both managerial and leadership skills.

10.3 Leadership Styles

Project managers often have a preferred leadership style, depending on their personal attributes. Some are charismatic, naturally inspiring others with their strong and confident personalities. Others may prefer a collaborative and teaching leadership style, earning respect through demonstrated technical expertise. While certain styles may feel more natural to an individual, relying on a single style is not always effective because of the diversity of teams and varying project situations. Instead, a project manager's leadership style is a skill that can be developed and should be *adapted* based on team members' personalities and the project manager's capabilities.

A project manager's leadership style is a skill that can be developed and should be adapted based on team members' personalities and the project manager's capabilities.

Managers must understand that a different leadership styles will either negatively or positively impact personnel motivation and morale.

Additionally, different styles may be appropriate in different situations. The right leadership style should be chosen based on the environment and always with the intent of moving the team toward project success. While some styles may come more naturally to an individual because of a project manager's personality or culture, an effective leader must be capable of more than one style. This adaptive leadership style considers project conditions, team characteristics, and client objectives.

This section identifies some common leadership styles, and several others are defined by PMI. Note that it often requires self-awareness and a deliberate effort to be effective with each style and to know when to seek training to develop these skills. The most common leadership styles are transactional, teaching, autocratic, laissez-faire, and servant. Each has benefits and challenges based on the project situation and the team's characteristics.

Transactional

A structured leadership process is considered *transactional.* In this case, there are clear expectations, processes, and direct feedback to the team. Cyclical feedback is a core element of this leadership style and is meant to continuously guide the team through rewards and penalties to achieve project objectives. This approach could work well for projects with a visible team structure and predictability so the team understands the expectations and the goal.

This hands-on approach may be most appropriate for newer engineers or those unfamiliar with the specific project environment. It can also be valuable for employees who prefer structure and feedback. This style would be most challenging when the project must frequently adapt to new requirements or schedules and the expectations become muddled, which could cause frustration among the team.

Teaching

A *teaching* leadership style supports the growth of team members, focusing on teaching new skills and supporting career advancement. Through a teaching leadership style, the project manager will seek to grow the team's capabilities as they learn through structured processes designed to refine a skillset; however, the team members are also encouraged to improve their skills independently.

While this approach has long-term advantages in strengthening team members, there must be a commitment from the engineering team and the organization to invest in this growth. The project manager will likely also learn from this style as it encourages open discussions and exploration into the *why* of processes. However, this style can be time consuming and may not be appropriate when a project demands rapid production with little time for exploratory learning.

Autocratic

An *autocratic* leadership style is authoritative in nature and is characterized by a leader who typically prefers a strict hierarchy and directs the team. This approach provides little room for the team to give input or work through a creative solution – instead, the team is expected to execute project work as directed by the manager. In a creative setting, such as engineering, many team members may not respond well to this leadership style, at least not for an extended period.

Indeed, this approach could be appropriate when it is necessary to be efficient and deliver results quickly under short deadlines. For example, a late change in design may not leave room for discussion as the team needs to make edits to the design quickly. Still, this is best reserved for temporary and critical actions or used when onboarding new employees who need (and appreciate) concrete direction.

Laissez-faire

A *laissez-faire* leadership style is the most hands-off approach and can seem almost void of involvement and direct leadership, but it does provide a unique situation for project team member growth. There is little interaction between a project manager and the engineering team under this approach. Instead, the team is encouraged to operate autonomously and learn through their actions. Some engineers on a project team may prefer this style if they are self-driven, experienced, and confident in their abilities (but they are not in the project manager role).

However, this approach may not work well for new engineers who require and expect guidance from a project manager, which would likely lead to confusion, frustration, and lack of productivity by the team. This style also may not be appreciated by team members who value recognition or feedback, or when personality conflicts need resolution. In the worst case, laissez-faire leadership can lead to unsupervised friction within a team, lack of equity, and poor project outcomes.

Servant

The *servant* leadership style shares some similarities to laissez-faire in that it grants autonomy to the project team, but there are crucial distinctions between the two styles. The primary difference is that while a servant leader allows the team significant operational freedom, this leader is still actively involved. As a leader who serves the team members, the manager frequently communicates with the team to understand the project status and needs. The project manager then seeks ways to serve the team by removing obstacles and providing access to the necessary resources. For example, the project manager may focus most on communicating the team's progress and inquiries during meetings, leaving the team to stay focused on the production.

This style is effective for a mature project team with mutual trust and in cases where exploration, creativity, and innovation are welcomed. With a disciplined team, this leadership style can be effective even with aggressive design schedules but relies more on the individualism of team members and less on the direct management of the leader.

10.3.1 Choosing a Leadership Style

Several other leadership styles exist, ranging from a directive to a noninterventionist, and some use charisma (or fear), reward (or penalties), and other motivators to yield results. Many engineering teams will appreciate freedom in creativity and time allocated to test new solutions, which is best accommodated by communicating project requirements, schedules, and budgets. Again, the expectation is that a project manager can operate in different leadership styles throughout the project based on current project conditions and team needs.

For example, in the early stages of a project, a project manager has the opportunity to use a teaching leadership style, allowing the team to explore various solutions and learn by actively engaging in tasks. However, a more directive, autocratic approach may become necessary as a project approaches its deadline. At this stage, the project manager may need to provide explicit instructions to ensure all deliverables are finalized on time. As the team matures and better comprehends the project's objectives, operations, and client expectations, the project manager can adopt a servant leadership style, respecting the team's earned autonomy. Furthermore, a project manager overseeing several concurrent projects with different teams may need to adapt their leadership style to meet the unique needs of each team. A self-aware and proactive leader consistently adapts their leadership style to fit the project's demands, underpinned by a deep respect for the team.

A self-aware and proactive leader consistently adapts their leadership style to fit the project's demands, underpinned by a deep respect for the team.

10.4 Leadership Proficiencies

Project managers must continuously cultivate and refine their leadership proficiencies to best serve their team and project stakeholders. However, leadership skills are not limited to managerial roles or specific positional authorities; any project team member can practice and exhibit them. Leadership skills are essential to managing scope, quality, cost, schedule, risk, communication, and other project operations. Leadership practices and norms have changed over time, and engineers should monitor industry practices while remaining aware of disparate stakeholder cultures, preferences, and perspectives. This section summarizes core proficiencies.

Leadership skills are not limited to managerial roles or specific positional authorities; any project team member can practice and exhibit them.

Listening and Communication Skills

Much of a project manager's responsibilities revolve around communication. Influential leaders excel in both verbal and written communication. They can convey ideas, listen actively, and adapt their communication style to suit different audiences. Civil engineering leaders must understand how to communicate complex topics to technical and nontechnical audiences.

Leaders provide constructive feedback and coaching to help team members grow and develop their skills. More information is described in Chapter 9, Section 9.4.

Empathy and Emotional Intelligence

Strong leaders understand and empathize with the emotions of team members. They use emotional intelligence to build relationships, show empathy, and foster a positive environment for the project team and stakeholders. Proficient leaders inspire and motivate their teams. They recognize achievements, provide positive feedback, and create a sense of purpose and enthusiasm. For example, it is important to celebrate project milestones with the team, especially with demanding schedules and complex requirements. Leaders with cross-cultural competence can work effectively with individuals from diverse backgrounds, respecting and leveraging cultural differences.

Adaptability

Effective leaders are adaptable and open to change. They can adjust to evolving circumstances, embrace innovation, and guide their teams through transitions. Most importantly, effective leaders may need to change their own behaviors or expectations to exercise adaptability. Many civil engineering projects will experience unforeseen conditions as new information is acquired through site investigations or project requirements may change in response to regulatory policies or other external factors. As project challenges emerge, leaders address conflicts by facilitating constructive conversations, identifying root causes, and implementing solutions to maintain a productive work environment.

Decision-Making

Leadership involves making informed and timely decisions. Proficient leaders assess situations, consider available information, and make choices that align with organizational goals. Engineering projects often have multiple viable solutions with different trade-offs associated with cost, development time, safety, risk, sustainability, and client value. While project managers are accountable for success (or failures), they rely on the project team to execute the work. Trust and delegation are essential leadership skills. Proficient leaders assign tasks appropriately, trust their team members' capabilities, and provide support as needed. Note that evaluating and properly supporting individual team members is a critical element of delegation, and managers need to be proactive in identifying the need for and providing training or other resources to support team members who may otherwise fail when a delegation does not align with their skill set.

Ethical Leadership

Engineers must identify ethical situations, understand how to assess related conditions and act in ways that "hold paramount the safety, health, and welfare of the public" as defined by the National Society of Professional Engineers (NSPE) code of ethics. They model ethical behavior, make moral decisions, and set an example for their team and organization. Proficient leaders build and maintain a solid professional network to connect with peers, mentors, and industry contacts to gain insights and support their goals and client objectives.

Risk Management and Resilience

Leaders identify and respond to risks effectively. They assess potential threats, develop contingency plans, and minimize negative impacts on projects and operations. Successful leaders excel in handling crises and unforeseen challenges. This is a continuous responsibility in any project as external factors and project requirements shift. Risk management is about engineers remaining composed, making informed decisions, and leading their teams through adversity. Resilient leaders bounce back from setbacks and challenges and maintain a positive outlook while learning from failures. Refer to Chapter 8 for more information on risk management.

Strategic Thinking

Proficient leaders have a strategic mindset. They can formulate and execute long-term plans, aligning their actions with the organization's vision and goals. This means that engineers think beyond project-centric decisions and understand the interdependencies across projects and programs. Proficient leaders are also skilled negotiators who can secure favorable agreements, resolve conflicts, and advocate for their team or organization's interests. Leaders efficiently manage their time and prioritize tasks.

Humility

Humility is the quality of being modest, open to feedback, and willing to admit one's mistakes. This means project managers should recognize and celebrate solutions from their team members and welcome novel perspectives, especially from new engineers. Humble leaders are approachable, value the contributions of their team members, and create an inclusive and collaborative work environment. They are not driven by ego but by the team's and the organization's collective success. Note that leaders must often balance humility with the need to maintain authority and the team's respect, which can be a delicate nuance in practice.

Innovation and Creativity

Effective leaders encourage innovation and creativity within their teams. They intentionally create an environment that welcomes new ideas and supports experimentation. This is especially important for the engineering industry, which continuously evolves as new technologies and crises demand innovative solutions. From a business perspective, innovation has value as a driver of competition.

People Skills and Team Building

Leaders strive to build cohesive and high-performing teams. Ideally, they select and embrace diverse talents, foster collaboration, and create an environment where each team member feels inclusive and can thrive. To achieve this, leaders may need to overcome their own biases and must be proficient in promoting a harmonious work environment by mediating disputes and finding common ground.

10.4.1 Behaviors in Leadership and Management

Successfully leading a project team requires the engineer to practice self-awareness and develop a basic understanding of leadership versus management. Even within each of these broad categories, there are different styles and communication approaches, which may be more appropriate depending on the situation and composition of a particular team. Table 10.1 provides some examples of leadership versus management activities and associated communication practices, all of which the engineer may be engaged with at different times throughout a project.

TABLE 10.1 **Leading versus managing.**

When leading, an engineer is. . .	When managing, an engineer is. . .
• **Inspiring, motivating, and empowering the team.** Example: Before a difficult client presentation, the project manager maintains an optimistic attitude, reassures the team that they are well prepared, and then sets the tone by greeting the client confidently and introducing the team. Communication note: These types of leadership activities often use uplifting messaging for the purpose of encouragement rather than focusing on the discussion of facts and data. However, quantifiable evidence can certainly be used to support aspirations.	• **Directing the team, allocating resources, and assigning work to team members.** Example: The project manager decides that during the client presentation, Engineer A will be responsible for discussing the site layout, and Engineer B will be discussing the budget. Communication note: It can be helpful to explain the reasoning behind decisions to avoid confusion and ensure transparency, e.g., the client is very picky about their site, and Engineer A has the most experience with this type of layout; Engineer B has not presented on budgets yet, and this will be a good learning opportunity.
• **Establishing strategies and innovating new solutions to non-project-specific problems.** Example: The team lead spends time identifying ways to increase the team's internal visibility and to connect the team's work program to corporate strategic goals. Communication note: It can be helpful to explain how this use of the project manager's time can benefit the team, for example, by positioning them to be assigned more complex projects that may be interesting to work on but that could also lead to bonuses or promotion and other growth opportunities for team members.	• **Facilitating the team's ability to achieve project results on time and budget.** Example: The team's progress is constantly delayed by the slow responses of another department, so the project manager sets up a series of meetings with functional managers in that department to troubleshoot the workflow process and find ways to improve response time. Communication note: It can be helpful to let the team know these meetings are taking place; this shows the team that their manager is proactive about working to support their progress and can also help prevent future surprises if the team needs to make a process change to work more collaboratively with the other division.

(Continued)

TABLE 10.1 *(Continued)*

When leading, an engineer is. . .	When managing, an engineer is. . .
• **Focused on vision and future development.** Example: The project manager invites the team to participate in a brainstorming exercise to see how the less-intensive times can be used strategically for professional development and explore what continuing education opportunities might be valuable to team members for their own growth and to upskill to win new types of work. Communication note: It can be helpful to explain that the goal in visioning and brainstorming exercises is to focus on open creativity and generating ideas; it is often not essential to identify strictly if or how these ideas will be carried out.	• **Focused on process, reporting, scheduling, and other administrative responsibilities.** Example: The project manager team lead requires summer vacation requests to be made at least one month in advance and only approves team members to be out of the office on non-overlapping weeks. Communication note: Explaining the rationale behind the vacation policy can be helpful. For example, the policy ensures that the team has plenty of time to prepare for absences and understand schedule and cost implications.

Understanding these differences can be particularly challenging because, while they are distinct activities, there is often overlap between leadership and management behaviors – and a project manager may engage in one or both roles at different times or simultaneously. Also, the degree to which a project manager successfully translates these skills into team communications can materially impact team morale, productivity, and overall workplace engagement.

For example, selecting the appropriate work for each team member is a management task that considers project cost implications based on personnel skills and efficiency. The corresponding leadership responsibility involves inspiring, motivating, and explaining project goals to the team members so they are aligned with objectives and find ways to correlate their personal career interests.

Communicating the value and purpose of each task provides greater context about the project objectives and may motivate personnel who might otherwise be apathetic to the assignment. In this scenario, embracing only the management role while failing to connect through personal leadership will create a potentially unpleasant work environment if the work assignments portray favoritism or discrimination, and result in reduced employee engagement, increased stress, dissension, or resignation. The project manager should be aware and empathetic to what motivates and interests team members and aim to align work accordingly. Does a team member have interests beyond a commonly assigned (and potentially repetitive) task? Is there an opportunity for a team member to learn a new skill? In other words, a successful project manager needs to be careful that their communication doesn't consistently portray an autocratic voice: "You will do this work because I told you to do it."

Management and leadership share common aspects and similar underlying goals, but the terms are not synonymous, and the associated behaviors are not identical. Both management and leadership traits and behaviors are required for successful projects, and they are used based on context, project situations, and team member personalities. Table 10.2 provides more examples of categorizing management and leadership aspects.

Management and leadership share common aspects and similar underlying goals, but the terms are not synonymous, nor are the associated behaviors.

TABLE 10.2 Categorizing management and leadership aspects.

Aspect	Project Management	Leadership
Role	Established by a position based on documented credentials to focus on defined processes to achieve results.	Does not rely on a formal position or role and is focused on alignment, motivation, and inspiration toward shared vision and goals.
Decision-Making	Data-based decisions using known models and formulas to optimize scope, schedule, cost, and resource management.	Considers broader objectives beyond the project and considers the long-term success for client, organization, and team members.
Accountability	Designates roles and responsibilities to team members based on hierarchical project structures.	As the project manager, recognizes they are ultimately accountable for performance; as a team member, recognizes individual responsibilities to support other team members and the project manager.

(Continued)

TABLE 10.2 *(Continued)*

Aspect	Project Management	Leadership
Solutions	Monitors project operations to determine issues and challenges resolved through root cause analysis and refinements in management processes.	Promotes and fosters innovation with dedicated support and resources to promote organizational goals and team interests.
Interaction	Establishes meeting and client-engineer communication practices based on project coordination requirements.	Considers personal aspects that influence team members' interests and performance with open and transparent communication.
Task Execution	Emphasizes work completion based on defined project tasks and adherence to planned scope, schedule, and cost.	Emphasizes team development and fosters a positive, inclusive work culture supporting professional growth.

10.5 Forms of Power

Many project managers (especially those who have recently transitioned from a technical production role) find themselves in a challenging position because they are accountable for project success but have limited positional power over organizational resources. A project manager will not have absolute power to acquire resources or direct team members and stakeholders. For example, project managers often rely on team members with supervisors in other departments – the availability and priorities of these team members may be governed by their direct supervisor (functional managers). Similarly, project managers often rely on corporate support, such as legal and accounting teams, with no direct authority over corporate schedules and priorities.

In its many forms, power often serves as a driving force behind decision-making, problem-solving, and effective project management. Power can come from a formal position held in an organization, but it can also stem from an individual's expertise, knowledge, charisma, network, and other sources

> Power can come from a formal position held in an organization, but it can also stem from an individual's expertise, knowledge, charisma, network, and other sources that enable that individual to influence others.

that enable that individual to influence others. This section references some of the commonly recognized forms of power.

Positional

An organization, whether public or private, can assign power to individuals through their labeled *position* (sometimes called legitimate power). While the position in an organization sometimes seems arbitrary, there could be policies based on titles. For example, the difference between a vice president and a senior vice president may correlate to a maximum signatory contract value. In this case, an individual may not be authorized (by internal policy) to serve as a project manager based on the contractual size. The position of an individual may also give them the power of resources, which could influence a project's efficiency and production capabilities. At the project level, the assigned project role could limit some responsibilities or participation in project activities.

Charismatic and Personal

Charisma can influence people to act, but this power relies heavily on an individual's traits. This power comes from a personality that inspires through strong communication, passion, trust, and confidence (sometimes called referent power). Similarly, power from *personality* comes from someone who is considered a role model to their peers. These forms of power recognize that human factors can influence decisions, sometimes even more than positional authority. For example, a charismatic leader can inspire and garner support from team members through encouragement and by setting an example of a strong work ethic. Even without formal authority, both charisma and personal power can effectively influence stakeholders to promote project success.

Information and Expertise

Power from information and expertise are two distinct forms of power, but they share the principle that knowledge can influence action. For example, *informational* power could come from an individual understanding the context or underlying motivations influencing stakeholder actions. This might include knowledge about a client's preferences. Power from *expertise* references deep technical knowledge, such as when an individual realizes they are the only one with experience with a product's complex technical specifications required for the project. In both cases, the individual's position is arbitrary, and instead, the knowledge they hold establishes their power.

Network

Sometimes, a person's power comes from their *network*, either through professional or personal connections. This power could be well-deserved based on a reputation of solid project performance or technical expertise. This could also come from charisma and kindness, providing a voluntarily supportive network. In other cases, network-based power may be solely a result of who the shared connections are or even nepotism. In some ways, power from networks is an extension of charismatic and personal power. Still, it relates to how relationships can establish a robust personnel network that can influence project resources or the actions of stakeholders.

Coercive or Reward

When an individual can impose action-based penalties, this is called *coercive* power. In an engineering management setting, this should be reserved for addressing noncompliance issues or safety violations. Team members should not constantly fear negative consequences but realize that frequent issues in compliance and performance can result in reassignment or termination.

The alternative action is *reward*-based incentivization, where behavior is encouraged because the individual holding power can provide meaningful incentives, such as salary bonuses or promotions. Although rewards typically serve as a positive motivator, surpassing the effectiveness of coercive power, their impact hinges on equitable distribution and consistency in acknowledging superior performance. If not managed fairly, rewards can inadvertently become a source of demotivation for others. In both cases, these forms of power are meant to establish intrinsic motivation as individuals seek personal growth and feel accomplished for their efforts.

10.5.1 Abuse of Power

It is imperative that power is wielded ethically, aligning with the project's positive objectives and promoting the professional growth of team members. The misuse of power can lead to severe and enduring consequences, impacting not only the company's culture and team morale but also the reputation of individuals and

It is imperative that power is wielded ethically, aligning with the project's positive objectives and promoting the professional growth of team members.

the organization as a whole. Engineers should remain mindful of two critical aspects:

1. *Self-reflection*. Engineers should evaluate their use of power to ensure it adheres to ethical and socially acceptable standards. It's important to be vigilant and avoid behaviors that are, or may be perceived as, unethical or inappropriate.
2. *Observation and intervention*. Equally important is recognizing and responding to instances where power is being wielded inappropriately, whether one is directly involved or a bystander. This awareness requires a keen understanding of power dynamics, which may not always be immediately evident. Note that intervention of any kind should be carefully handled so as not to further undermine a vulnerable party.

For instance, within the context of civil engineering project management, subtle abuses of power, or attempts to claim it, may manifest in several ways:

1. Selective Communication
   - This abuse of power involves manipulating the flow of information within a team and controlling who receives or contributes to vital communications. An example of this could be the deliberate scheduling of key project meetings at times known to be inconvenient for certain team members. For instance, a project manager might schedule an important meeting late in the day, fully aware that a particular team member, who may have other commitments such as parental duties, cannot attend. Such actions effectively exclude certain individuals from crucial discussions and decision-making processes based on non-merit factors.
2. Gaslighting and Undermining
   - Gaslighting involves manipulating perceptions and creating self-doubt in others. This action of undermining can take the form of subtle comments questioning a colleague's work or ideas, especially in public settings. In civil engineering, gaslighting could involve undermining the credibility of a colleague's design or engineering solutions, leading to self-doubt. Undermining may manifest through comments, such as openly questioning or challenging a design option during a meeting without constructive intent. Note that team members without legitimate sources of power may use gaslighting and undermining

in an attempt to gain power by making themselves look more valuable in comparison to their colleagues. These may also be devices used to deliberately harm others over whom a person does not have any other power.

3. Sabotage via Withholding
   - Within a project team, someone with power may intentionally withhold critical information or resources, hindering project progress. For instance, a team member failing to share essential geotechnical data with the design team promptly can affect the site feasibility or cost implications. This could also create undue stress for other team members as they are forced to rework designs or rush their operations after receiving late information. Higher-level management should oversee information and resource allocation to prevent this. Again, team members without other sources of power may use withholding to create power and significance for themselves as gatekeepers of certain information.
4. Microaggression
   - This involves directing a stream of backhanded compliments, passive-aggressive comments, or even direct insults toward team members based on their physical, behavioral, performance, or other attributes. In the civil engineering context, microaggressions can occur when project members use a harsh tone in engineering design review comments (e.g., "You should know better," "Why is this still not right?" or "This is wrong, fix it."). This is especially damaging when the review notes are visible to a wider audience and the designer is known. These comments can create a hostile work environment and affect team morale, potentially leading to project delays and staff turnover. Engaging in direct and private conversations with team members regarding recurring technical issues is advisable, employing a tone that fosters learning and personal growth.
5. Selective Recognition or Enforcement
   - As a leader, it's crucial to acknowledge the contributions of all team members fairly. Even seemingly minor instances of recognition can carry significance. For instance, a project manager publicly and enthusiastically celebrates one team member's work service anniversary while merely forwarding a template email note to another team member. Likewise, individuals with the authority to enforce rules may engage in selectivity reprimanding some individuals for

noncompliance while ignoring others. This differential treatment can lead to resentment and reduced collaboration, or even HR complaints.

6. Unbalanced Work Distribution
   - Within a civil engineering project, managers must distribute workloads fairly, ensuring that all team members have opportunities to contribute meaningfully. Overburdening or underutilizing specific team members can lead to project delays and decreased morale.

Engineers must remain vigilant in their project environments, actively identifying, reacting to, and, when necessary, reporting these behaviors to human resources. Self-assessment by managers is crucial, as subordinates may hesitate to provide candid feedback due to power dynamics. Do any of the examples of abuse of power sound familiar? New team members, in particular, may struggle to discern whom to confide in, potentially resulting in unreported abuses and cause personnel to seek alternative employment, which can harm the firm's overall talent pool. Engineers who may be the recipients of abuses of power or who find their values and culture misaligned with their project team or organization should feel empowered to seek support from their human resources department or even explore external opportunities rather than remain in a potentially hostile environment. Addressing power-related challenges hinges on recognizing and taking appropriate action against abuse, even if it requires individuals in positions of power to do so.

Self-assessment is crucial, as subordinates may hesitate to provide candid feedback due to power dynamics.

10.6 Team Charter

Given the variabilities in leadership styles and team member responsibilities, personalities, and preferences, a document such as the *team charter* effectively communicates expectations. The team charter (or similar documents such as team ground rules, teaming agreements, team covenant, or working agreements) includes information facilitating collaboration. The document often includes team roles,

The team charter includes information facilitating collaboration.

communication plans, conflict resolution mechanisms, decision-making processes, resource support, and other information. While some of this content is included in the project plan (described in Chapter 6), the charter is a team-centric resource used to account for the team's individualism. Unlike a project plan that could be shared with other stakeholders and focuses primarily on operating metrics, the team charter is expected to be internal or even team-only.

While the team charter can be developed with an informal style, all team members must understand and accept the conditions. This way, the team can trust each other's adherence to these "ground rules." For example, each team member might identify their typical working hours, and other team members should be able to trust that the availability of personnel will be as documented. As a team-centric document, the charter can be as detailed or simple as the team needs. Ultimately, the team charter is a collaborative artifact that reflects mutual understanding and respect between the project manager and team members, ensuring everyone's expectations are considered.

Standard information in a team charter is listed in this section, including team values, project goals, team member roles, operating hours, communication, resources and standards, budget and schedule details, and a project manager's operating manual, as outlined below. Note that all team charters should be reviewed for inherent biases and ensure that content treats all team members fairly.

The content of a team charter can vary but benefits from content such as:

1. Team Values
2. Project Goals
3. Team Member Roles
4. Operating Hours
5. Communication Preferences
6. Resources and Standards
7. Budget and Schedule Summary
8. Project Manager's Operating Manual

This section includes summary information about each of these topics, which would be tailored to the project and team.

1. Team Values

Values guide the team's actions, understanding that many decisions require tradeoffs that should be measured against the defined values. For example, the team may dictate that work-life balance is a foundational value and that there should be no pressure or expectation that the team will work outside regular business hours. If this is an accepted value, the team must recognize that schedule variances must be closely monitored and that there is enough schedule reserve to accommodate disruptions or inefficiencies. Note that values related to work commitment will need to align with firm's policies and individual employment contracts. Other examples include a focus on environmental sustainability, frequent communication, respect, quality, and professionalism. With each value, there should be a description of how the value influences decisions and the expected tradeoffs or resulting operating procedures. An exhaustive list of values would be challenging to accommodate and may dilute the importance of core values, so a shorter, focused list will better serve the project team.

2. Project Goals

Project technical requirements are likely documented in the contract, proposal, and project plan; however, project goals and success criteria may be less apparent. As part of the team charter, these goals and success criteria will also be team-centric and somewhat tailored to the larger project goals. For example, the team may establish goals that focus on continuous improvement in quality based on past performance (of other projects). Personal goals may also be associated with staff looking to learn new skills or hold a new role on the current project. For example, an engineer may aim to become more proficient in traffic engineering by working with other experienced team members who are often responsible for that scope of work. Establishing goals and communicating objectives reinforces the project's broader purpose, transcending a task-centric technical approach.

3. Team Member Roles

This section would provide general guidance on the responsibilities of each team member. While a project plan will include high-level assignments and operations, often with organizational charts, this charter section may focus only on one part of the team. For example, with three or four engineers working on the project's stormwater design, each person may focus on different aspects, such as hydrology, hydraulic modeling, narratives, system design, or the person responsible for preparing the plan sheets. This might also include notes about availability, such as identifying when some personnel must work on other projects.

4. Operating Hours

There are several reasons why team members may not operate on the same schedules. Team members in different geographies could be in different time zones, there could be obligations for other projects, or personal responsibilities may dictate when an individual can be available. Sometimes, personal preferences may dictate when a team has decided to establish core hours where everyone is expected to be available. These conditions and preferences are best shared through the team charter to establish a mutual understanding and identify potential scheduling concerns. For example, a team may elect to have reoccurring meetings, which should be scheduled when the team has agreed to be available. This charter section may also list overtime interest or availability and upcoming scheduled time off.

5. Communication Preferences

The communication plan of a team charter may focus mostly on communication preferences, such as understanding what forms of communication the team prefers and for what kind of information. For example, the team may use messaging applications for rapid informal discussion, prefer daily meetings, or operate best with formal emails and task lists. These policies may be extended to agree on the expected number of internal meetings, a time limit for meetings, expectations for distributing agendas, meeting format, and others. The communication plan may also inform how or when team members should communicate with external stakeholders. In all cases, the team's communication should reflect best practices, as outlined in Chapter 9.

6. Resources and Standards

When project requirements do not dictate operational decisions, the team may establish standard operations for performing work. This team charter section may identify standard practices that the team should follow (such as plan sheet formatting or color assignments for utilities), software (and version) to be used by the team, methodologies (such as a selected hydrologic model), and others. Similarly, the team may gather helpful resources, such as training materials, sample content, and summaries of applicable regulatory requirements (e.g., rainfall depths, road geometric requirements, unique design parameters, etc.).

7. Budget and Schedule Summary

If necessary, the team charter can include additional details beyond the project plan to communicate the planned schedule and budget, along with periodic updates. For example, the project plan may allocate a fixed price to a large task, such as developing a geotechnical report, but the team charter might provide additional detail regarding the assumptions of budget and time for each team member. Similarly, the schedule details may identify internal precursor activities required to meet the milestones established in the project plan. For example, the team may agree to print a "pencils down" date (stoppage of major design edits) a week before any deadline to avoid issues caused by late changes.

8. Project Manager's Operating Manual

A team charter might also include specific information about the project manager's operations (or this could be a separate document) that describes their preferences. For example, how does the project manager prefer to communicate (email, in-person, chats), do they have a strong preference for the graphic design of the plans, how often do they expect the team to provide progress reports (and in what format), what are their preferred tools, and what is the standard leadership style. This type of information, and other short biographies of the team, can go a long way to fostering teamwork and help each team member better understand the perspectives of their peers. This is especially important when new teams are developed, and preconceived notions about skills and preferences could be incorrect.

10.7 Stakeholder Management

Identifying, monitoring, and engaging stakeholders is known as *stakeholder management*. This does not imply that the stakeholder's personnel are managed; rather, stakeholder management focuses on processes, tools, and techniques to manage stakeholder relationships and communication.

All stakeholders have a relationship with the project that can be articulated regarding legal rights and obligations. For example, an engineer engaged by an owner to conduct preliminary studies on the owner's property will typically have a contractual right to be on the property with the contract defining respective rights and obligations of the owner and the engineer regarding the safety and wellbeing of the engineer while conducting work on the property. On the other hand, an engineer engaged by a contract buyer of a property may have no right to be on the property (and be deemed a trespasser) unless the contract buyer obtains sufficient permission for the engineer to be on the seller's property. These examples highlight the need to understand stakeholder relationships in the context of legal rights, which influences project operations.

Even with low formal power, a stakeholder with a high interest in the project can affect the operations.

Even with low formal power, a stakeholder with a high interest in the project can affect the operations. For example, community members and associations can influence early project planning efforts during public hearings by persuading the jurisdictional authority that a project is not aligned with the public interest. Conversely, some stakeholders may have significant authority over project operations but show little interest and prefer to be uninvolved. For instance, a department of transportation has authority over many project design elements but would show little interest unless there are concerns regarding the safety or mobility of travelers.

Those stakeholders necessary to the project's success are prioritized as key stakeholders. Some stakeholders may be critical throughout the entire lifecycle of a project (such as an owner), while other stakeholders may be critical only at certain junctures (such as an architect designing the initial improvements of a project). In many cases, the criticality of a stakeholder is determined by their power and interest in the project. Failure to identify critical stakeholders impairs the entire risk assessment process and exposes the project to unanticipated risks that could challenge the project's feasibility (as indicated in Chapter 8).

10.7.1 Strategies to Identify Stakeholders

The first step is to identify stakeholders relevant to the project. Stakeholders extend beyond the project team and clients and include external organizations and individuals, some of whom may support the project and others who may oppose the project's objectives.

The project design team, client, and functional managers are examples of internal stakeholders. The internal list extends to additional tiers, such as recognizing the client's organization, the project's users (tenants, customers, drivers, etc.), and the engineering firm's corporate structures (accounting, communications, marketing, etc.).

External stakeholders include the community, government agencies (except when they are the client), media, and other non-partnering design firms (competitors). The list of stakeholders can be extensive for civil infrastructure and real estate development projects based on geographic locations, local politics, governing agencies, federal requirements, etc. An organization should maintain a standard list of stakeholders to be referenced with new projects so stakeholder identification is more efficient and consistent across projects. Chapter 1 includes a map of stakeholders identified from the client's perspective by contractual (internal) and noncontractual (external) stakeholder relationships. Still, the project-specific stakeholder register (i.e., list) would include details regarding the organization or individual names. Figure 10.2 provides a small sample of stakeholders categorized as internal and external.

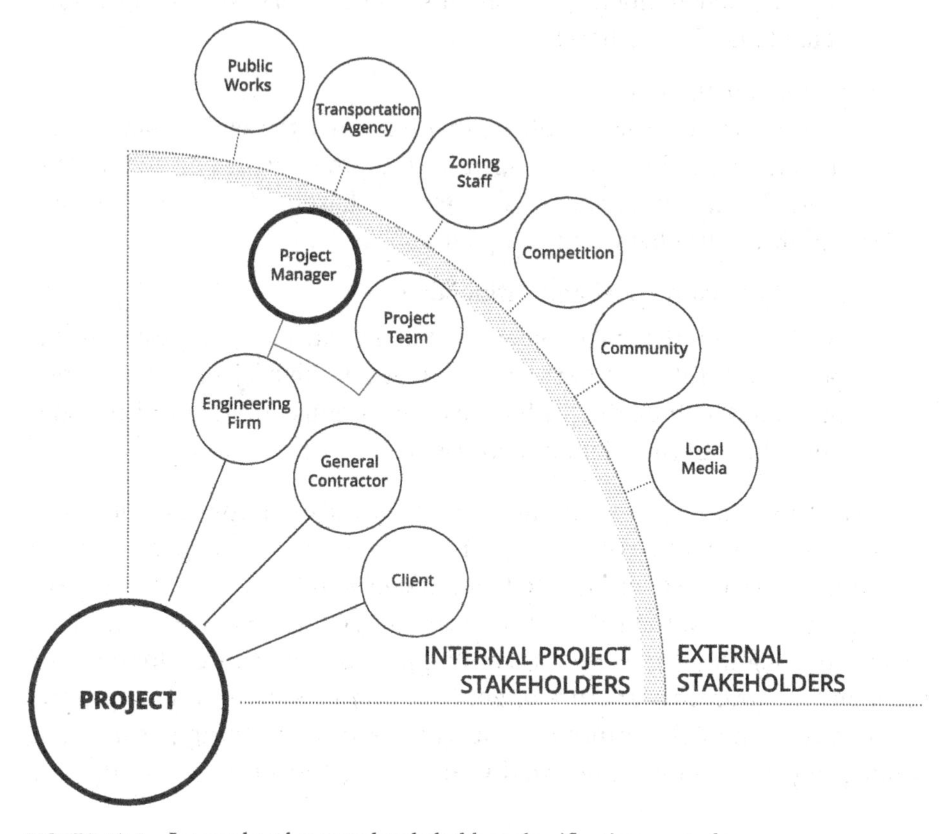

FIGURE 10.2 Internal and external stakeholders classification example.

A stakeholder register is a common way to document the identified stakeholders. In the context of risk assessment, there is also a benefit to prioritizing stakeholder management. Stakeholder identification often begins with organizational process assets or industry knowledge about applicable stakeholders as identified with prior projects. Expert judgment also plays a significant role, but the general process benefits from prompts that guide the project team in brainstorming the list of stakeholders. Such prompts include the following:

- Internal Stakeholders
  - The project manager, team members, and other employees of the organization are stakeholders common across all projects.
- Key Stakeholders
  - From the engineer's perspective, the client is the primary stakeholder. From the client's perspective, the end-user or tenants are the primary stakeholders (such as utility customers, airport operators, retail tenants, or others).
- External Stakeholders
  - This list includes stakeholders that influence the project but are not directly controlled or managed by the project team or client. This includes regulatory agencies, investors, shareholders, the local communities, suppliers, and competitors.
- Secondary Stakeholders and Influencers
  - At this level, the stakeholders will likely have low interest and low power but still maintain some influence in the project. For example, this could include the media, advocacy groups, or others with a relationship with the client or end-users.

Over time, the context of the project will also shape the roles and significance of stakeholders. For example, when a project evolves from vacant land to a construction site to an active stabilized project with tenants and customers, the role of the local government agencies will transition from issuing initial permits and approvals for the development of the land (i.e., granting land use or zoning approvals) to issuing approvals for construction of the improvements (i.e., issuing building permits and conducting inspections) and finally to issuing the ultimate occupancy and use of the improvements for the intended benefit of the tenants and the customers (i.e., issuing certificates of occupancy). Note, however, that though the types of approvals evolve, the role of government stakeholders remains one of approval authority. In contrast, the role and importance of community stakeholder may be more pronounced at the beginning of a project during initial feasibility reviews but reduced once construction has begun.

There are several tactics to identify the key values and risks with stakeholders. Internal stakeholders and clients will likely contribute to interviews, questionnaires, focus groups, and workshops. In this way, the engineering team could contribute a standard list of risks (obtained from organization process assets or developed for the project) and add or remove risks based on these tactics. Risk management techniques can be used later to prioritize the identified risks based on potential scenarios (more information is provided in later sections of this chapter).

The tactics with external and secondary stakeholders will be less direct, except for data gathering from the community and public meetings. For example, the project team can learn about tangible or perceived risks during public hearings for a rezoning project. Information may also be obtained from publicly available data, such as social media, forums, news, and other historical data that might inform stakeholder risk criteria.

Once stakeholders are identified, their key values must be determined and evaluated. Key value systems are the strategic compass of a stakeholder and consist of the primary motivation structure of the critical stakeholder within the enterprise. For example, the local government will likely be interested in long-term solutions that improve sustainability and community well-being and may be willing to invest more upfront to save on long-term costs. Key value systems drive decision-making and are essential when anticipating or forecasting outcomes. In some cases, key values are evident and clearly stated (e.g., a commercial real estate firm's interest in maximizing financial returns). At the same time, other scenarios present a tone of indifference where the stakeholder has no particular motivation associated with the project (e.g., a subcontractor may be focused only on delivering their requested services with little interest in project aesthetics, community benefits, or others). Although most stakeholders will focus on cost, schedule, scope, and quality, identifying key

Stakeholder identification becomes an integral process that uncovers various sources of risk and contributors to risk events, providing a foundation for a more robust and comprehensive risk management strategy.

values will determine the client's appetite for tradeoffs and inform stakeholder management.

With these methods, stakeholder identification becomes an integral process that uncovers various sources of risk and contributors to risk events, providing a foundation for a more robust and comprehensive risk management strategy. This process highlights immediate risks and reveals the dynamics of stakeholder influence, which can evolve and change throughout the project (and, therefore, affect the risk landscape). For instance, the early involvement of environmental consultants may identify potential ecological risks that could alter project timelines or require design adjustments. As the project progresses, the focus may shift to operational risks identified by future facility managers or safety concerns raised by future tenants. Understanding and mapping these shifting influences is essential to ensuring that the project adapts to changing risk profiles and that stakeholder concerns are addressed proactively, safeguarding the project's objectives and ensuring its successful delivery.

Several stakeholder characteristics should be monitored and managed, such as understanding stakeholder power, interest, influence, and engagement. Stakeholder interest is affected by (but distinct from) engagement. Stakeholder engagement is often monitored with a stakeholder engagement matrix, which lists several engagement levels and identifies a current and desired engagement level. This matrix is a resource for the project team to understand what actions may be required to transition from the current to a desired state, such as scheduling meetings, sharing information, connecting with the media, or providing publicly available resources (such as social media or websites of the project plans and details). Stakeholder engagement also informs risk management and mitigation strategies if any opposition (or supportive problem-solving) is identified. A sample stakeholder engagement matrix is shown in Figure 10.3.

Stakeholder interest and power are dynamic throughout the project and should be monitored by the team.

Stakeholder interest and power are dynamic throughout the project and should be monitored by

	Unaware	Resistant	Neutral	Supporting
Client				C|D
Transportation Agency		C ········→	D	
Adjacent Neighborhood	C ········	········	········	········→ D
County Staff			C ········	········→ D

C: Current
D: Desired

FIGURE 10.3 Sample stakeholder engagement matrix.

the team. A standard project management tool is a power-interest grid, which maps how various stakeholders could influence a project and what actions should be taken by the project team to mitigate risks. For example, suppose a developer plans to apply for rezoning of a property. In that case, there will likely be requirements for public outreach so the elected officials can gauge the perception of the project. Community members may not hold direct power to dictate the project scope; however, with enough interest and involvement, they could influence the elected officials and exert power over architectural, operational, and infrastructure design changes. As previously mentioned, this influence from the community is more prevalent in the early project phases, such as entitlement processes. It dissipates as the project moves toward final design, when it is primarily dictated by prescriptive technical requirements managed by public agencies. Figure 10.4 provides a sample power-interest grid for a civil infrastructure project.

In the example shown in Figure 10.4, the stakeholders have been mapped and are in various grids that represent different tactics for coordinating with each stakeholder. For a by-right development project, community members may show interest but do not have any direct power over the project scope or schedule; conversely, community members may have extensive power to influence elected officials and zoning review boards when a project requires rezoning or other entitlement approvals.

Public agencies, such as environmental and transportation, will have some power over permitting processes, regulations, design standards, and

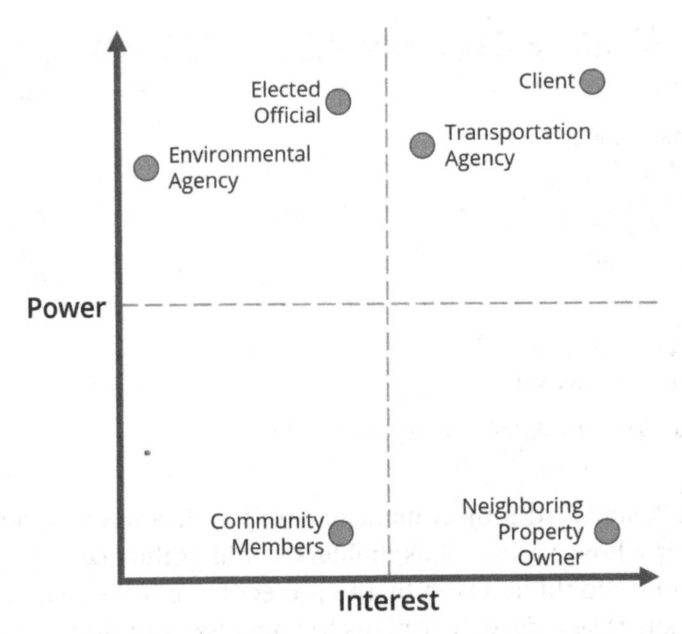

FIGURE 10.4 Sample power-interest grid for a civil infrastructure project.

plan approvals and would have variable levels of interest based on the project's potential impacts. An elected official would have significant power over some project elements, and the interest would vary based on community perception and involvement. The power interest grid (matrix or similar register) requires updates and monitoring throughout the project. Changes to project scope or other environmental factors could influence a stakeholder's interest or power. Suppose a new retail development project receives a comment from the transportation agency requesting improved intersection visibility. If the intersection sight distance provision requires an easement from a neighboring property owner, that stakeholder now has power over the project costs, schedule, and design. This shift is shown in Figure 10.5.

The shift of the Neighboring Property Owner's location in the power-interest grid shown in Figure 10.5 demonstrates the importance of continuous monitoring of stakeholders. This shift also demonstrates the importance of maintaining a professional and respectable relationship with all stakeholders, whose power and interest may change throughout the project life cycle.

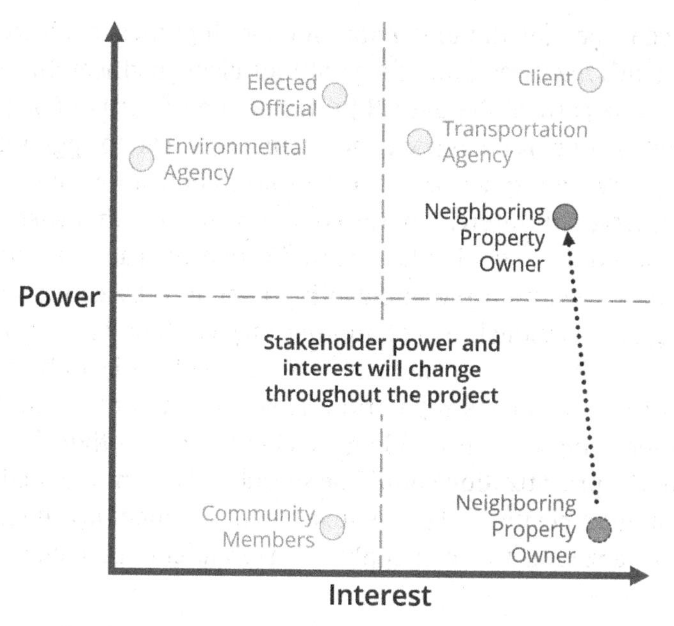

FIGURE 10.5 Updated power-interest grid for a civil infrastructure project.

Stakeholder Clients

Project managers should be aware that other stakeholders involved in the project could also be clients of the engineering firm or a desired future client.

From the broader perspective of the organization, the relationship between a client and engineer does not solely pertain to a project's primary client. Project managers should be aware that other stakeholders involved in the project could also be clients of the engineering firm or a desired future client. For instance, while working on a project for a commercial developer, it's common to interface with county staff or municipal officials. These interactions can be related to permits, inspections, regulations, or any number of collaborative tasks. Recognizing that an engineering firm may also seek or hold contracts with county agencies, the stakeholder should be considered an auxiliary client. Respecting these stakeholders as clients will support organizational goals and better serve the project objectives.

These overlapping client relationships pose a unique challenge. It's possible that a project manager enters a situation where the interests or demands of one client might conflict with another's. In such instances, the goal is to be respectful and understanding for all parties involved. Prioritizing one client's needs over another without clear justification or communication can jeopardize the overall project and erode trust. For example, a commercial client may encourage the engineering firm to aggressively pressure the county staff to act faster on a project application. While this may seem like it serves the interests of the commercial client that wishes to accelerate their schedule, it would likely be ineffective and cause undue strain on the relationship between the engineering firm and the county. Recklessly prioritizing one relationship over another may conflict with organizational objectives, even if it seems beneficial to the project, and can harm both the firm's and the individual's reputation. Note that circumstances will often require prioritizing one stakeholder or client over another; however, the reasons for the prioritization should be sound and communicated appropriately to maintain positive relationships. Chapter 9 includes suggestions for difficult conversations that can apply to stakeholder management.

10.7.2 Client-Engineer Relationship Management

The most time-intensive stakeholder relationship likely occurs between the client and the engineer. This relationship extends beyond individual projects and considers the programs and portfolios managed for a client (often with different project managers). Extending from pre-project planning until project closeout, this relationship is built around the engineer's intent to provide value to the client's organizational strategic objectives.

The most time-intensive stakeholder relationship likely occurs between the client and the engineer.

Several dimensions associated with client-engineering relationships will mature over time and mutual project success. In the initial stages of forming a client-engineer relationship, the parties learn each other's capabilities, objectives, and communication styles. Trust is just being established, so communicating context and core values may be limited between the parties. In this stage, the project manager likely responds reactively to client requests until they better understand the context of project decisions (while ultimately aiming to be proactive of client needs). Examples of the progression of the client-engineer relationships are described in this section.

Developing Client-Relevant Skills

While some projects and clients only require generalized services, engineers can better serve target clients with a tailored skillset. The engineering firm can invest in refining skillsets to better meet the client's needs, including esoteric requirements. For example, a client may have a frequent requirement to include solar panels or rainwater harvesting to meet their sustainability objectives. In this case, the engineering team can learn design practices and monitor emerging technologies, pricing, specialty contractors, and other aspects of the system. As tailored expertise grows and the engineer has successfully demonstrated competency in other engineering skills, the client-engineer relationship will benefit because the client may provide more context and information about their objectives and trust the technical advice of the engineer.

Reciprocal Financial Performance

Understandably, clients aim to minimize project development expenses and want to acquire services for the least possible cost to the project. However, this objective does not preclude a client from acknowledging that the engineering firm should be compensated fairly and that project value can be achieved in many ways. In the early stages of a client-engineer relationship, the client may be change-resistant to mitigate risk and variability in scope, schedule, and cost. As the engineer demonstrates that they understand the client's measure of value and priorities, the client may be willing to openly discuss changes that financially impact the project.

For example, an engineer who understands a client's priorities for early project delivery could make a case for additional costs associated with project acceleration, and the client would trust – but confirm – that any price premium will result in a return on investment (such as early completion and retail occupancy that provides the client with earlier revenue generation).

Predictive Market Research

Pre-positioning for projects will increase an engineering firm's opportunity to win a project because the team will have more time to prepare, better context on the project's history and objectives, and will have a team organized for the project. As the client-engineer relationship matures, a client may be more willing to reveal strategic objectives to help the engineering firm better prepare for upcoming projects. As the client provides more context, the engineering team can better understand project and program priorities and communicate tradeoffs to inform decision-making. At one of the highest relationship maturity levels, the client may even illicit advice from the engineering firm as objectives are established.

For example, an engineering firm may begin its relationship with a retail development client by responding to a request for proposal (RFP). As the client-engineer relationship matures, the client may discuss planned RFPs or eventually seek input from the engineer to identify potential sites suited for the client's typical retail program.

Advocacy

Engineers and clients may advocate for one another when they have common strategic objectives and mutual respect. This often begins with an engineering firm demonstrating its technical skillset, project management acumen, and comprehension of the client's core objectives and values. Internally, an engineer can advocate for a client and begin to prioritize resources to best support the client's projects. In some cases, an engineering firm can help the client foster new relationships, apply for grants, submit projects for industry awards, and support the client's brand recognition. Similarly, the client may advocate for the engineering firm, offering to serve as a reference when the engineering firm submits proposals for new work with other clients. Genuine advocacy requires respect and trust between the client and engineer, which is earned with demonstrated project success.

10.8 Continuous Improvement

No process is perfect, but systems can evolve for the better through incremental improvements. Embracing this mindset involves engaging every member of an organization in identifying and implementing improvements, from corporate leaders to those in the design studio or out in the field. Those directly involved in day-to-day operations are often best positioned to innovate and uncover areas ready for enhancement.

No process is perfect, but systems can evolve for the better through incremental improvements.

Organizations can deliver superior services and products as processes evolve and people are inspired to learn. While corrective actions are essential, they tend to recur if the root causes aren't addressed. Thus, a commitment to continuous improvement requires an investment in project personnel who are encouraged to pinpoint improvement opportunities and feel supported in honing their skills. Achieving this level of improvement demands multidisciplinary collaboration to identify, measure, and communicate the necessary changes. Implementing performance metrics is crucial to tracking progress and ensuring data informs the organization's continuous improvement journey.

In the realm of project management, continuous improvement stands as a core principle. These concepts of continuous improvement are derived from the Japanese word *kaizen*. This term means "change for the better" and is recognized as a mindset that has proven benefits to management objectives. Its success is often quantified by reduced waste and increased efficiency. In a civil engineering project context, waste encompasses rework time, unproductive tasks (e.g., unnecessary meetings), low-value but time-consuming processes (e.g., document generation), resource-heavy design solutions, and inefficient production methods.

Quality management (Chapter 5, Section 5.7) plays a pivotal role in recognizing these sources of waste and introducing new processes that enhance the current project and future endeavors. Indeed, this philosophy of continuous improvement is best implemented beyond project-specific objectives and must be part of organizational culture and vision. The highest levels of organizational leadership must factor in long-term benefits, understanding that achieving these goals requires both an investment in their people and a focus on tangible and intangible organizational values.

This philosophy of continuous improvement is best implemented beyond project-specific objectives and must be part of organizational culture and vision.

> In Appendix A, Part 10 of Millbrook Logistics Park, the narrative offers a retrospective analysis of the project several years post-completion of engineering designs and during the construction phase. This review delves into the impact of various forms of power on project operations. It highlights the importance of adjusting leadership styles to suit the needs of stakeholders and project conditions, as recognized by the engineers involved.

APPENDIX A

Millbrook Logistics Park

Narrative Developmental Editor: Jennifer L. Holt

APPENDIX OUTLINE

Management Essentials for Civil Engineers: A Practical Guide to Business, Communication, Ethics, and Risk, First Edition. Cody A. Pennetti, C. Kat Grimsley, and Brian M. Grindall.

This narrative, Millbrook Logistics Park, provides anecdotal experiences portraying the nuanced challenges of implementing project management and leadership skills with a scenario for each corresponding chapter the book's technical content. The scenario progresses through the major milestones of a large project from pre-project planning until project completion. The narrative structure mirrors real-world situations, revealing insights that may not be apparent through fact-based exposition. By adopting a story-like format, these scenarios describe the application of project management topics to practice.

Each part of the scenario delves into the perspectives of the client and the engineering team, providing a holistic view of the project dynamics. From the client's viewpoint, the narrative explores differing priorities and objectives, including evaluations of how success is quantified and how project objectives are communicated. The engineers' narratives offer a contrasting experience of a novice engineer and a seasoned project manager, highlighting how early choices and interactions among stakeholders can lead to a cascade of ethical and legal challenges as the project evolves.

By following the project from its initial planning stages to a final retrospective, each part of the scenario serves as a pedagogical tool and a nuanced exploration of a professional journey in the engineering world.

* * *

TerraHaven Land Development, a prominent national real estate development firm, is considering acquiring and developing property in the County of Millbrook. Millbrook, located outside a bustling metropolitan area of Harbor View City and near a highway interchange, has remained a mostly rural community despite the proximity to nearby urban development areas (as conceptually shown in Figure A.1). Local farmers and small business owners have intentionally buffered themselves from the rapid development of the surrounding city of Harbor View. However, in recent years, the County's prime location has piqued the interest of real estate developers. As a result, land prices have increased exponentially, along with the growth of the adjacent metro area.

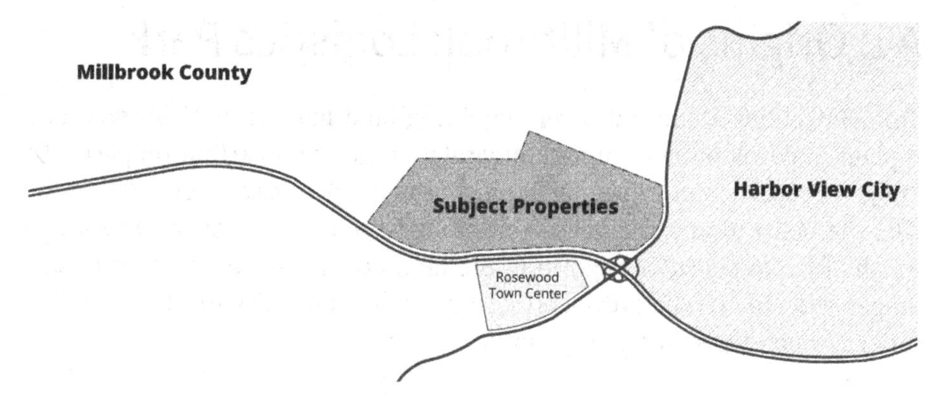

FIGURE A.1

The main parties referenced in these narratives are:

- **TerraHaven Land Development, LLC** – national real estate development company (the client)
  John Branson – Senior Development Manager
- **Meadow Law Partners PLLC** – Land Use Attorney
  Sarah Chen – Attorney
- **ApexTech Engineering, LLC** – National Engineering Firm
  Maya Mendez – Civil Engineer, Project Manager
  Veronica Statton – Civil Engineer, Designer
- **Avrum Planners, LLC** – Local Land-Planning Firm
  Amar Singh – Owner, Senior Planner

A1: Origins of Millbrook Logistics Park

TerraHaven has identified seven adjoining land parcels in Millbrook, and it plans to develop them into an integrated industrial distribution park (the Subject Properties in Figure A1.1). After years of passively evaluating the site, the political and financial conditions were deemed favorable enough for the firm to consider beginning the land acquisition process. Prior due diligence on the site had revealed several challenges, but with the right conditions, there was significant potential.

* * *

April 30, Year 1

Sarah Chen, the lead attorney at Meadow Law Partners PLLC, was catching up with TerraHaven VP, John Branson, at a locally owned coffee shop in Rosewood Town Center. John and Sarah had made a habit of meeting in Rosewood to talk casually about the region's development opportunities. Built a few years prior, the new town center had transformed an older suburban region of Millbrook. Historically, Millbrook's citizens opposed this kind of development due to looming concerns that the aging infrastructure couldn't support the population increase and general opposition to the resulting increase in traffic and noise. However, the rising popularity of Rosewood within the community was clear evidence that perceptions of new growth might be shifting. The mixed-use project had initially faced opposition in the early planning stages. Still, active negotiations with the community had shaped Rosewood into an attractive development with high-end finishes, vibrant plazas, and a substantial diversity of tenants.

> This region has potential for development but also carries risk with new work. When rezoning is required, the community perception will often influence project scope and schedule.

"You know, John," Sarah started, with a hint of enthusiasm in her tone, "new developments like this are becoming more popular for the younger generations of Millbrook." Sarah and John sat outside in one of the well-designed plazas, enjoying the spring weather. She paused to sip her coffee and take in the surrounding activity, noting several new window signs for high-end retail tenants advertising "coming soon." John nodded and followed her gaze around the open plaza, observing the decent-sized crowd milling about and acknowledging how much things had changed in the last few years. "The new members of the Board of Supervisors," she continued, "the recent promotions of a few key zoning staff; now is the right time to move forward on this." John knew she was referring to the potential industrial site, which he had discussed with Meadow Law several times over the past few years.

> The surrounding context and external environmental variables play a pivotal role in the outcome of a project. Identical project parameters might lead to success or failure contingent upon the prevailing political, social, and economic climates beyond the project's immediate control.

John looked down at his coffee, contemplating Sarah's inferences but knowing he agreed. In reality, John's team had been carefully evaluating this site for several years as they tracked demographics, comparable sales, and tenant activity. John had already made the decision to move forward. Still, he was curious about Sarah's perspective on the potential for development based on what she's been hearing. He glanced up and prompted her to offer her opinion, "You know who isn't ready to move though? Ridderfield. Oloport. To make this development happen, we need their properties in addition to the others." Sarah maintained her gaze but sighed; she knew John was referring to the two landowners with a reputation for aggressively defending their farmland from interested developers. After a moment, she turned, "John, you know those two can't fight the changes happening around here forever. Heck, I think I even saw Ridderfield out here at Rosewood the other day with his grandson. They know Millbrook is evolving with the times, and they can't expect their farms to be here in another 10 years. Their kids aren't interested in keeping them running anyway; they'll sell."

> The developer may not openly communicate all information to other stakeholders.

John kept quiet for a while as Sarah waited expectantly. Finally, he broke the silence with a simple, "Alright, let's do it."

Sarah smiled with a slightly triumphant gleam. "Yeah? You're ready for Millbrook Industrial Park?"

"Millbrook *Logistics* Park," John corrected with a small smile. He didn't want to appear too eager right off the cusp, but he had to admit he was always the most energetic at the start of a new project. "Have your office make sure you're still on retainer, Sarah. We'll want you in on this. Also, let's see if we can get Option contracts on all the different lots we need."

Sarah was pleased. She was happy for John and recognized that this would be a complex project, which meant her team would be involved as land-use attorneys for TerraHaven. Bringing in a significant contract for Meadow Law would look good for her. "Great, I'll have my team write one up and send it over. Thanks for involving me in this." She lifted her drink to John and took a sip before quickly shifting into planning mode. Knowing John would also need engineering services, she instantly thought of Maya Mendez, a professional colleague she'd worked with. "Have you thought about who you'll use for the planning and engineering side? Maya has done some good work for you, right? I think she's still with ApexTech Engineering."

Professional relationships are important. For a private development project, the selection of consultants is often influenced by an established network and the perceived qualifications.

John chuckled, "You mean, have I thought about it in the maybe 30 seconds or so since I told you we're moving forward?" She shrugged, "No time like the present. Besides, I know you've *actually* been ready to move on this project for a while now" He smiled at being called out, but paused a moment before giving his answer. He again wanted to hear her opinion, "ApexTech has the BOA with us, sure, but they've always focused on Harbor View City. I'm not sure if they've ever done any work in Millbrook County."

"Well, who has?" Sarah countered. "There's been no development out here in ages." John raised a brow at her comment and gestured to their surroundings. "Ok, you're right," she conceded with a roll of her eyes. "*This* was obviously just built. But I heard it was just some small, local engineering firm. You'll be better off with Maya, and we both know that."

Sarah and John recognize that Rosewood Town Center sets a precedent for future development and that at least one development team has successfully built a new project in this region.

John nodded in agreement, "You're right. Maya's been amazing and has a great track record of getting things done efficiently. And honestly, I don't feel like shopping around for a new firm and spending months developing an RFP , reviewing proposals, reviewing contracts," he trailed off with a heavy sigh, already imagining the tedious procurement process. "I'll go ahead and connect with her, get her up to speed. Let's set up a meeting for later this month so we can corroborate on our next steps."

> As a private developer, TerraHaven can sole source a project and work directly with ApexTech without a formal RFP (unless TerraHaven's internal processes dictate the need for one).

* * *

August 15, Year 1

Several months later, Maya Mendez was in her AppexTech office finishing an online meeting when one of her junior engineers, Veronica Statton, tapped on her door. "Hey, Maya," she greeted, peeking around the open door. "I was just wondering if we are going to ride together to that TerraHaven thing today?"

Maya looked at her blankly, trying to decompress from the stressful meeting to processing the new conversation. Blinking slowly for a beat, it eventually clicked. About a month ago, she had a conference call with TerraHaven for an active project in Harbor View City. During the call, she had promised John Branson a quick meeting to review a potential new project he had in mind in Millbrook County. For over 10 years, Maya has served as John's point of contact for TerraHaven. That relationship and the success of prior projects bolstered Maya's career and garnered several promotions at ApexTech Engineering. While beneficial to her career, the promotions had inevitably placed more responsibility on her shoulders, and it had gotten to the point where she had begun needing to delegate far more work to her team than she was used to (or comfortable with) to keep her from being spread too thin.

> This signifies that TerraHaven is a key client for ApexTech and personally for Maya.

Case in point, overlooking a client meeting, even a casual one with John, was unusual for her. She pulled up her calendar, and a quick scan of her day revealed several meetings overlapping one another. She took a deep breath.

Engineers are often pressured to accommodate multiple client demands, and time delegation is required to avoid overbooking. Maya notes that this particular meeting was communicated informally during a phone call and deprioritized compared to other meetings that were formally scheduled.

The project manager, Maya, recognizes some risk in sending a young engineer to the meeting and provides cautionary advice to avoid potential issues.

"I'm afraid we'll have to cancel that one, Veronica. I'm double, well, triple booked this afternoon. I'm pushing it as it is today, so John will have to wait. It was nothing set in stone anyway." She returned to her emails, trying to reorient her day, but she noticed Veronica still standing in her doorway. "Was there anything else?" Veronica seemed to hesitate before fully entering the office.

"Well, if this is just sort of a courtesy meeting, I was thinking maybe I could go and fill in for you." Maya mulled it over. Veronica only had a few years of experience and wasn't a project manager, but she had learned a lot in a short time. Allowing Veronica to go would give her more experience and help lighten the workload. "Alright," she decided. "But remember, this is an unofficial meeting. We don't have a contract signed for any new work, so just take notes and listen to what he has to say. John's a nice guy, but don't let him pressure you into committing to anything prematurely. He can be laser-focused when he gets going on a new project." Veronica nodded enthusiastically, pleased about the opportunity to go solo. Veronica shifted the plans she had held under her arm as she prepared to head to the meeting with John. Maya noticed, and her brow furrowed in confusion. "What are those for? I didn't ask for any prints or anything, did I?"

"Oh, um no, but I had heard you mention TerraHaven wanted to talk about Millbrook, so I went ahead and pulled some GIS maps together so we'd have some clear talking points at the meeting."

Maya placed her head in her hands, and Veronica thought she had overstepped a line. Trying to hide her frustration, Maya looked up and questioned, "And how long did all that take?" Veronica sensed the mood shift, and her previous confidence faded a bit. "Not that long, really, just a few days. Millbrook doesn't have a great mapping or record system, so I had to do a bit of digging around." She took a deep breath, waiting for Veronica's response.

Maya sighed heavily after hearing this. She had received a lot of pressure from higher-ups at ApexTech about keeping utilization rates high, which meant her team needed to stay active on *billable* projects, not client favors. Maya was hopeful this could be the next big project for ApexTech, but John had a history of asking for favors with projects that never evolved into anything tangible. Still, she had to admit she admired Veronica's tenacity and forethought. She noticed Veronica fidget slightly during the silence and decided to let go of it.

From the engineer's perspective, this work effort carries risk. Maya's team must balance investments in business development and time spent on profitable work.

"All right, well, I appreciate the extra leg work, but let's try to stay on the same page from here on out. Remember, we don't even have a contract yet, so let's prioritize other projects until this one officially moves forward . . . assuming it does," she finished doubtfully. Veronica relaxed a bit, then tilted her head in thought, "But, don't we have that BOA with TerraHaven?" She was referencing a basic ordering agreement (BOA) between the two firms.

Maya paused a moment, "Well, it's complicated. Yes, that BOA helps facilitate new work, but it's not like we can bill work without a signed contract and formal scope." Veronica nodded to show she understood. "Anyway," Maya checked her watch, "I guess you're good to go. Sarah Chen, from Meadow Law, will probably be there too, and they'll want to bounce some ideas around. But remember, you're just there to share your plans and to take notes."

"Got it." Veronica responded with conviction as she turned to leave.

"Good luck!" Maya called out before the door closed.

* * *

When Veronica arrived at the TerraHaven offices, the receptionist guided her to a large conference room where two individuals, she presumed to be John and Sarah, were seated. Spread through the conference room were various charts, paperwork, and food debris – John and Sarah had been working on land acquisition contracts earlier in the day and likely through lunch. While Veronica had worked on the engineering tasks on multiple projects with TerraHaven before, she had never met John in person. Her knowledge of Sarah was just as limited, knowing that she was Maya's professional friend. She stood up a little straighter and entered the room to greet them.

"Hello!" she started enthusiastically, "I'm Veronica Statton, I work with Maya at ApexTech." John and Sarah looked up at Veronica and smiled at her, standing to

This is a sign of a communication issue. John understands that this meeting is an important discussion, but Maya understood this as a casual conversation and felt Veronica could handle it.

shake her hand. John not so subtly glanced around her toward the door, expecting Maya to enter any moment. When she didn't, he seemed a bit confused.

"Good to meet you, Veronica. Uh, is Maya stuck in traffic?"

"Well, no. Actually, Maya just sent me today, her loyal trainee." John gave a sympathetic chuckle at Veronica's response, but he was a bit unsettled that Maya had chosen not to attend such an important planning meeting herself.

"I hope everything is alright," Sarah chimed in, sensing John's unease.

"Oh, everything is fine. Maya just had too many things on the docket for today, so she's letting me handle this meeting on my own." John and Sarah snuck a hesitant glance at each other.

"How long have you been with ApexTech, Veronica?" John inquired, wanting both to gauge her experience level, as well as to get an indication of how worthwhile this meeting was about to be.

Veronica is being transparent here about her limited experience.

"Almost two years now. I joined the firm after graduating with my civil engineering degree but interned even before that," Veronica replied. John still seemed unsettled but slid a few stacks of paper out of the way and motioned to an empty chair, "Why don't you join us." Veronica thanked him and set her things down before sitting across from them.

"I assume Maya's brought you up to speed on our Millbrook endeavor?" John began.

"Yes, she's mentioned the basics of the project, but I'm not sure I know all the details. That's why I've also brought some maps along, so we have a reference of the area to really clarify things." Veronica unrolled the base maps she had brought, which compiled publicly available GIS data for the area that TerraHaven had been studying. John looked more than pleased to see the plans Veronica had brought (Figure A1.1).

"Great! Maya must have read my mind. I love a good map to brainstorm with." Veronica hesitated but decided not to take credit for the idea and

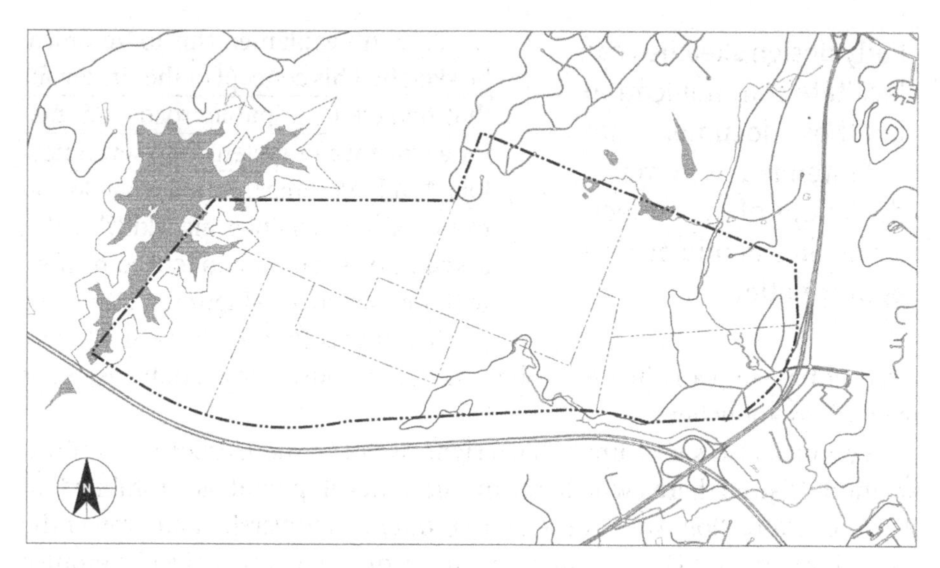

FIGURE A1.1

merely smiled in response. John helped to unroll the remaining papers, which showed different layers of information, like the property lines, roads, streams, and other mapped data. John was familiar with the site by now, albeit mainly from a financial perspective and simple layouts. As he took in everything before him, he picked up a red marker and started to trace the key properties on the map while explaining some of the project's conditions to Veronica:

1. We need a rezoning of all the parcels from the current agricultural use to an industrial distribution use ("Rezoning") that includes a development plan.
2. We need site plan approval and a permit for land disturbance consistent with the intended development plan for the project site infrastructure and rough grading of the anticipated lots.
3. Concurrent with the site plan permit process, all existing properties will need to be consolidated into a single lot and then subdivided into six or seven new lots – each with an option to operate independently of the other (i.e., each lot will enjoy independent access to public rights of way to new roads, independent satisfaction of all stormwater requirements and direct access to all necessary utilities).

> Early design sketches can facilitate communication, but these documents are not intended as formal depictions of the project scope and should be used appropriatley.

As John explained this to Veronica, he sketched his concept of the site layout that depicted the public roads, an area for a stormwater system, and seven new lots. "Well, Veronica, Sarah, what do you think of my engineering skill?" John asked as he sat back and looked pleased with his site layout (Figure A1.2).

Sarah was first to chime in: "Don't forget John, you can't just level the site. The resource protection easement regulations apply here."

Veronica also spoke up: "That's right, we're in the Harbor Creek Preservation District here, which means new development is prohibited in Resource Protection Areas, or RPA, to promote biodiversity and protect the natural waterways. This is usually 100 feet from a stream or river." Veronica then pointed to a dashed line on the west side of the plans near a lake that extended into the site, "And this is where the County has mapped this site's RPA, but they will require a formal survey and mapping." Sarah looked over at Veronica as if acknowledging her thoroughness.

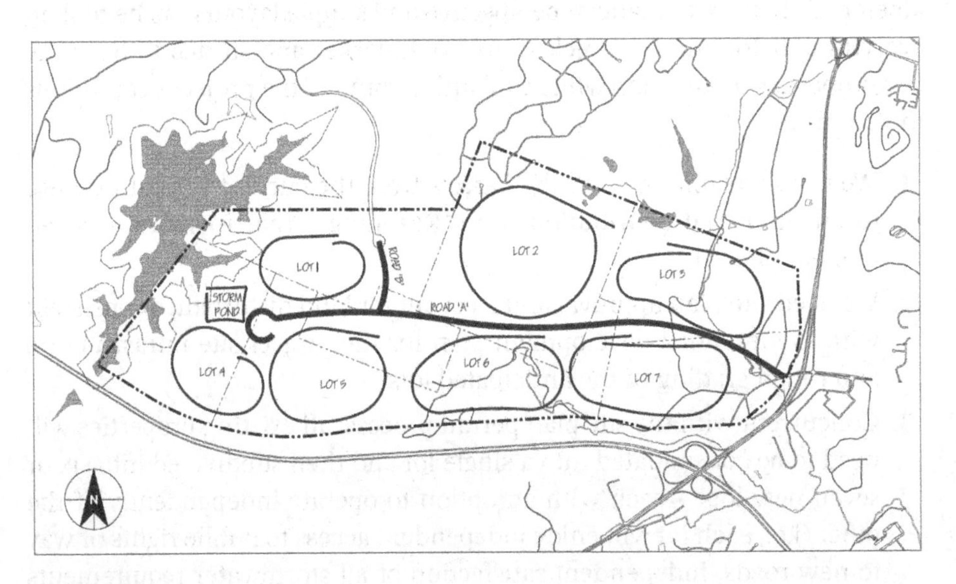

FIGURE A1.2

John looked back at his sketch to see if the RPA had any sort of impact on his plan. "Well, we need to follow the rules, of course. It's farmland, so it's mostly cleared as far as I know. And we figured we'd need to avoid the western lake already. It sounds like you know your way around Millbrook County, Veronica." Veronica smiled in response; she was pleased that her initial research into the County regulations had proven helpful for this discussion. John continued, "OK, Veronica. How long do you think it will take before we get shovels in the ground?" Veronica was quick to respond, "Well, we're pretty efficient, and we can usually have site plan approvals in about six months."

Although Veronica is eager to provide an answer that will satisfy the client, schedules are complex and should be carefully evaluated. The context of this discussion is informal, but erroneous information could create issues if the client intends to use this for scheduling.

At this point, Sarah felt like she could step in to facilitate and noted, "John, why don't we let Veronica go back and talk to Maya to answer that? We also still have to get these sites under contract. Plus, don't you still owe ApexTech a contract before they do any more favors?" Veronica appreciated that Sarah was helping her talk through the next steps. Veronica nodded in agreement, "Yes, I will take this back to Maya and talk more with her."

"Great! Why don't you take these plans back and work with Maya on a proposal so we can get started." John carefully rolled up the plans and returned them to Veronica. As she packed up her bags to leave, Veronica said, "Well, it was great meeting you both. I look forward to working with you."

A1 Review Questions

1. How might John's concerns about development risks influence the project planning for the new development? What challenges and risks have been identified by the team? How might TerraHaven mitigate these risks?
2. How has the community perception changed recently, and how might it influence a rezoning process that transitions the Subject Properties from farmland to industrial uses? What economic impact could the new development have for the local community.

3. How is the professional relationship and contractual relationship best defined for TerraHaven, Meadow Law, and ApexTech? How can TerraHaven ensure a productive partnership with ApexTech?
4. Based on what's been indicated so far, what is a reasonable timeline to expect for the project, which would include procurement, conceptual designs, rezoning, and then final design and permitting?
5. What are some early sustainability goals that could be considered, how do these coincide with the resource protection area requirements, and how might they best be introduced to the project?
6. Veronica acted proactively to prepare for the meeting with TerraHaven – why might Maya be concerned and appreciative of these early actions, specifically with how Veronica communicated during the meeting with John and Sarah?

A2: The Request

A few months ago, Veronica (civil engineer at ApexTech LLC) met with John (the client at the development firm TerraHaven) and Sarah (the land-use attorney at Meadow Law). Things had gone well, and Veronica thought she'd handled the meeting in a way that would have pleased her supervisor, Maya. Veronica expected that this introductory meeting of a new TerraHaven project would mean that ApexTech would receive a contract to do the work. Veronica felt she did well to represent ApexTech during the meeting, taking ample notes about TerraHaven's project objectives and providing some initial but informal guidance for the new project. After the meeting, Veronica filled Maya in on everything and felt sure a contract between TerraHaven and ApexTech would be completed soon.

During the initial meeting, John sketched some concept designs on the base maps Veronica had developed and brought to the meeting and asked for copies of them. ApexTech does not have a contract for this work yet, so Maya wasn't thrilled about handing over free resources; however, she relented. Veronica figured it was primarily general information publicly available to anyone willing to track it down, so it wasn't that official or proprietary to begin with.

After keeping in touch with John over the past few months and providing more exhibits and notes for the Millbrook Logistics Park, Maya was antsy that nothing had officially moved forward. She had anticipated that TerraHaven might get stuck in negotiations with brokers and landowners, but her team's ongoing "favors" felt like a type of pro bono work, and her bosses were leaning on her to keep her team's focus on clients with actual billable work.

> Small favors can add up over time. Without a contract or billing mechanism, the engineering firm must treat this as an investment.

* * *

October 27, Year 1

Maya sighed heavily and leaned back from her desk, massaging her temples.

"Um, bad timing?" Veronica stood in the doorway of Maya's office at ApexTech Engineering, holding a brown paper bag with the logo of a local restaurant on the front.

"No, I definitely need a break. Come on in." She shifted some papers on her desk to make room for her lunch. "Why don't you join me, I just got off the phone with John and have some frustrating news." Veronica sat down opposite Maya, nervous about how bad the news might be.

"What's wrong? I thought work on the Millbrook project was going well?" Her contact with John had been productive, but Maya had asked Veronica to be cautious of her time to avoid spending too much effort on a project without a contract.

> Although ApexTech has a BOA, this work is being treated by both sides as business development or complimentary services to demonstrate an interest in the future project.

"That's the problem. We're about to have to put even more work into it."

"Did we get the contract?" Veronica sat up straighter, more interested in the positive spin the conversation seemed to have taken. But she noticed Maya shake her head before she could ask any follow-up questions.

"There's been a new development. I thought the delay was due to the negotiations to acquire the parcels we need, but apparently, John failed to mention that all future projects at TerraHaven can't be sole-sourced. TerraHaven now requires John to develop an RFP, and then send it to three firms. Obviously, we'll be one of them, but now I need to write a proposal and estimate pricing for a project we've technically already started working on." Veronica heard the stress increase in her voice as Maya explained everything. "I shouldn't have been so lenient in letting you work on this. John is a friend, but he has a habit of pushing his favors just a bit too far. It means there's a risk we might not get this project. This mess is just as much on me, though, I guess."

> In the private sector, this decision is determined by the client's company. It's not required or formalized like the public sector, but a private company may elect a process like this to better evaluate their options.

"But it seemed like we basically had this project locked in. His emails definitely seemed like we were dealing in tangibles and not hypotheticals." It was difficult for Veronica not to feel like she had been a little taken advantage of.

> Communication can infer a situation or agreement, but without a contract, there is nothing binding the engineering firm's work to a billable service for the client.

"No point in dwelling," Maya continued with a deep breath. "I'll need help getting this proposal written up as quickly as possible. And until we have a signed contract, all side work for this thing stops. Understood?" Veronica nodded, "I'll help however I can."

Veronica left Maya's office and was finally able to get to her lunch, though she had suddenly lost a bit of her appetite.

* * *

John hung up the phone with a bit of frustration. He'd just gotten off a call with Maya, who wasn't pleased about the surprise that an RFP would be needed for Milbrook Logistics Park. Really, John wasn't happy, either. With all the sites under Option Contracts, John had paid a hefty fee to have the unilateral right to purchase the land within a fixed period of time – and the clock was ticking. The idea of stopping the due diligence work in order to write up an RFP felt like it would slow down his momentum on everything.

John intended to work with ApexTech; they'd had great results in the past, but the new TerraHaven policy that precludes sole sourcing had kept him from getting a contract signed. The policy was a risk-mitigation technique so TerraHaven could see comparable qualifications and pricing with any new project. John had hoped that because of his seniority at TerraHaven, the company would give him the reins to do what he felt was best.

> This implies that John was just as surprised as Maya about the new company policy. His communication and work with ApexTech likely would have been different if he had known TerraHaven would require an RFP.

Unfortunately, the policy was a requirement for all new projects. John planned to develop the RFP as quickly as possible. He was sure that once the three proposals were in, it would be an easy decision to award ApexTech the work. So he concluded that, since they'd end up with Maya and her team anyway, it would be OK to go ahead and use the resources that Veronica had brought to their initial meeting and any other information they'd exchanged over the past few months. Even though she was a newer engineer, Veronica seemed more than capable, so he was sure her sources had no significant

> John is taking on some risk here by relying on complimentary work that has not been vetted. Often, a client will formally procure services that inform the RFP and scope.

errors. This would also help move things along more efficiently. He didn't want to do this alone, but he knew he couldn't use Maya for help. That evening, after contemplating his options, he decided to give Sarah a call.

* * *

Sarah Chen of Meadow Law Partners had nearly finished wrapping up her final work emails for the day when her phone rang.

"Sarah Chen," she answered, slightly impatient. So much for getting home a bit early, she thought.

"Sarah, it's John. Have you got a minute?" He sounded a bit out of sorts.

"Sure, John. What's going on?" He briefly explained the situation with the new TarraHaven policy and the RFP. He wondered if she could help him develop the RFP content to take some of his burden. He figured she could help since TerraHaven already had a contract with Meadow Law Partners.

"Listen, I know this type of thing isn't your cup of tea, but unfortunately, I can't do much to help you out here, John. Unless you're willing to redraw our contract to include the extra work?"

This conversation implies that John is not confident in his development of an RFP, but Sarah has indicated that she does not have this contract in place to provide this service to support John.

John sighed heavily. "We're already doing so much on Millbrook together. Couldn't you just do me this one favor?"

Sarah laughed, "Engineers might gloss over the fine print of a contract, but for a lawyer, we follow what's written. You know you'll be hard-pressed to find a lawyer willing to do any favor for free, present company included. Sorry, John."

This quip from the attorney highlights how one company might approach business development differently from another. ApexTech was willing to provide free engineering services to John with the expectation of winning the project, but Meadow Law Partners has a more reserved approach.

"Yeah, I hear you," John responded sullenly. "The details and wording of these RFPs always get me. But I can manage on my own. Thanks anyway, Sarah."

"Good luck," she returned before hanging up.

* * *

December 2, Year 1

John worked to develop the RFP and benefited from using the materials he'd received from Veronica (Figure A2.1). He sent it out to ApexTech and two other engineering firms he recently connected with.

Request for Proposal

Millbrook Logistics Park

Civil Engineering and Consulting Services

December 2nd

Dear Julia Whitman,

TerraHaven LLC is pleased to invite ApexTech Engineering LLC (Offeror) to submit a proposal for the Millbrook Logistics Park project's civil engineering and consulting services. We have reviewed your company's portfolio and expect that your expertise aligns with the specific requirements of this project. We are excited to explore the possibility of working together to bring this project to fruition.

Project Overview:

- Project Name: Millbrook Logistics Park
- Location: Millbrook County
- Proposal Submission Deadline: January 5th

Competition Intended: TerraHaven intends that this Request for Proposal (RFP) permits competition. All comments, questions, or requests for clarification, including any notification regarding restrictions to competition, must be received by TerraHaven no later than ten (10) days before the date set for receipt of offers. Offerors may not rely on oral explanations, clarifications, or changes to the solicitation. All explanations, clarifications, or changes will be issued in written form. TerraHaven will not be bound by any oral statements.

All inquiries and any changes to the requirements of this solicitation shall be answered by the issue of written addenda to the solicitation. It shall be the responsibility of the Offeror to acknowledge all addenda by signing and returning a copy of all addenda with the offer submission or by separate acknowledgment of each addendum by number and date, in writing. Offerors are advised to contact this office to confirm the number of addenda five (5) days before the date established for the offer due date. All addenda will be issued no later than five (5) days before the offer due date.

FIGURE A2.1

I. Background and Project Description

Millbrook Logistics Park is a strategic development project initiated by TerraHaven, a prominent real estate development company with a proven track record of delivering high-quality, sustainable, and innovative projects across the region. TerraHaven specializes in creating dynamic spaces that cater to the evolving needs of businesses and communities, and Millbrook Logistics Park is no exception.

Millbrook Logistics Park is envisioned as a modern, state-of-the-art logistics and industrial hub that will serve as a catalyst for economic growth and job creation within fMillbrook County. Located on 251 acres, this project occupies a significant portion of land, making it a pivotal development.

The project requires several orchestrated phases to achieve construction permits. The first is consolidating seven existing lots, followed by rezoning and subdivision to accommodate parcels that will be leased to future tenants. TerraHaven will develop the shared infrastructure for Millbrook Logistics Park, which includes new roads (to be transferred to the Department of Transportation), utility systems (water, communication, electric, sewage), and the requisite stormwater conveyance and management systems.

TerraHaven has prepared the following base map to be used for information only:

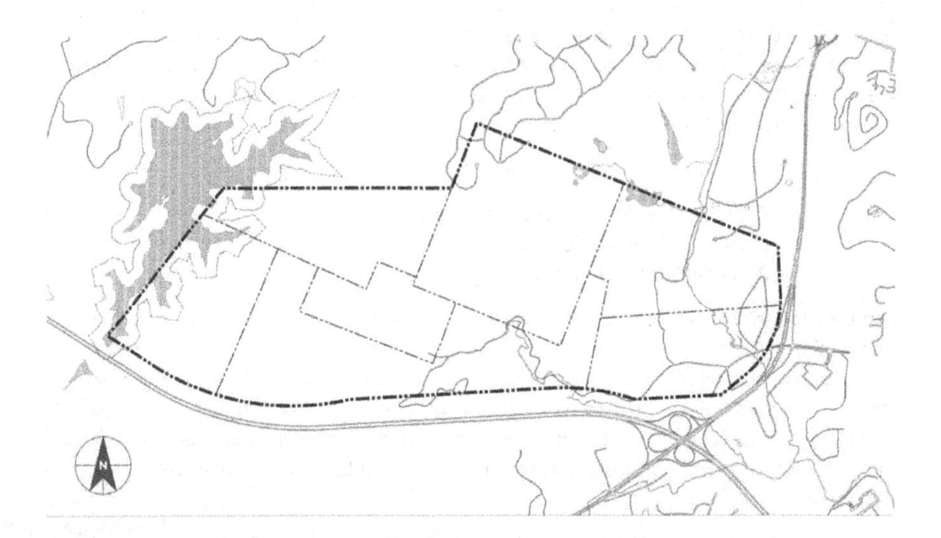

FIGURE A2.1 *(Continued)*

II. Scope of Work:

TerraHaven intends to develop the Millbrook Logistics Park as a state-of-the-art logistics and industrial park. The project requires extensive surveying, civil engineering, planning, and rezoning work to prepare the site for construction.

We require comprehensive services that include, but are not limited to, the following:

1. **Preliminary Site Analysis & Concept Plan:** Conduct a detailed analysis of the project site, identifying any potential challenges and opportunities. TerraHaven will provide a topographic and planimetric survey acquired under a separate contract. This survey data will be provided to the Offeror. Provide at least three concept plans to TerraHaven that provide six (6) or seven (7) new lots between ten (10) to thirty (30) acres each.
2. **Consolidation, Subdivision, and Rezoning:** Submit the necessary survey plats to consolidate the existing seven lots. Navigate the subdivision and rezoning process, ensuring compliance with all local regulations and zoning requirements.
3. **Site Plan:** Develop a comprehensive site plan that includes grading, drainage, utilities, roadways, and landscaping. The site plan shall accommodate future development of the lots but the scope shall only include the primary infrastructure.
4. **Environmental Assessment:** Conduct an environmental impact assessment and provide recommendations for mitigation measures.
5. **Project Management:** Offer project management services to oversee the entire process from conception to completion, including meetings and coordination efforts as required.

III. Schedule

Proposals will be evaluated, and an offeror will be selected within sixty (60) days of receipt of proposals deemed compliant with the RFP. Notice to proceed (NTP) is expected within thirty (30) days of selection. TerraHaven expects the offeror to submit a schedule for the rezoning process. Site plan approval is expected in nine (9) months.

IV. Proposal Submission Requirements:

Please submit your proposal in accordance with the following guidelines:

1. **Cover Letter**: Provide an introductory cover letter summarizing your company's qualifications and expressing your interest in the project.
2. **Company Profile:** Include a detailed overview of your company, highlighting relevant experience in civil engineering and rezoning projects in Millbrook County to demonstrate your qualifications.

FIGURE A2.1 *(Continued)*

TERRAHAVEN
LAND DEVELOPMENT

3. **Project Team:** Describe the key team members assigned to this project, including their qualifications and experience. The offeror shall note the intended use of sub-consultants.
4. **Approach and Methodology:** Outline your approach to the project, including the steps you will take to meet the project's objectives.
5. **Demonstration of Experience:** Offeror must show through entity and key personnel experience a track record of providing consulting services of similar scope. Offeror must provide evidence of organization and financial capacity to deliver the proposed services. Bidders shall identify three (3) similar consulting projects completed or in the process of completion, comparable to the scope of services. For each project, the bidder shall identify: (a) the Project name, (b) the Location of the project, (c) a description of the project, including work performed and total acreage, (d) the period of performance, (e) project commencement and completion dates, and (f) the proposed project team's role in the listed projects.
6. **Budget and Timeline**: Present a detailed cost estimate and timeline, including resource estimates for each of the outlined task items of the project. TerraHaven is specifically interested in a timeline of the rezoning process and acknowledgment by the offeror that site plan approval will be obtained within nine (9) months.
7. **No Conflicts of Interest:** The Offeror must make a statement of no knowledge of any potential conflicts of interest with the TerraHaven project.
8. **Proof of Insurance:** Offerors must be fully licensed for this type of work required by this solicitation no later than the date the proposals are due. The project will require the consultant to provide proof of the following insurance coverages:
   a. Commercial General Liability insurance having limits of at least 1 million dollars per occurrence, 2 million dollars aggregate;
   b. Umbrella Liability insurance of at least 2 million dollars;
   c. Commercial Auto (including owned, leased, non-owned and hired) having limits of at least $500,000; and
   d. Workers Compensation Insurance at statutory amounts.

V. Evaluation Criteria

After determining compliance with the requirements of this RFP, TerraHaven shall conduct its evaluation of the technical merit of the proposals. Each proposal received as a result of this RFP shall be subject to the same review and evaluation process. The following criteria will be used in the evaluation of submitted proposals scored out of 100 points, where the lowest fee is allocated the total point value (20 points) and other offeror score are calculated as a comparison in price (20 points x Lowest Price / Offeror Price).

FIGURE A2.1 *(Continued)*

TERRAHAVEN
LAND DEVELOPMENT

Criteria	Weight
Approach and Methodology	30
Demonstrated Experience	30
Key Personnel	10
Project Fee	20
Project Timeline	10

The evaluation criteria contained herein shall be scored by TerraHaven based on the stated weight factors for each evaluation criteria. Based on the initial review of proposals, TerraHaven may invite, without cost to itself, ranking finalists to make a presentation of their proposal and their capabilities as a further consideration in the selection process.

VI. Proposal Format Requirements

Title Page: Each proposal shall begin with a Title page. It should display the words "RFP Millbrook Logistics Park." It should also have the company's name, project manager name, title, business address, email address, and telephone number of the person authorized to obligate the company.

Table of Contents: The proposal shall contain a "TABLE OF CONTENTS" with page numbers indicated.

Proposal: The Offeror shall present their offer on double-spaced typed pages. Proposals must address each area covered under the scope and evaluation criteria in the order provided in this RFP. Proposals shall not exceed forty (40) pages.

Please submit your proposal by **January 5th** to TerraHaven, with attention to John Branson. If you have any questions or require further information, please contact John Branson.

TerraHaven looks forward to receiving your proposal and potentially partnering with your esteemed company on the Millbrook Logistics Park project. We believe that your expertise will be instrumental in the successful realization of this exciting venture.

FIGURE A2.1 *(Continued)*

A2 Review Questions

1. What are the potential risks and ethical considerations associated with sharing information and resources without a formal contract, as depicted in the interaction between Veronica and John?
2. How does TerraHaven's policy shift regarding the requirement of an RFP impact the existing working relationship between ApexTech and TerraHaven, and what strategies could ApexTech employ to safeguard its interests in such situations?
3. Consider the professional and ethical implications of John's decision to use the materials provided by Veronica for the RFP, given the lack of a formal agreement between TerraHaven and ApexTech, and the potential conflicts it creates with ApexTech's policies and expectations.

A3: The Proposal

Several weeks ago, Maya, a civil engineering project manager at ApexTech Engineering LLC, received a call from John, a commercial land developer working for TerraHaven. Maya's staff engineer, Veronica, has been working with John over the last few months to help him investigate a potential new project: Millbrook Logistics Park. To date, the work efforts by ApexTech have been tracked as a business development investment; there has been no contract between ApexTech and TerraHaven.

Based on their support for John's initial site investigation, Maya expected that ApexTech would be sole-sourced; however, Maya was surprised when John explained that TerraHaven had a new policy that required a request for proposal (RFP) to be sent to at least three engineering firms to receive competitive bids for any new projects. After some quick work, John developed an RFP and sent it to Maya, expecting a formal proposal for the new project. ApexTech was now responsible for developing a proposal based on the published RFP.

* * *

December 3, Year 1

Maya had hoped to receive the RFP sooner, so when she received an email from John in early December, she prioritized it. Though the timing could have been better, preparing a proposal while dealing with the holiday season would be tight. She opened the attached document.

> Given the holiday season, Veronica recognizes that the noted duration to work on the proposal is often pressured.

TerraHaven is pleased to invite ApexTech Engineering to submit a proposal for civil engineering and consulting services for the Millbrook Logistics Park project.

"Why are they always pleased to send an RFP?" She quipped. She also saw that the proposal was due in early January, giving her a month to develop a proposal during the holiday season. Annoyance aside, she knew she would need help. Maya messaged Veronica to join her in her office. After a moment, Veronica appeared in the doorway,

"Hey, what's going on?" Veronica questioned.

"We finally got that RFP from TerraHaven," Maya explained. She noticed Veronica's face perk up.

"Oh, good. I'm glad we can finally get this moving forward officially. I feel like we've already put so much work into it. Is there anything I can help you out with?" Veronica asked.

"I'm glad you still have a strong interest in this because I'd like you to be really involved here. I'm being pulled in too many directions at the moment, and I could use some extra help. You've done well working with John and Sarah so far. You've managed everything that's been asked of you efficiently and skillfully. So I was wondering, how do you feel about leading this proposal?"

Veronica appreciated the recognition but was surprised about the request and hesitated. Maya, reading her mood shift, continued quickly, "I know you're more comfortable with the technical side of things, but you're already familiar with this project. I think it'll be a good learning experience for you to write a proposal. Besides, you'll have support from other departments to write content related to transportation, stormwater management, environmental, and anything else we need. Also, the marketing staff will do most of the work; it's a lot of copying and pasting from other ApexTech proposals, just marketing lingo about how great ApexTech is, you know? And I'll work closely with you on schedule and pricing." Veronica seemed to relax a bit as Maya explained.

> This exchange shows that Maya recognizes Veronica's contributions and expects that much of this proposal development will be procedural, something that could be quickly compiled from existing ApexTech content.

"I've never really worked on a proposal before," Veronica responded as she scanned the RFP that Maya had open. "But it doesn't sound too overwhelming. What sort of timeline are we looking at?" She knew John had been slow on getting the RFP out and wanted to know if that would make things more difficult for them.

"It's due early next month. ApexTech's policy is to put pencils down two days before a proposal's due date,

> This is company policy that has reduced the proposal preparation duration. While the requirement is not as strict as a proposal submission deadline, Maya and Veronica must keep this date in mind when planning the proposal activities.

reserving that last day for critical errors, and then we like to submit a day early. We have a few rounds of internal review before all that, though. Still, that should be plenty of time to get things sorted. I'll offer you any help I can, of course; it's just that I have deadlines on several other projects, and I can't push this ahead of those, as much as I'd want to." Veronica nodded, but was worried when she heard the January due date. Trying to get everything and everyone pulled together that close to the holidays would be difficult. She had just started making a mental to-do list when Maya huffed loudly and pulled Veronica's focus back to her. Maya stared at her computer screen with an irksome look on her face,

"Well, that's just crossing the line," she mumbled.

"What's wrong?" Veronica asked, concerned.

"Oh, John's just using our *free* work in his RFP. He's gone and used the map you pulled together for that initial meeting. I know you made that of your own initiative, but you did it on ApexTech time, so it wasn't just a free resource for him to take." Maya shook her head and continued to scan through the proposal. Veronica was a bit taken aback that John used her work, and said as much to Maya. "Yes, it is frustrating," Maya answered back. "I'd be more upset, but I'm confident we'll get this contract, so we'll be paid for our work in due time."

> This raises some professional and ethical questions. John could or should have used content developed by their company or formally procured the services for developing RFP content.

"Well, I hope there aren't any more surprises." Veronica was growing more stressed about the short time she'd have on the proposal, even with Maya's leadership.

"Hm, I wouldn't hold your breath, unfortunately," Maya replied. "They're really pushing it on the schedule for site plan approval. Six months for rezoning and site plan approval? That just seems unnecessarily aggressive."

"It is?" Veronica questioned timidly. She recalled the first meeting with John, where the schedule had been discussed and she had told John that six months was typical, but she decided it might be

> In an earlier meeting (A1) Veronica verbally estimated the project duration. It was a verbal exchange and Veronica made her level of experience clear, but John has decided to use that information in a formal document.

better not to mention that conversation to Maya since she hadn't meant for her schedule estimate to end up in writing on an RFP. "But didn't Greenleaf Cafe in Harbor View City only take us about six months?" she offered.

"That was a different case. The project was a whole lot smaller, we didn't need a rezoning, which can double or triple the duration and add a lot of complexity. Plus, we know the city's process and their planning and engineering staff," Maya explained offhandedly, still scanning the screen in front of her. Maya continued, "All this, and they want a firm fixed price from us."

A firm fixed price is risky for the seller, ApexTech Engineering, especially when the scope is ambiguous or if there are concerns (such as the proposed schedule).

She looked at Veronica and saw the concern on her face. "Hey, listen, you can do this. I'll connect you with the marketing team, and things should go smoothly. I'll handle the cost estimate. And as I mentioned, I'll review everything before the proposal is sent so don't feel like you're on your own on this, OK?" Veronica nodded hesitantly and smiled weakly in return. "I'm here if you need me," Maya added. "My first step is to follow protocol and get an official 'go or no-go' decision so we can track our time and effort on the proposal. It might take a few days to get a meeting scheduled. Given our years of prior work with TerraHaven, I'm sure that we'll receive a 'go' on this. Let me know if you have any issues spring up. I know you're the best person for this."

Most firms require a process to determine if the pursuit is worth investing time and resources with a go/no-go decision. This process establishes a budget for the pursuit and documents the necessary resources, such as support from a marketing team.

"Thanks." Veronica smiled more genuinely, but as she turned to leave, she was already thinking she probably wouldn't have a very relaxing holiday.

* * *

December 9, Year 1

Several more days passed before Maya finally met with the ApexTech marketing team and some other staff when she was approved to work on the proposal. The go/no-go review had gone well for the most part, but there had been some hesitancy because ApexTech hadn't worked in Millbrook County before. However, given the long history with TerraHaven and because ApexTech's transportation department had coincidently just won a basic ordering agreement (BOA) with the Millbrook County Department of Transportation, ApexTech figured it was worth the investment to pursue the Millbrook Logistic Park project.

> A benefit of a go/no-go discussion is identifying risks, including potential weaknesses. Although ApexTech has a strong relationship with TerraHaven, there was concern about working in a new region.

> This is a potential benefit to the TerraHaven proposal, but the BOA award is recent, and there no tasks have been completed yet. This means it will not be possible to demonstrate experience in the County.

Veronica was excited to get going on the proposal and the anticipated project design. Maya called Veronica into her office so they could lay out a solid plan to mitigate any potential trouble spots for the proposal. She informed Veronica that they could proceed with the proposal development and relayed some concerns from the go/no-go review meeting.

"Is it usual to have issues with a proposal like this?" Veronica questioned. This experience was a new test for her skills, and she wasn't sure if, by accepting this task, perhaps she'd bitten off more than she could chew.

"I wouldn't say these are serious concerns. We're just doing our due diligence here so that we present ourselves as the best option for this project. Essentially, we're trying to curb any potentially negative view of ApexTech." Veronica nodded as Maya continued:, "For example, a key point here is that our firm hasn't worked in Millbrook County before, so we want to make sure we focus more on our work experience in the surrounding area but reference similar project types. Our transportation department just won a project in Millbrook, but we don't have any project examples to show yet."

There is often a benefit with teaming agreements, which are best established before an RFP is issued, and with teams with an existing professional relationship.

"Would it maybe help to get in touch with a local engineering firm, someone familiar with working in the area?" Veronica questioned, looking up from making a note on her laptop. She wanted every possible advantage to land the contract, realizing that all of her work so far would be pointless if ApexTech were not selected.

"I had thought of that, yes. I got the contact information for the land planner who led the Rosewood Town Center site design, but I've held off getting in touch."

"Really? But that seems like such a great partnership for us. They'd have more knowledge of the area than we would, which would save us a bunch of research time. Not that I'm not willing to, of course." Veronica added on quickly.

This is a common consideration for teaming arrangements. An engineering firm often needs to decide if there's a value in bringing on an external partner because it strengthens the team's qualifications, but it requires sharing some of the revenue.

Maya smiled, "I know. I've been impressed with your engagement on this so far. You're doing a great job." Veronica returned her smile, clearly pleased. "It's just that it comes down to a revenue issue." Maya continued. "If we bring in outside people, it means less revenue for us. Also, I'm unfamiliar with this team, and the thought of cold-calling an unknown firm for help doesn't sit well." Veronica agreed but privately thought there was still an advantage to working with someone local if they wanted to win this project.

"Alright," Veronica summarized, "so we highlight our work in the surrounding area. What else do we need to focus on?"

"It seems like we'll need to submit a few RFIs to clarify a some things. They don't say it, but I know they'll need traffic engineering, and ApexTech doesn't provide those services, so we need to understand what's expected." Veronica noted down the details of the request for information (RFI) while Maya continued. "And there's mention of some environmental engineering services needed, but it seems pretty vague as to who's responsible for that. Honestly, I'm hoping we don't have to do much on that front. Environmental scope is a bit outside of my wheelhouse, but we do have ApexTech staff that can help if needed so that we could show an environmental engineer on our team."

She scanned through the remainder of the RFP again for any last thoughts. "Oh, one more thing," Veronica stopped writing and looked up expectantly. "I'm a bit uneasy about the effort of the rezoning work; it can be unpredictable on a local level like this, so our scope needs to be solid."

Veronica nodded but questioned, "Solid?" as she made a few more additions to her notes. Maya answered, "Yes, meaning that we need to define our expectations, such as how many zoning submissions, how many public meetings, how many iterations, how many internal meetings, and so on. The zoning work is usually hourly, or time and materials, because nobody knows how long it will take to receive approval. But, John wants this all as a firm fixed price."

"Would it be faster for me just to call John to get the answers we need?" Veronica suggested. Maya shook her head.

"I'm afraid not. These RFP rules state we need to send questions in writing, and then TerraHaven issues a written response to everyone. I know that will slow things down, but we need to play this by the book." Veronica double-checked the date in the corner of her screen and felt her anxiety rise a notch. She shut her computer, took a deep breath, and rose to leave, "OK, I'll draft the RFIs and send them to you, and then I'll get back to the rest of the proposal so we can have a good draft as soon as possible." Maya thanked her and returned to her other work. Veronica turned to leave and was getting concerned all these delays were going to make things even more difficult as people started taking off for the holidays.

> These conditions are more prevalent with public projects, but a private client may choose to control information exchanges with similar processes.

* * *

December 15, Year 1

A few days later, Maya received the RFI responses from TerraHaven she'd been waiting for. Veronica's nerves had rubbed off on her a bit, and the long wait had left her restless to get the proposal submitted. She quickly scanned over the addendum that answered the questions (Figure A3.1).

ADDENDUM #1 – DECEMBER 15

Millbrook Logistics Park

Civil Engineering and Consulting Services

The Proposer is to acknowledge receipt of this Addendum #1, including all attachments in its Proposal by so indicating in the proposal that the addendum has been received. Proposals submitted without acknowledgment of receipt of this addendum may be considered non-conforming.

The following Questions have been received by TerraHaven. Responses are being provided in accordance with the terms of the RFP. Respondents are directed to take note in its review of the documents of the following questions and TerraHaven responses as they affect work or details in other areas not specifically referenced here.

#1 Questions & Answers

Question	Answer
Are traffic engineering services required with this scope of work?	No. TerraHaven will manage a separate contract for requisite Traffic Engineering services. Relevant information will be provided to the offerors. The offeror is expected to coordinate with TerraHaven's consultant.
Are environmental engineering services required with this scope of work?	TerraHaven expects the offeror to handle the environmental engineering services required for site plan approval.
There is no schedule or defined scope for the consolidation, subdivision, and rezoning. What is the expected schedule and scope for the engineering firm?	TerraHaven expects the offeror to describe the schedule and scope of work in their proposal of services.
Has TerraHaven evaluated unsuitable soil conditions for the site? Do we need to include geotechnical engineering?	No. TerraHaven will manage a separate contract for geotechnical engineering services. Relevant information will be provided to the offerors. The offeror is expected to coordinate with TerraHaven's other consultants.
Does TerraHaven own all subject properties?	TerraHaven will acquire rights to develop the properties.

FIGURE A3.1

This questions and the answers (Figure A3.1) are important. For example, ApexTech did not submit the unsuitable soils question, indicating that a competitor may know more about the County's review requirements (or it could be a misleading question). The answer to the question provides no substantive information and may indicate that the client is also unfamiliar with this scope.

Maya noted that traffic engineering services were under a separate contract, so they were off the hook for that. Unfortunately, the answer to their rezoning scope and the environmental question was still frustratingly ambiguous. She shook her head at the non-answers and picked up her phone to call Veronica and fill her in. Maya also studied the last two questions, which didn't come from ApexTech. It meant that TerraHaven sent the RFP to other engineering firms with serious interest and questions. Maya had hoped for more information on the schedule and was unfamiliar with the "unsuitable soil conditions" question that was asked. She called Veronica in to get her to do some research on the matter.

Here, Maya uses the RFI responses to consider what other relevant content should be included in the proposal.

As Maya filled Veronica in on the RFI responses, Veronica again suggested they contact the land planner responsible for Rosewood Town Center to help answer some of the remaining questions about rezoning and unsuitable soils. Maya conceded, primarily because she felt backed into a corner due to her unfamiliarity with the environmental and potential geotechnical issues and the overall zoning scope ambiguity.

She pulled up the contact info for Avrum Planners, LLC, and made the call. The phone rang several times before a man picked up,

"Amar Singh," said the voice on the other line.

"Mr. Singh, hi, this is Maya Mendez from ApexTech Engineering. I'm sorry to call you out of the blue like this, but am I correct that you were the planner responsible for the Rosewood Town Center?"

"Well," Amar chuckled, "not me alone, but yes, my team here at Avrum Planners designed Rosewood. What can I do for you, Ms. Mendez?"

"ApexTech is submitting a proposal to TerraHaven Development for a new logistics park in Millbrook County. Rosewood turned out wonderfully, and I was thinking you might want to join our team for the new project. We specialize in site-civil engineering and could use a great land planner. With your knowledge and history of the area, we'd create a strong team and pull ahead ahead of any competition. What do you think?" Maya was intentionally reserved about noting ApexTech's lack of experience and portraying the request as a desperate plea.

This is a bit more direct than is typical for a teaming discussion but represents some of the pressure on ApexTech to feel qualified for the work.

"I'm so happy that you like Rosewood, but regrettably, we've already been contacted by another engineering firm on that particular project," he explained.

Maya was disconcerted when she heard this. Now it was more evident that her competition was interested in winning this. She realized she hadn't thought the competition would be so earnest but remained optimistic. "I see. Well, not to seem too presumptuous, but ApexTech and TerraHaven have been working together for years. John is a personal friend of mine, and it seems more than likely that ApexTech will be selected. I think a partnership here would be beneficial to us both." She hoped she was coming across as persuasive enough to change his mind on working with her. With his experience, it would certainly make things a lot easier if he did.

This is critical information for ApexTech, as they understand that there is interest from competitors and that they have a strong teaming partner with an exclusivity arrangement.

"You make a compelling argument, but I'm afraid we already have a gentleman's agreement with the other firm. I appreciate your interest in Avrum Planners, though. Perhaps we might be able to work together in the future," Amar offered.

"Right," Maya tried not to sound too disappointed. "Thank you for your time Mr. Singh. I will go ahead and keep you in mind in the future; I appreciate that. Have a nice day. Happy holidays."

"Ah yes, of course, you have a good day as well, Ms. Mendez."

She hung up the phone, more than a little defeated. Amar has been professional, but the rejection still stung a bit. Maya called Veronica again to tell her she didn't have much help to offer on the rezoning or the unsuitable soils issues. This meant the next step was to proceed with what they had and provide the current draft to marketing as a Gold Team review.

> Gold team represents one of the final stages of internal review for a proposal; no significant changes are expected after this review occurs.

Thankfully, the review of the draft proposal occurred in just a day, so at least something was going smoothly. The ApexTech proposal review team's concerns continued to focus on their gaps in local knowledge and a lack of detail regarding the approach to rezoning and environmental services, which wasn't surprising. However, with little time for any significant changes to the proposal, Maya felt skeptical about being able to adjust any content to sound more knowledgeable about the Millbrook area.

As a last resort, and knowing she was bending the rules a little, Maya decided to call John directly. She wanted to understand what was expected for environmental services and understand if John really thought that six months was feasible for rezoning and site plan approval. John was willing to have a conversation with Maya to help her out.

> In public projects, this could disqualify a bidder. Here, this could still be seen as an unethical practice.

Regarding the ambiguous environmental scope, John referred to the meeting he had had with Veronica, where she mentioned a resource protection area (RPA) designation. John conceded that he didn't know how long rezoning or site plan approval would take or anything about unsuitable soils, so he told Maya to use her professional judgment and the firm's experience to determine something reasonable if the six-month duration wasn't feasible, and they could work through it more later.

> This type of verbal direction introduces risk, as ApexTech uses this direction to propose schedule and pricing that is then documented in the proposal.

Maya was relieved to hear this. Her team was familiar with RPA requirements, and now she had something to expand upon. The rezoning and site plan process now seemed to be something she could generally describe, schedule, and price. Still, she wanted to gather as much information about Millbrook County as she could so she'd feel better about a price and schedule.

As Maya prepared to wrap things up before the holidays, she wrote out some final notes in an email to Veronica to prepare the proposal for submission to TerraHaven. She trusted that Veronica could finish the proposal on her own by coordinating with the marketing team and she felt the proposal was almost complete in her mind. They'd give it one more review before sending it out, but her work seemed done for now. Maya was ready for a nice break.

* * *

December 23, Year 1

Veronica spent a few weeks working hard with other department managers and the ApexTech marketing team to develop a professional proposal. Veronica felt pleased with her work, going above and beyond the "copying and pasting" from other proposals. She used her knowledge of the project, and John's interest, and was careful to think about overcoming their challenges with a lack of experience in Millbrook County. It was a lot of work and an unfamiliar task for Veronica, but she was learning a lot and feeling confident in her work.

Veronica had just been about to call a friend to make dinner plans for a much-needed catch-up when her inbox notified her of a new email from Maya. Veronica was exhausted, but with a sigh, she opened it. She didn't think Maya realized that perhaps she had wanted to spend the remainder of the month relaxing and not occupied with this seemingly never-ending proposal. The message gave an overview of how Maya wanted her to polish up the final document:

Thank you again for all your hard work on this proposal. I know it's a new skill for you, but you're doing amazingly well and preparing a great-looking proposal. I know I'd choose us!

I have just a few remaining comments from the Gold Team Review. I would do this myself, but I have some family obligations for the next few days, and you're the most familiar with the content. Here are my notes:

1. Add a few notes about our environmental services on our prior RPA-related work.
2. Make a note to exclude surveying and traffic engineering (this will be contracted out by TerraHaven per their notes, but I'd like to reiterate it).
3. Be sure to describe in our approach that we can commit to a 9-month site plan approval, but clarify that this will be measured after zoning approval.
4. Our experience in Millbrook is lacking, so focus more on the surrounding areas where we've worked and our general expertise with industrial development sites and rezoning. The transportation department recently won a BOA with Millbrook County, which we can note, but we haven't done much there, so we have limited detail.
5. Overall, the rezoning process needs more details. I see you pulled the guidelines from the Millbrook County website. Still, it would go a long way if we could be more detail-oriented in our description – maybe try contacting the County office directly and discuss this process (but do not give specifics on the project, just speak generally to the need).

We'll need these changes made to the proposal in the next few days, and that will take us from Gold to White Glove review, where we're ready for submission to TerraHaven. I don't need to re-review the final version if you can make these edits. Email me about what you hear from the County on the rezoning process so I can fill in a few blanks on pricing. And email me if you need me, and I'll try to get back to you as soon as possible.

Thank you again — Happy Holidays!
-Maya

As she read through the email, Veronica decided to nix her evening plans and get as much done as possible; she made her way through Maya's first comments without too much effort. It was late by the time she'd made it to Maya's fourth comment, and she decided the marketing team would be best at describing the new BOA contract from the transportation department. She made a note for them to add it to the Demonstration of Experience section.

For the fifth and last comment, she was a bit concerned with how she could determine more information so late in the proposal effort. Figuring that she could at least send an introductory email to County staff, she checked the Millbrook County website for the contact information. Unfortunately, a large, red header at the top of the website stated that the County offices were closed until January 2. This wouldn't give her enough time to make contact and update the proposal before it was due to TerraHaven.

Veronica sent a message to the County asking for their best schedule estimate on rezoning and plan approvals. She then sent an email to Maya letting her know that the County contact was probably a dead end, figuring Maya would be able to develop some cost and schedule estimates from similarly sized projects completed by ApexTech.

January 3, Year 1

Veronica managed a few days' rest but realized if these were the types of deadlines Maya had to deal with regularly, it wasn't a wonder why she seemed stretched so thin all the time. When Veronica returned to the office, she noticed an email from the County in her inbox.

> Veronica,
>
> Thank you for contacting Millbrook County Office of Engineering. We cannot guarantee any review times, but in most cases, a site plan (SP) is approved around three months after the first submission, depending on the size, complexity, and accuracy of the plan.
>
> If your project does not include any new buildings you can apply for a rough grading plan (RGP), which takes closer to six weeks.
>
> Thank you and Happy New Year!
> -Carl
> Engineer 2
> Millbrook County Office of Engineering

Although late, Veronica appreciated the email from the County and sent a “thank you” to follow up. She then forwarded the information to Maya but acknowledged that it only provided some guidance on estimating the site plan schedule and didn’t answer the question about rezoning. Maya expected the site plan design effort to take about five to six months, and based on the County’s three-month estimate, she figured her initial estimate of nine months for the site plan was still reasonable. Maya felt uneasy about the rezoning schedule, but since they were out of time, she made her best estimate that it would also take nine months.

With everyone else back in the office in the first week of January, the proposal was reviewed one last time with the final edits on cost and schedule and given official approval to be sent to TerraHaven on the 4th of January. John let Maya know it had been received and that a decision would be made in about 60 days.

A3 Review Questions

1. How does ApexTech’s company policy, particularly the rule of ‘pencils down’ two days before a proposal due date, impact the proposal preparation process, and what are the implications of such policies in a time-sensitive project?
2. How can ApexTech mitigate the challenges of preparing a proposal during the holiday season, considering the potential for reduced staff availability and other seasonal disruptions?
3. Analyze TerraHaven’s ethical considerations using ApexTech’s preliminary work in their RFP. What are the professional boundaries regarding the use of shared information in business dealings without a formal contract?
4. Discuss the potential benefits and drawbacks of partnering with a local firm for proposal preparation, as contemplated by Veronica and Maya, in the context of revenue implications and local expertise.
5. Considering the tight timeline and Maya’s instruction to Veronica to handle the proposal with minimal oversight, what strategies should Veronica employ to ensure the proposal’s completeness and quality while adhering to ApexTech’s standards and deadlines?

A4: The Contract

In early January, ApexTech Engineering submitted a proposal to provide professional engineering and consulting services for the new Millbrook Logistics Park proposed for development by TerraHaven. Although TerraHaven is a private company, a recent company policy change required that TerraHaven develop an official RFP and distribute it to at least three engineering firms. John Branson at TerraHaven distributed the RFP to multiple firms, including ApexTech.

ApexTech has a longstanding professional relationship with TerraHaven, led by the ApexTech project manager Maya Mendez. Maya has worked with John for years, and more recently, Maya has allocated additional responsibilities to a staff member on her team, Veronica. It was Veronica who authored most of the proposal and provided preliminary (and informal) engineering work to TerraHaven. The RFP noted that an award selection for qualified bidders, including ApexTech, would be made within 60 days of the proposal submission. Maya and Veronica have been waiting anxiously to hear from John.

* * *

May 2, Year 2

Maya wasn't surprised when the 60-day response time John mentioned had come and gone – it had been four months since their submission of the proposal for Millbrook Logistics Park. She knew she would hear back eventually and had plenty of other work to keep her occupied. Veronica, however, felt she had a personal stake in the project because it was her first proposal, and she was quite anxious to see things come to fruition. Maya had told her things often move slower than originally advertised, and that she should try not to focus too much energy on it. Still, that hadn't stopped Veronica from casually mentioning the topic at least once a week.

"So," Veronica started. Maya smiled slightly, knowing what was coming. It seemed to be Veronica's habit of dropping in a mention of the Millbrook Logistics Park after lunch towards the end of each week. "I was just curious if you'd heard anything from John yet?"

"Not yet, but you have to remember there's a lot of moving parts here, so it's not going to be as cut and dry as you'd think. My guess is that they're negotiating with the purchase agreements, but John hasn't told me anything."

"They would tell us if we *hadn't* been chosen, wouldn't they?" Veronica inquired.

"Well, don't get me wrong, I'm still fairly confident that . . ." but a phone call cut Maya off before she could finish. She glanced down at her phone, and a look of mild surprise crossed her face when she saw John's name on the screen, "Well, speaking of," she trailed off as she picked up and put the call on speaker, "John, good to hear from you." At the mention of his name, Veronica instantly perked up, and Maya had to stifle a laugh. John picked up on it, "I hope I'm not interrupting anything. Sounds like you're working on something fun," he joked.

"No, no, Veronica and I were just talking about you. I hope you've called for more than just a friendly chat," she insinuated. It was John's turn to laugh in response.

"I hope you two are saying all good things about me. I didn't mean to keep you in suspense for so long, but we had to smooth a few things out first before making a final decision."

"Which would be?" Maya pushed.

"TerraHaven has chosen ApexTech to handle the Millbrook project."

"Yes!" Veronica mouthed and leaned back in her chair. She looked more relaxed than she had in the past few months since submitting the proposal. Maya smirked slightly, "I'm glad to hear it. It'll be good to be working together again officially."

"Well, that's the good news," John continued. Maya noticed Veronica's eyes narrow slightly and girded herself for whatever was coming next. "The good news?" She repeated.

"You were the most *reasonable* candidate, but upper management raised a few red flags about your lack of familiarity with Millbrook County."

This verbal exchange is an example of a debrief, where a client provides feedback from the submitted proposal. These can be formal or casual, as indicated here.

Maya subtly understood "reasonable" to most likely mean their competition had probably lost out due to being too expensive. "And so the bad news would be?" she questioned.

"We decided it would be in our best interest to bring on a local firm to split the contract. They will be in charge of all the rezoning and subdivision and consolidation work, and ApexTech will develop the site plan."

Maya recalled her conversation with Amar Singh about already working with another firm for the project and putting two and two together. "A local firm? Let me guess, Avrum Planners? "

"Yes, actually," he answered. She heard the slight surprise in his voice but decided not to mention her call with Mr. Singh to John. Maya continued, "They're a good choice. I think they did a great job with the Rosewood site."

"That was our general idea. I understand this might not be ideal, having to pair up with a new consultant, but we think it's what's best for the development."

He was right, she wasn't thrilled. Splitting the work meant splitting the revenue, but she understood that the job would probably go smoother working with someone who knew the local jurisdiction more than they did. "Right, of course," she responded slowly, already trying to find a way to capture a bit more work to account for the loss of rezoning scope. Remembering her call to Amar gave her an idea. "I understand the need to coordinate with a local firm, which means we should work out an hourly coordination budget for all the back and forth we'll inevitably need."

> This is a reasonable ask because there is effort involved in coordinating with a new stakeholder, which ApexTech could not have anticipated.

John readily agreed, "I think that's something we can go ahead and do."

"Also," she pushed on, thinking quickly, "since ApexTech won't be in control of the rezoning design anymore, we should reserve the right to revise our budget based on anything you decide to change with Avrum Planners and because we don't have any control over their schedule." John was quiet for a moment but hesitantly agreed.

> This is risk management. ApexTech recognizes that they should not commit to a schedule that begins outside their control or to a change in scope that could arise from Avrum Planners.

As she listened to the conversation, Veronica couldn't help but be impressed with how deftly Maya shifted the situation to negotiate additional work and fee to address the surprise change when suddenly she heard her name being dropped.

"You mentioned speaking with Veronica. I'm gonna assume she's there with you?" John asked.

After a nod from Maya, Veronica leaned forward slightly to be heard. "Yes, I'm here. I'm so glad you decided to go ahead and use us for the project. We've put a lot of work into it so far, and I'm excited to get going."

"Well speaking of hard work, I understand you were the main writer on the proposal that was sent in, and I just wanted to say that my team and I were impressed with it. I know these things can be a bit tedious, but it was well organized and detailed."

Veronica couldn't help but blush a little bit, "Thank you. I definitely had help, but I appreciate it."

Maya and John chatted for a few more moments, then hung up. She congratulated Veronica on her strong proposal and said that they couldn't have gotten the contract without her.

"I know things got a bit dicey around the holidays, conflicting schedules, and everything, but you really came through, and I wasn't the only one who noticed." Veronica grinned again, "Thank you, that means a lot."

* * *

After the call with John, as Maya had explained to Veronica, things proceeded slowly because the ApexTech scope of work needed to be modified, and the contract needed to be drafted and executed based on the revised scope and fee. Maya received a contract with comments from the ApexTech internal counsel (Figure A4.1) for her to review prior to signature and distribution to John.

This contract is modified from the Engineers Joint Contract Documents Committee (EJCDC) E520, used with permission. Not for reproduction.

A4 Review Questions

1. Analyze the role of professional relationships in project management, as exemplified by the longstanding relationship between ApexTech and TerraHaven, and the collaboration between Maya and John. How do these relationships impact the selection process and project execution?
2. Based on the scenario where ApexTech is chosen for the Millbrook project but with a split contract, how should a project manager approach negotiations for revising the scope and budget of a project? Consider the strategies used by Maya in negotiating additional work and fees to address the surprise change in the project scope.
3. Reflect on the impact of delays and changes in project timelines, as seen in the extended response time from TerraHaven. How should project managers like Maya and team members like Veronica manage expectations and adjust their planning in response to such uncertainties in civil engineering projects?

SHORT FORM OF AGREEMENT
BETWEEN OWNER AND ENGINEER
FOR PROFESSIONAL SERVICES

↓DRAFTING NOTE TO APEXTECH ENGINEERING: THE EMBEDDED COMMENTS REFLECT ISSUES FOR CONSIDERATION BY APEXTECH ENGINEERING AND ARE NOT INTENDED TO BE DISTRIBUTED TO OWNER.

THIS IS AN AGREEMENT effective as of ("Effective Date") between **TerraHaven Land Development LLC** ("Owner") and **ApexTech Engineering, LLC**. ("Engineer").

Owner's Project, of which Engineer's services under this Agreement are a part, is generally identified as follows: **Millbrook Logistics Park** ("Project").

Engineer's services under this Agreement are generally identified as follows:

1. ~~Preliminary Site Analysis & Concept Plan: Conduct a detailed analysis of the project site, identifying any potential challenges and opportunities. TerraHaven will provide a topographic and planimetric survey acquired under a separate contract. This survey data will be provided to the Offeror. Provide at least three concept plans to TerraHaven that provide six (6) or seven (7) new lots between ten (10) to thirty (30) acres each.~~ ***←DRAFTING NOTE TO APEXTECH ENGINEERING: PER NEGOTIATIONS WITH OWNER – THIS COMPONENT OF THE PROPOSED SCOPE HAS BEEN REMOVED. PLEASE CONFIRM.***
2. ~~Consolidation, Subdivision, and Rezoning: Submit the necessary survey plats to consolidate the existing seven lots. Navigate the subdivision and rezoning process, ensuring compliance with all local regulations and zoning requirements.~~ ***←DRAFTING NOTE TO APEXTECH ENGINEERING: PER NEGOTIATIONS WITH OWNER – THIS COMPONENT OF THE PROPOSED SCOPE HAS BEEN REMOVED. PLEASE CONFIRM.***
3. **Site Plan**: Develop a comprehensive site plan that includes grading, drainage, utilities, roadways, and landscaping. The site plan shall accommodate future development of the lots but the scope shall only include the primary infrastructure.
4. **Environmental Assessment**: Conduct an environmental impact assessment and provide recommendations for mitigation measures.
5. **Project Management**: Offer project management services to oversee the entire process from conception to completion, including meetings and coordination efforts as required.
6. **Coordination with Other Consultants**: Miscellaneous consulting services for the Owner's consultants to advise in civil engineering matters.

Collectively, the foregoing are referred to as the "Services".

Owner and Engineer further agree as follows:

Page 1

FIGURE A4.1

1.01 *Basic Agreement and Period of Service*

A. Engineer shall provide or furnish the Services set forth in this Agreement. If authorized by Owner, or if required because of changes in the Project, Engineer shall furnish services in addition to those set forth above ("Additional Services") pursuant to an amendment to this Agreement approved by Engineer and Owner.

B. Engineer shall complete its Services within the following specific time period: ***18 Months*** of [ ] ***←DRAFTING NOTE TO APEXTECH ENGINEERING: RECOMMEND SPECIFYING WHEN THIS TIME PERIOD BEGINS BY REFERENCE TO A FORMAL NOTICE TO PROCEED ACCOMPANIED BY DOCUMENTS NEEDED TO COMMENCE WORK***; provided, however, Engineer shall use commercially reasonable good faith efforts to accelerate the performance of Services in a manner so as to complete the Services within 9 Months of ***←DRAFTING NOTE TO APEXTECH ENGINEERING: THIS PROVISION REFLECTS THE GOAL OF ACCELERATING SERVICES TO EXPEDITE COMPLETION; HOWEVER, THIS IS SILENT ON REMEDIES FOR FAILURE TO ACHIEVE THE GOAL (SUCH AS LIQUIDATED DAMAGES) OR ADDITIONAL COMPENSATION FOR SUCCESS IN ACHIEVING THE GOAL. PLEASE ADVISE.***

C. If, through no fault of Engineer, such periods of time or dates are changed, or the orderly and continuous progress of Engineer's Services is impaired, or Engineer's Services are delayed or suspended, then the time for completion of Engineer's Services, and the rates and amounts of Engineer's compensation, shall be adjusted equitably. ***←DRAFTING NOTE TO APEXTECH ENGINEERING: CONSIDER SPECIFYING ANY RATES AND/OR ADJUSTMENTS FOR ANY PARTICULAR FORESEEABLE CHALLENGES IN DELIVERING SERVICES PURSUATN TO THE SCHEDULE.***

2.01 *Payment Procedures*

A. *Invoices:* Engineer shall prepare invoices in accordance with its standard invoicing practices and submit the invoices to Owner on a monthly basis. Invoices are due and payable within 30 days of receipt. ~~If Owner fails to make any payment due Engineer for Services, Additional Services, and expenses within 30 days after receipt of Engineer's invoice, then (1) the amounts due Engineer will be increased at the rate of 1.0% per month (or the maximum rate of interest permitted by law, if less) from said thirtieth day, and (2) in addition~~ ***←DRAFTING NOTE TO APEXTECH ENGINEERING: OWNER DELETED THE LATE PAYMENT PROVISIONS. PLEASE ADVISE IF THIS NEEDS TO BE RESTORED.*** Engineer may, after giving seven days written notice to Owner, suspend Services under this Agreement until Engineer has been paid in full all amounts due for Services, Additional Services, expenses, and other related charges. Owner waives any and all claims against Engineer for any such suspension.

B. *Payment:* As compensation for Engineer providing or furnishing Services and Additional Services, Owner shall pay Engineer as set forth in Paragraphs 2.01, 2.02.1 AND 2.02.2 (Services), and 2.03 (Additional Services). If Owner disputes an invoice, either as to amount or entitlement, then Owner shall promptly advise Engineer in writing of the specific basis for doing so, may withhold only that portion so disputed, and must pay the undisputed portion.

↓DRAFTING NOTE TO APEXTECH ENGINEERING: NEED TO CONFIRM THE BUSINESS AGREEMENT AS TO WHETHER THE COMPENSATION WILL BE (1) PAID IN A LUMP SUM FORMAT (I.E., ALL PAID AT ONCE), (2) PAID AS SERVICES ARE PERFORMED OVER TIME, OR (3) A COMBINATION OF THE TWO. THE FORM CONTRACT PROVIDES TWO ALTERNATIVE COMPENSATION PROVISIONS IN SECTION 2.02 BELOW.

EJCDC® E-520, Short Form of Agreement Between Owner and Engineer for Professional Services.

Page 2

FIGURE A4.1 (*Continued*)

2.02.1 *Basis of Payment— Site Plan (Scope Item 3) and Environmental Assessment (Scope Item 4)*

A. With respect to Site Plan (Scope Item 3) and Environmental Assessment (Scope Item 4), Owner shall pay Engineer for Services as follows:

1. A Lump Sum amount of **$630,000**.

2. In addition to the Lump Sum amount, reimbursement for the following expenses: **Printing, shipping, and travel (mileage).**

B. The portion of the compensation amount for Section 2.02.1 billed monthly for Engineer's Services will be based upon Engineer's estimate of the percentage of the total Services actually completed during the billing period.

[and]

2.02.2 *Basis of Payment— A. Project Management (Scope Item 5) and Coordination with Other Consultants (Scope Item 6)*

A. Project Management (Scope Item 5) and Coordination with Other Consultants (Scope Item 6), Owner shall pay Engineer for Services as follows:

1. An amount equal to the cumulative hours charged to the Project by each class of Engineer's employees times standard hourly rates for each applicable billing class, plus reimbursement of expenses incurred in connection with providing the Services and Engineer's consultants' charges, if any.

2. Engineer's Standard Hourly Rates are attached as Appendix 1.

3. The total compensation for Section 2.02.2 Services and reimbursable expenses shall not exceed **$62,000** without Owner's prior written approval. ***←DRAFTING NOTE TO APEXTECH ENGINEERING: THIS REVISION REFLECTS NEGOTIATIONS WITH OWNER. PLEASE ADVISE IF THIS IS NOT THE FINAL RESOLUTION ON THE ISSUE OF TIME AND MATERIAL ESTIMATES.***

[or]

2.03 *Additional Services:* For Additional Services, Owner shall pay Engineer an amount equal to the cumulative hours charged in providing the Additional Services by each class of Engineer's employees, times standard hourly rates for each applicable billing class; plus reimbursement of expenses incurred in connection with providing the Additional Services and Engineer's consultants' charges, if any. Engineer's standard hourly rates are attached as Appendix 1.

3.01 *Termination*

A. The obligation to continue performance under this Agreement may be terminated:

1. For cause,

FIGURE A4.1 *(Continued)*

a. By either party upon 30 days written notice in the event of substantial failure by the other party to perform in accordance with the Agreement's terms through no fault of the terminating party. Failure to pay Engineer for its services is a substantial failure to perform and a basis for termination.

b. By Engineer:

1) upon seven days written notice if Owner demands that Engineer furnish or perform services contrary to Engineer's responsibilities as a licensed professional; or

2) upon seven days written notice if the Engineer's Services are delayed for more than 90 days for reasons beyond Engineer's control, or as the result of the presence at the Site of undisclosed Constituents of Concern, as set forth in Paragraph 5.01.I. ***←DRAFTING NOTE TO APEXTECH ENGINEERING: PLEASE CONFIRM A 90-DAY LAG IN DELIVERY IS APPROPRIATE FOR THIS ENGAGEMENT / THIS PROJECT***

c. Engineer shall have no liability to Owner on account of a termination for cause by Engineer.

d. Notwithstanding the foregoing, this Agreement will not terminate as a result of a substantial failure under Paragraph 3.01.A.1.a if the party receiving such notice begins, within seven days of receipt of such notice, to correct its substantial failure to perform and proceeds diligently to cure such failure within no more than 30 days of receipt of notice; provided, however, that if and to the extent such substantial failure cannot be reasonably cured within such 30 day period, and if such party has diligently attempted to cure the same and thereafter continues diligently to cure the same, then the cure period provided for herein shall extend up to, but in no case more than, 60 days after the date of receipt of the notice. ***←DRAFTING NOTE TO APEXTECH ENGINEERING: PLEASE CONFIRM A 30-DAY CURE PERIOD IS SUFFICIENT FOR THE NATURE OF SERVICES COVERED IN THIS ENGAGEMENT / THIS PROJECT***

2. For convenience, by Owner effective upon Engineer's receipt of written notice from Owner.

B. In the event of any termination under Paragraph 3.01, Engineer will be entitled to invoice Owner and to receive full payment for all Services and Additional Services performed or furnished in accordance with this Agreement, plus reimbursement of expenses incurred through the effective date of termination in connection with providing the Services and Additional Services, and Engineer's consultants' charges, if any – such payment to be due and payable within ten (10) days of receipt of written demand.

4.01 *Successors, Assigns, and Beneficiaries*

A. Owner and Engineer are hereby bound and the successors, executors, administrators, and legal representatives of Owner and Engineer (and to the extent permitted by Paragraph 4.01.B the assigns of Owner and Engineer) are hereby bound to the other party to this Agreement and to the successors, executors, administrators, and legal representatives (and said assigns) of such other party, in respect of all covenants, agreements, and obligations of this Agreement.

B. Neither Owner nor Engineer may assign, sublet, or transfer any rights under or interest (including, but without limitation, money that is due or may become due) in this Agreement without the written consent of the other party, except to the extent that any assignment, subletting, or transfer is mandated by law.

FIGURE A4.1 *(Continued)*

Unless specifically stated to the contrary in any written consent to an assignment, no assignment will release or discharge the assignor from any duty or responsibility under this Agreement.

C. Unless expressly provided otherwise, nothing in this Agreement shall be construed to create, impose, or give rise to any duty owed by Owner or Engineer to any Constructor, other third-party individual or entity, or to any surety for or employee of any of them. All duties and responsibilities undertaken pursuant to this Agreement will be for the sole and exclusive benefit of Owner and Engineer and not for the benefit of any other party.

5.01 *General Considerations*

A. The standard of care for all professional engineering and related services performed or furnished by Engineer under this Agreement will be the care and skill ordinarily used by members of the subject profession practicing under similar circumstances at the same time and in the same locality. Engineer makes no warranties, express or implied, under this Agreement or otherwise, in connection with any services performed or furnished by Engineer. Subject to the foregoing standard of care, Engineer and its consultants may use or rely upon design elements and information ordinarily or customarily furnished by others, including, but not limited to, specialty contractors, manufacturers, suppliers, and the publishers of technical standards.

B. Engineer shall not at any time supervise, direct, control, or have authority over any Constructor's work, nor shall Engineer have authority over or be responsible for the means, methods, techniques, sequences, or procedures of construction selected or used by any Constructor, or the safety precautions and programs incident thereto, for security or safety at the Project site, nor for any failure of a Constructor to comply with laws and regulations applicable to such Constructor's furnishing and performing of its work. Engineer shall not be responsible for the acts or omissions of any Constructor.

C. Engineer neither guarantees the performance of any Constructor nor assumes responsibility for any Constructor's failure to furnish and perform its work.

D. Engineer's opinions (if any) of probable construction cost are to be made on the basis of Engineer's experience, qualifications, and general familiarity with the construction industry. However, because Engineer has no control over the cost of labor, materials, equipment, or services furnished by others, or over contractors' methods of determining prices, or over competitive bidding or market conditions, Engineer cannot and does not guarantee that proposals, bids, or actual construction cost will not vary from opinions of probable construction cost prepared by Engineer. If Owner requires greater assurance as to probable construction cost, then Owner agrees to obtain an independent cost estimate.

E. Engineer shall not be responsible for any decision made regarding the construction contract requirements, or any application, interpretation, clarification, or modification of the construction contract documents other than those made by Engineer or its consultants.

F. All documents prepared or furnished by Engineer are instruments of service, and Engineer retains an ownership and property interest (including the copyright and the right of reuse) in such documents, whether or not the Project is completed. Owner shall have a limited license to use the documents on the Project, extensions of the Project, and for related uses of the Owner, subject to receipt by Engineer of full payment due and owing for all Services and Additional Services relating to preparation of the documents and subject to the following limitations:

FIGURE A4.1 (*Continued*)

1. Owner acknowledges that such documents are not intended or represented to be suitable for use on the Project unless completed by Engineer, or for use or reuse by Owner or others on extensions of the Project, on any other project, or for any other use or purpose, without written verification or adaptation by Engineer;

2. any such use or reuse, or any modification of the documents, without written verification, completion, or adaptation by Engineer, as appropriate for the specific purpose intended, will be at Owner's sole risk and without liability or legal exposure to Engineer or to its officers, directors, members, partners, agents, employees, and consultants;

3. Owner shall indemnify and hold harmless Engineer and its officers, directors, members, partners, agents, employees, and consultants from all claims, damages, losses, and expenses, including attorneys' fees, arising out of or resulting from any use, reuse, or modification of the documents without written verification, completion, or adaptation by Engineer; and

4. such limited license to Owner shall not create any rights in third parties.

G. Owner and Engineer may transmit, and shall accept, Project-related correspondence, documents, text, data, drawings, information, and graphics, in electronic media or digital format, either directly, or through access to a secure Project website, in accordance with a mutually agreeable protocol.

H. To the fullest extent permitted by law, Owner and Engineer (1) waive against each other, and the other's employees, officers, directors, members, agents, insurers, partners, and consultants, any and all claims for or entitlement to special, incidental, indirect, or consequential damages arising out of, resulting from, or in any way related to this Agreement or the Project, and (2) agree that Engineer's total liability to Owner under this Agreement shall be limited to $100,000 or the total amount of compensation received by Engineer, whichever is greater. ***←DRAFTING NOTE TO APEXTECH ENGINEERING: PLEASE CONSIDER THIS CAP ON DAMAGES IF NEGOTIATIONS PROCEED WITH RESPECT TO LIQUIDATED DAMAGES FOR FAILURE TO ACHIEVE CERTAIN MILESTONES IN THE PERFORMANCE OF SERVICES UNDER SECTION 1.01(B).***

I. The parties acknowledge that Engineer's Services do not include any services related to unknown or undisclosed Constituents of Concern. If Engineer or any other party encounters, uncovers, or reveals an unknown or undisclosed Constituent of Concern, then Engineer may, at its option and without liability for consequential or any other damages, suspend performance of Services on the portion of the Project affected thereby until such portion of the Project is no longer affected, or terminate this Agreement for cause if it is not practical to continue providing Services.

J. Owner and Engineer agree to negotiate each dispute between them in good faith during the 30 days after notice of dispute. If negotiations are unsuccessful in resolving the dispute, then the dispute shall be mediated. If mediation is unsuccessful, then the parties may exercise their rights at law.

K. This Agreement is to be governed by the law of the state in which the Project is located.

L. Engineer's Services and Additional Services do not include: (1) serving as a "municipal advisor" for purposes of the registration requirements of Section 975 of the Dodd-Frank Wall Street Reform and Consumer Protection Act (2010) or the municipal advisor registration rules issued by the Securities and Exchange Commission; (2) advising Owner, or any municipal entity or other person or entity, regarding municipal financial products or the issuance of municipal securities, including advice with respect to the

FIGURE A4.1 *(Continued)*

structure, timing, terms, or other similar matters concerning such products or issuances; (3) providing surety bonding or insurance-related advice, recommendations, counseling, or research, or enforcement of construction insurance or surety bonding requirements; or (4) providing legal advice or representation. ***←DRAFTING NOTE TO APEXTECH ENGINEERING: PLEASE ADVISE WHETHER ANY OTHER SPECIFIC SERVICES SHOULD BE EXCLUDED FROM THE ENGAGEMENT – ESPECIALLY TO THE EXTENT SUCH SERVICES ARE RELATED TO WHAT IS EXPRESSLY INCLUDED IN THE DEFINITION OF SERVICES.***

6.01 *Total Agreement*

A. This Agreement (including any expressly incorporated attachments), constitutes the entire agreement between Owner and Engineer and supersedes all prior written or oral understandings. This Agreement may only be amended, supplemented, modified, or canceled by a written instrument duly executed by Owner and Engineer.

7.01 *Definitions*

A. *Constructor*—Any person or entity (not including the Engineer, its employees, agents, representatives, and consultants), performing or supporting construction activities relating to the Project, including but not limited to contractors, subcontractors, suppliers, Owner's work forces, utility companies, construction managers, testing firms, shippers, and truckers, and the employees, agents, and representatives of any or all of them.

B. *Constituent of Concern*—Asbestos, petroleum, radioactive material, polychlorinated biphenyls (PCBs), hazardous waste, and any substance, product, waste, or other material of any nature whatsoever that is or becomes listed, regulated, or addressed pursuant to (a) the Comprehensive Environmental Response, Compensation and Liability Act, 42 U.S.C. §§9601 et seq. ("CERCLA"); (b) the Hazardous Materials Transportation Act, 49 U.S.C. §§5101 et seq.; (c) the Resource Conservation and Recovery Act, 42 U.S.C. §§6901 et seq. ("RCRA"); (d) the Toxic Substances Control Act, 15 U.S.C. §§2601 et seq.; (e) the Clean Water Act, 33 U.S.C. §§1251 et seq.; (f) the Clean Air Act, 42 U.S.C. §§7401 et seq.; or (g) any other federal, State, or local statute, law, rule, regulation, ordinance, resolution, code, order, or decree regulating, relating to, or imposing liability or standards of conduct concerning, any hazardous, toxic, or dangerous waste, substance, or material.

Attachments: Appendix 1, Engineer's Standard Hourly Rates.

FIGURE A4.1 *(Continued)*

IN WITNESS WHEREOF, the parties hereto have executed this Agreement, the Effective Date of which is indicated on page 1.

Owner: **TerraHaven Land Development LLC**	Engineer: ApexTech Engineering LLC
By: [ ]	By: [ ]
Print name: [ ]	Print name: [ ]
Title: [ ]	Title: [ ]
Date Signed: [ ]	Date Signed: [ ]

FIGURE A4.1 (*Continued*)

This is **Attachment 1, Engineer's Standard Hourly Rates,** referred to in and part of the Short Form of Agreement between Owner and Engineer for Professional Services.

Engineer's Standard Hourly Rates

A. *Standard Hourly Rates:*

1. Standard Hourly Rates are set forth in this Attachment 1 and include salaries and wages paid to personnel in each billing class plus the cost of customary and statutory benefits, general and administrative overhead, non-project operating costs, and operating margin or profit.

2. The Standard Hourly Rates apply only as specified in Paragraphs 2.01, 2.02.1, 2.02.2, and 2.03, and are subject to annual review and adjustment.

B. *Schedule of Hourly Rates:*

Billing Class	**Rate**
Principal	$ 260/hour
Project Manager	$ 225/hour
Senior Engineer III	$ 180/hour
Senior Engineer II	$ 160/hour
Senior Engineer I	$ 140/hour
Staff Engineer III	$ 110/hour
Staff Engineer II	$ 100/hour
Staff Engineer I	$ 90/hour

FIGURE A4.1 (*Continued*)

A5: Scope Ambiguity

ApexTech Engineering LLC is under contract to provide engineering and consulting services for a land developer, TerraHaven, for the new Millbrook Logistics Park project. The engineering project manager, Maya, works with her staff engineer, Veronica, during the project's early site design and rezoning work.

When TerraHaven awarded the project to ApexTech, the client, John Branson, surprised Maya by announcing a forced teaming arrangement. After ApexTech initially submitted a full-service proposal for the project, TerraHaven elected to split the scope of work into two parts: A local firm, Avrum Planners, would complete the initial planning, rezoning, and survey work, with the expectation that ApexTech would provide the final engineering services.

ApexTech accepted this condition but added scope and fee for coordination efforts with Avrum Planners. This scope would include miscellaneous meetings and coordination efforts to facilitate the site design.

* * *

November 11, Year 2

After the success of the Millbrook proposal, Maya was pleased to see that Veronica had been doing well with her other assignments and keeping up with the rapid pace of the Millbrook work. Veronica clearly felt more comfortable now that she had more experience and was establishing a rapport with the client and the other team members. Veronica appreciated the autonomy on the project tasks but made a point to stop by Maya's office at least once a week to catch up.

Maya and Veronica had multiple other projects to keep them occupied, but Maya liked keeping up to date on Millbrook Logistics Park. As an aside, Maya saw Veronica's potential for advancement and wanted to help mentor her where she could. Each of their informal meetings looked a little different. Still, since the project was nearing the milestone of completing the first submission of the rezoning application, Maya was curious to inquire more about the project. Veronica had just finished briefing Maya on the latest developments on some of her other projects.

"Things are going well with our other projects, but that reminds me, what about the Millbrook Logistics Park? Everything still moving efficiently on that?"

Veronica nodded, "I think so. I know we're just coordinating everything during the rezoning phase, but working with Amar is great. Watching the land planning evolve is exciting, especially knowing I've got a hand in it."

"That's great. I have to say, it's rare for everything to go as planned on these sorts of things, so I'm glad you haven't run into any major issues." She watched Veronica's face tighten a little.

"Well, it's going well for the most part," Veronica amended. Maya sighed. She had noticed Veronica was in the habit of taking on too much and glossing over exactly how heavy her workload was. *I couldn't imagine where she'd learned that from*, she mused to herself.

"John told me that one of the original landowners, Oloport, had only agreed to sell on the condition that the woods around the western part of the site are preserved. Still, I think John has been looking for ways to still find value with that part of the site." The name *Oloport* rang a bell for Maya. She remembered John mentioning an owner's early resistance to sell and some resultant conditions.

"What do you see as your best move?" Maya questioned, wanting to see the solution Veronica would settle on.

"I don't know all the details of John's agreement with Oloport, but I think everyone can benefit if the western part of the site is preserved. It's part ecological, part personal for Oloport; the land has been in his family for quite a while, and I think there's some nesting site there for a few different bird species, nothing endangered or anything, but Oloport wants to protect what's there. I know John first envisioned it as a developable area, but there would probably be other environmental challenges in that part of the site, and we could still use it as required open space per the County ordinance. I think preservation of that would also please the County and community, showing that we're considering the importance of the land and our environmental impact."

Maya was impressed. "I think that's a smart move, and going that route will most likely give you the result you want." Veronica looked pleased at the compliment. "John might be reluctant because he sees it as developable land, but it sounds like there could be some challenges on that part of the site. I think he has a requirement to provide a certain percentage of open space, so avoiding that area could satisfy that requirement, too." Maya continued, "Is that the only issue you seem to be having? It doesn't seem too major outside our coordination scope."

Maya is suggesting some ideas, but ultimately, there are tradeoffs that need to be considered with all stakeholders.

"Well, I guess it's not a big deal. But honestly, there's just been a lot more back and forth than I anticipated." She tried to shrug it off, but Maya seemed concerned by hearing this.

"A fair amount of communication between everyone is to be expected. What in particular has you bothered?" Maya questioned.

"Not bothered per se. Amar is knowledgeable about the local area, which is a big help, and he is very focused on land planning. But he's a talker. And when it comes down to the more technical side of things, I feel like I need to give him a lot of guidance." Maya pressed for an example.

> The coordination scope of work is established as time and materials. This means that the time ApexTech spends working with Amar and the Avrum Planners team is paid for by the client, which may need to be communicated if it seems excessive.

"Well, at first, I was mostly listening in on the meetings between TerraHaven and Avrum Planners." Veronica elaborated. "Then a certain question arose about geometric road design, and I could tell Amar was unfamiliar, so I stepped in. But when they realized I'm also well versed in the road design requirements, I started getting requests to review their designs. Of course, I was happy to help, but now Amar is asking if I can review and maybe redesign a few things to keep things progressing. It's just a lot of added back and forth about the overall layout and checking and editing the Avrum team's work." She sighed heavily, "I'm glad to be so involved in the project here, but I have other projects to keep me busy right now, and this is more than I expected we'd be doing; I just feel spread pretty thin, is all."

Maya shook her head. "I had a feeling this might happen. Is John aware of this? I know he has a habit of casually asking for a few too many favors outside the contract scope of work."

"I'm mostly coordinating directly with Amar, so I'm not sure what he's been telling John. But they're paying us hourly, right? So aren't we just meant to do the work that's needed?" Veronica assumed.

> In a time and materials (i.e., hourly) cost structure, all time spent on the project task is billed to the client, but the work must be within scope. The 'not to exceed' condition means there's an upset limit to how much can be billed to the client.

"You're partially right, but you need to remember the 'not to exceed' condition that was written into the contract so we stay within budget. TerraHaven specified that condition in contract negotiations to keep their costs under control, so I'm slightly annoyed the Avrum team is trying to overstep a bit."

"Should we speak with John about it?" Veronica seemed worried.

"I'm not exactly sure what their financials look like, but I think if we discussed expanding the scope and adjusting the budget, he might push back a bit – this phase of work can be ambiguous in terms of who is responsible for what. Still, I wouldn't stress too much about that aspect. I'm more concerned about the work you're unofficially doing on behalf of Avrum, especially from a general ownership and liability standpoint." Maya decided to pose a follow-up question to Veronica to see if she was following her thought process. "What negative impacts do you think this scope push could result in?"

> This distinction is made to identify the risks associated with cost management from the contracted scope and the associated liability.

"Oh, um," Veronica paused and thought for a moment. "Well, you mentioned ownership and liability. If we're doing this preliminary engineering work that TerraHaven is paying Avrum for, then there's confusion on who's actually responsible for each aspect of this project and the resulting payment for that work. Also, if we're providing extra design advice and planning, and it's not being checked, then are any potential errors blamed on us or Avrum? In which case, are we responsible, or are they?"

Maya nodded, pleased with Veronica's line of thinking on the issue. "With our role as coordinator, it seems like we'd be responsible for ensuring everything runs smoothly, especially with so many moving parts. But without clear guidelines on what's expected of each party involved, being too flexible can cause more harm than good. Client satisfaction is very important, but we need to be true to the scope of work."

"I'm guessing all the pre-contract work we did with TerraHaven kind of blurred the lines before we even officially started," Veronica noted.

> ApexTech provided early consulting services as a form of business development before a contract was established. This is identified here as another risk.

"You're not wrong. We had assumed the engineering work would go to our firm and were acting as such; but after John shifted the rezoning over to Avrum, you and I should have discussed how our work parameters were changing. I thought I clarified the scope with John when we revised the contract. I can't stress enough the importance of concrete, well-written scope," Maya elaborated. "When things are left ambiguous, more often than not, the workload tends to tip to one side because of lack of communication."

"So, for example," Veronica ventured for clarification, "not just 'meetings as needed,' but explicitly stating how many meetings, the duration, and the expected attendees."

"Exactly," Maya agreed. "Sometimes with this sort of work, certain aspects can be difficult to quantify, especially in terms of budget. So it's important to have measurable tasks." Maya asked Veronica to take a moment and pull up the contract with TerraHaven and focus on the original scope statement. "After looking at it again, let me know if you think it's worded in such a way as to imply that ApexTech is responsible for preliminary engineering work."

> Maya is referencing the contract here because it dictates the scope of work to be provided by ApexTech.

Veronica took a moment to read the section Maya had requested and glanced up, "You were right about how the scope is written. It says our focus will be solely on communication and awareness of the rezoning efforts by Avrum Planners. Nothing alludes to us handling any design or engineering efforts during the rezoning phase."

"So this means . . ." Maya prodded.

"It means," Veronica sighed heavily, "I've already pushed the scope of our work by helping Amar." Maya could see Veronica was frustrated by her unintended misstep.

"Listen, you went above and beyond here, again, and I of all people can't fault you for that. But I don't want to see you doing extra work for free, especially if the credit will go toward the other firm. Or worse, any potential mistakes landing on us."

"So what should happen now?" Veronica questioned, still worried.

"Well, as I see it we have two real options here. One, we submit a change order to modify the scope of our work to include some preliminary engineering, which will likely increase the budget; or two, we'll reiterate that ApexTech is not responsible for design work during the rezoning process and we will be pulling back on our involvement from here on out."

Veronica listened attentively. "What do you think is our best option?"

"What do *you* think is our best option?" Maya countered. Veronica paused to think over the pros and cons of both choices.

A5 Review Questions

1. Analyze the original scope of work agreed upon between ApexTech and TerraHaven. What aspects of the project did ApexTech initially anticipate managing, and how did these expectations change with the involvement of Avrum Planners?
2. Discuss the implications of the 'not to exceed' condition in the contract between ApexTech and TerraHaven. How does this condition affect ApexTech's approach to managing the project, especially with unexpected requests from TerraHaven or Avrum Planners?
3. Evaluate the relationship between ApexTech and the various stakeholders (TerraHaven, Avrum Planners, and landowners). What strategies could ApexTech employ to manage expectations and communications among these groups effectively?
4. Identify potential risks that could arise from ApexTech's expanded involvement in rezoning, planning, and preliminary engineering work, which was initially not part of their contract. How might these risks impact the project, and what mitigation strategies could be employed?
5. Discuss the ethical implications of ApexTech potentially doing extra work without additional compensation. How should Maya and Veronica navigate this situation while maintaining professional integrity?
6. Examine the balance between maintaining client satisfaction and preventing scope creep. How can ApexTech ensure client satisfaction without compromising their own resources and project boundaries?

A6: Schedule Acceleration

As Millbrook Logistics Park enters into the third year since the origin of the project, the team has done well to navigate the complexities of the rezoning process and some initial community opposition to the project based on traffic and environmental impact concerns.

John Branson, the client at a private development firm, TerraHaven, has negotiated acquisition purchase and sale agreements (PSAs) for each of the parcels needed to create the Millbrook Logistics Park. Per the terms of these agreements, TerraHaven has already paid a deposit to the current owners and will begin to pay nonrefundable fees to extend the contracts if needed while the team pursues the necessary rezoning. John has been working with his brokerage team for several months now to solicit tenant interest in order to pre-lease the new lots.

Unsurprisingly to John, there is strong tenant interest – but what he had not predicted was that the vast majority of national creditworthy tenants would insist on an earlier occupancy date due to the evolution of logistics business cycles and the timing of site delivery relative to the busy holiday shipping season. Given his broker's increasing warnings about a possible softening of market conditions, accelerating the project timeline seemed highly advisable, especially if the tenants would be willing to pay a premium for expedited construction and delivery.

As the project nears a milestone of rezoning approval, which feels certain at this point, John is looking for creative ideas from his design team (Amar at Avrum Planners, Maya and Veronica at ApexTech Engineering, and Sarah at Meadow Law) for earlier design completion. While John has not communicated all PSA and tenant details with his design team, he is motivated to encourage them to finish their work faster.

* * *

May 29, Year 3

With things well underway, John had called a meeting with Maya, Veronica, Amar, and Sarah. After a few back-and-forth pleasantries, John announced that TerraHaven's broker had informed him that the site was getting a lot of interest from potential tenants.

"That's positive news," Veronica spoke up. The first major project she'd had a hand in was finally taking shape, and she was excited to see everything come to completion.

"Well, we've had a few inquiries about different available parcels," John responded. "We're negotiating letters of intent (LOIs) for a few parcels; but I wouldn't get too excited. It's still tentative, since the rezoning has been taking so long."

Sarah rolled her eyes a bit, "Oh, don't let him downplay anything. These are serious tenants looking to build; we're all very excited."

John couldn't help but smirk a bit, glad to see everyone was pleased by the news. "Yes, this is obviously good to hear, but as I mentioned, things could be moving a bit faster if we wanted to keep these parties interested. What does our schedule look like, Amar?"

"As we've discussed," Amar explained, "we anticipated 18 months for rezoning, leaving us with about 6 months or so of work, and then it transitions to ApexTech for final engineering, with maybe another 9 months after that before we'll have a site permit. Though I suppose we could manage to shave off a few weeks, if we kept things moving efficiently after the rezoning is finished."

Overly aggressive acceleration should be discussed first with the team to determine what is reasonable.

John shook his head, " 'Weeks' isn't good enough. I'm thinking if we all buckle down we can push the design work ahead by 6 months." At this, Maya and Veronica cast a quick glance at each other. Amar tried to point out how 6 months would be impossible from his side since the rest of the rezoning schedule was out of their hands. As Amar continued to explain this to John, Maya leaned over for an aside with Veronica.

"He can't be serious, 6 months?" Sure, Maya wanted to keep the ball rolling on this, but removing one-third of their timeline would just mean added stress to her and her team. Not to mention she would have to reprioritize her workload by shifting more focus and resources away from other projects to this one.

Optimism can spur creativity in strategy, but it's essential to consider what's realistic and which team members are impacted by these decisions.

"I don't know," Veronica countered. "I know it'll be tough, but I think I can make it work. I'm already familiar with everything, and all this waiting around doesn't seem very productive." If the client wanted to speed things along, she'd be willing to put in the hours to make it work.

Maya was quick to point out that an acceleration of this magnitude would put an unnecessary strain not only on Veronica but on the rest of the team as well. "I don't doubt you could handle the extra workload, but you shouldn't overburden yourself here because of an unrealistic timeline. Besides," she continued before Veronica could argue. "This big change would have some heavy cost and risk implications."

> Schedule accelerations are possible but often come with a tradeoff of risk, cost, or both as described in Chapter 6.

Veronica hadn't thought of that but insisted she wouldn't mind the overtime work.

"No, I know there has to be a better way around this that doesn't involve so much new risk and pressure." She and Veronica started brainstorming, as Amar finished explaining to John there was nothing he could do to speed things up on his end.

John sighed heavily, "You've just got to look at your watch, not your calendar." Getting slightly annoyed, Amar started to answer back but noticed Maya and Veronica deep in conversation and turned to them for help. "You two seem like you're putting your heads together for something, any *realistic* ideas?"

Veronica noticed his emphasis on "realistic" and quickly pushed ahead when she saw John hadn't missed it, either. Trying to keep things constructive, she offered up her own idea.

"It might save us some time if we move ahead early, we could begin the final site plan engineering *now*. I think any sort of potential plan changes from zoning would probably be minimal at this point; and even if something did come up, I'm sure we could handle any alterations."

John's mood lightened at this idea, "That's what I like to hear. No point waiting around for a bit of red tape." He turned to Maya, "You've got yourself quite the protege here Maya, I like her drive."

Maya could see Veronica enjoyed the praise but wanted to offer some caution to the idea. "A concurrent design strategy is creative and would help reduce the overall schedule, but there are potential risks, John. I just want you to know that if something comes up between now and zoning approval, which results in any design changes, then ApexTech has the right to request a change order for any rework."

> Stakeholders need to understand the tradeoffs of schedule accelerations.

Hearing this caveat, John backstepped a bit. He didn't like the risk of potential rework that would cost more. Instead, he countered with his suggestion. "Now, I know Veronica here is incredibly dedicated to this project, but the workload shouldn't be all on her shoulders. Why don't you add a few extra team members to help her out? That way everything could be finished in no time. Come to think of it, I do remember once hearing a six-month timeline was always a possibility," he finished with a smile.

> This carries different risks regarding ramping up new staff and increasing communication channels, as described in Chapter 6.

> John is referring to an early verbal inquiry (A1) to Veronica from before many of the project elements were defined.

"Twice as many ovens can't bake a cake twice as fast, John," Maya quipped.

John let out a short laugh at her joke, "I know, I know. But Maya, you've worked miracles for me before and we've been able to easily shave off a quarter of our original schedule when we need to get things done."

Maya interjected, "Yes, for some projects we've been able to add more people to key tasks and get things done faster – for this project, we started with an aggressive schedule that already requires all the resources I can throw at it, so it just doesn't help here. Besides, a six-month timeline was never feasible, which is why it's not in the official contract schedule, as you know."

"Listen, I understand your restlessness here," she continued, "especially with interested parties looking to start building. But there are just some tasks that take more time, and things dependent on other outside factors we can't control." Still sensing his frustration, she followed through with a positive spin. "Look, we're all genuinely excited about this project, and I've got a great team lined up that would be able to step in early, if needed, to help Veronica. The Millbrook Logistics Park is one of the firm's top priorities, and we're definitely treating it as such, rest assured."

John nodded, "I'm glad to hear it. I'd just hate to lose any prospective tenants if it looks like things are dragging on for too long."

> This is John signaling his priorities for project success based on the current project conditions.

"I think Maya's got a good plan in place here," Sarah chimed in. "The logic of Veronica's idea is a good one, and any risk involved would be outweighed by the benefits of earlier delivery. If we do this, we could submit the site plan shortly after the rezoning approval."

Veronica couldn't help but note that Sarah's reassurances had helped to satisfy John and mostly pacify the situation. He agreed to the concurrent design strategy and asked for a modification to the contract that would provide a new site plan completion date for the accelerated timeline.

Maya added another note for assurances, "Keep in mind, John – as much as we will do to push this forward, we don't have any control over the speed of the County reviewers. We're basing a lot on what the County has told us with their average review times."

"I know, I know," John responded. "The County knows how much good this project will do for new jobs and new tax revenues. They should be just as motivated as we are." John seemed settled on the new direction, "Well, I really appreciate everyone coming together to find a good solution here," he finished. "You've got a good engineer on your hands, Maya; make sure you keep her around."

A6 Review Questions

1. What schedule modification technique does Veronica suggest when the discretionary dependency between the zoning documents and the site plan is changed to establish concurrent processes?
2. Evaluate the risks associated with Veronica's suggestion to proceed with the final site plan engineering before the rezoning process is complete. What potential issues could arise from this approach, and how might they impact the project's overall success?
3. John mentions that assigning more people to the project should get things done faster, but what are the challenges with this logic? Reflect on Maya's role as a leader in this scenario. How does she balance the need to satisfy the client's desires with the practical realities of project management? What qualities does she exhibit that make her an effective leader in this situation?
4. Analyze the conversation between John and the ApexTech team regarding accelerating the project timeline. How does Maya manage John's expectations without compromising the integrity and feasibility of the project?
5. If you were in Maya's position, would you modify the original contract to consider this change? If not, why? If so, how would you approach the modification of the contract to accommodate the new accelerated timeline? What factors would you consider regarding project scope, budget, and team capacity?

A7: Revenue Generation

Maya and Veronica have been hard at work over the last few months on the final engineering design for the Millbrook Logistics park for their client John Branson, at the private development firm, TerraHaven. The engineering project manager, Maya, continues to coach Veronica in different project management skills. As they approach a new milestone of the project, Maya feels it's a good opportunity to talk through cost management with Veronica.

* * *

August 15, Year 3

It was early afternoon when Maya called Veronica into her office.

"So, I'll just cut right to the chase," Maya began with a smile. "Things seem to be going well with the Millbrook Logistics Park. Would you be interested in diving into some cost management?"

Veronica's eyes lit up. "Absolutely! I have been interested to learn more about the financial side of our work."

As Veronica sat down, Maya turned her monitor around so they both could view the screen. A few reports and several spreadsheets were open. She began by explaining the budget and general billing structure. "Let's start with the task for the site plan," Maya said, pointing to a figure on the screen. "We are contracted to provide engineering and consulting services for $630,000 as a fixed price. This includes $510,000 for the site plan and permits and $120,000 for environmental work. Additionally, we have $40,000 for zoning coordination and $22,000 for meetings, both as a time and materials budget."

ApexTech has identified some tasks that are allocated as fixed price (or lump sum) or as time and materials, depending on the scope and agreed terms in the contract.

Veronica nodded as Maya moved down the list, explaining the breakdown for landscape planning, environmental and water resources, and the time-and-materials costs for zoning coordination and meetings. "We divided the $630,000 fixed-fee amount based on different tasks. This helps us budget and track each task, and also helps us when evaluating budgets split across other ApexTech departments.

John also agreed to a change order to accommodate his request for schedule compression, but that has not been processed yet so it will not show on our current invoice."

In A6, ApexTech was requested to modify the schedule to achieve an early site permit. The narrative identified that there could be a cost increase to accommodate the request.

Maya continued as she collected some paperwork, "For the invoices, John will only see the cumulative cost based on how the contract was established: one cost for the site plan, and one for the environmental engineering work."

Maya paused as Veronica jotted down a few notes before continuing. "Sometimes we break this down further for internal cost management. For example, the stormwater design and landscape architecture tasks were separated from the site engineering, providing $120,000 to our stormwater department, and $60,000 allocated to the landscape architecture department. We'll only see this additional breakdown in our internal cost report. Our invoice to John only includes the high-level items."

Maya continued to review each task budget and hierarchy with Veronica in greater detail:

- Site Plan: $510,000 fixed price (total)
    - Site Engineering: $330,000 fixed price
        - This covers all site engineering-related work to prepare the plan for submission to Millbrook County.
    - Stormwater Management: $120,000 fixed price
        - This covers the design of the stormwater pond and other systems required for stormwater management.
    - Landscape Plan: $60,000 fixed price
        - This work is performed by a different department in ApexTech and covers the landscape architecture for the site plan.
- Environmental: $120,000 fixed price
    - This scope includes the resource protection area (RPA) delineation and other environmental services.
- Zoning Coordination: $40,000 T&M
    - This is included to cover the effort to coordinate with Avrum Planners during the rezoning process.

- Meetings: $22,000 T&M
  - This covers all time associated with meetings and coordination.

"Now, this is the monthly Invoice Proof from accounting (Figure A7.1)," Maya elaborated. "I'm required to review it for accuracy and determine the billings before it goes back to accounting for final processing and then sent on to TerraHaven. By the time I receive this, there shouldn't be any surprises because I've been tracking our internal weekly expenditures."

INVOICE PROOF

Bill To: John Branson
TerraHaven Development LLC

Invoice #: ______
Invoice Date: ______
Due Date: ______
Client #: ______
Contract #: ______

FIXED PRICE BILLING

Task Description	ID	Contract Amount	Percent Complete	Amount Earned	Previously Billed	Current Amount
SITEPLAN	L001	510,000	-	-	204,000	-
ENVRIO.	L002	120,000	-	-	36,000	-

TIME & MATERIALS BILLING

Task Description	ID	Personnel	Hours	Rate	Amount
MEETINGS	T001				
		Engineer 3	12	110	1,320
		Engineer 8	5	195	975
				Total For T001:	2,295

TOTAL INVOICE DUE: -

FIGURE A7.1

Veronica nodded again as she looked over the invoice proof before asking, "I see the tasks listed here, but I don't see an invoiced amount for the fixed-feed billings. How is that determined exactly?"

Maya brought up a cost report and directed Veronica to take a look. "The part that TerraHaven doesn't see, but we need to calculate and invoice, is our internal cost report. This report helps us track our spending against what we can bill; it's crucial information to manage project costs."

Maya positioned the cost report next to the invoice proof and moved down the task list line by line. "For fixed prices, like the site plan work, the invoiced amount will need to consider both our expenditure and the amount of work completed." She then shifted their attention to the values listed in the cost report for the site plan task. Veronica noted that there was $35,000 listed as an unloaded cost and made the observation, "It doesn't look like we've spent very much on this task at all, considering the $330,000 budget we have."

> For fixed price work, the team must consider how much work has been completed as compared to the costs to complete that work.

Maya clarified, "That's our direct labor cost, the unloaded rate. But we need to consider overhead costs like rent, benefits, and, of course, a bit of profit. ApexTech aims for a 3.0 multiplier on direct costs."

> The multiplier goal will vary by organization. In this case, ApexTech has identified that a 3.0 multiplier will accommodate the firm's overhead costs and profit margins.

Veronica did some quick mental math, "So, we need to bill $105,000 to cover ApexTech costs?"

"Yes," Maya responded. "However, with a fixed price structure, we bill TerraHaven for the amount of work completed, not just how much it costs. John doesn't see how much we've spent on the task, nor does he really care. John cares about the value we've delivered."

"So, we're billing John based on the work completed, not just what we've spent?" asked Veronica.

"Exactly," Maya confirmed, pleased she was catching on quickly. "How far along would you say we are with the project?"

Veronica thought for a moment. "Well, that's sort of hard to say. I feel like we're getting close to submitting the plans to the County, but we still have some final details and calculations to finish. Also, we've heard from Amar that we should expect a few rounds of comments and coordination with the County. Optimistically, I'd like to say we're at about 60–70% completion. But considering the project's novelty, I'd realistically say closer to about 50%, which would be $165,000 for the site engineering, and still a good amount more than the work has cost."

Maya smiled approvingly. "That's a good number. We're ahead of our expenditure, and our cost performance index is looking great. This pace is consistent with our previous billing too, for all tasks associated with the site plan work." Maya pointed to the invoice proof that referenced the previous amount billed before continuing, "I feel like John would agree with us being at, or beyond, 50% completion with the work."

Veronica felt the need to clarify the billing conditions and asked, "But, isn't it strange that we can charge him more than what it actually costs to do the work? Don't we already have some profit built into that multiplier?"

"It's great that you're thinking about this, and I know John would appreciate a discount on our service," Maya replied. "However," she continued, "TerraHaven agreed to a fixed price structure because there is less risk for them. They can predict their own costs, and we would be hard-pressed to ask for more money even if it costs us more to do the work."

Veronica nodded, working out her grasp of the thought process. "So, we invoice based on the amount of work completed, or the client's recognized value of the work. And if it costs us less to do the work, then we improve our profit. But if it costs us more, TerraHaven doesn't pay more, and we potentially *lose* money?"

"Yes, you've got it," Maya confirmed. Veronica smiled, glad she was keeping up with these new concepts fairly easily. "For example," Maya elaborated, "let's take a look at the landscape design costs. They tend to put in the most effort right before a plan submission after the civil infrastructure layout is near final." Maya referenced their unloaded costs to date at $11,000 before continuing. "So, $11,000 with our 3.0 multiplier means their cost is $33,000, which is above half their budget. I'd say landscape is more like 30% through their work, even though your engineering work is close to 50% complete."

Veronica spoke up, "Alright, I follow you. So, we might have a few tasks that are doing better than others, and we'll be tracking costs for all of them, along with how much work has been completed. Then, we'll use that data to determine what to invoice TerraHaven and to track our performance."

"Right again," Maya noted positively before continuing. "My responsibility as the project manager is to work with other task managers, or functional managers, to understand how much work has been completed and how much it has cost." Maya sat back in her chair as Veronica reviewed the cost report and the invoice proof. After a moment, Maya continued, "Your site design tasks and the stormwater team are doing well, which helps the project do well. I've been working with our landscape architecture department to understand why their costs are so high and help them get back on track."

"Got it." Veronica was now focused on the costs and invoice for the meeting time. "It says that we've spent close to 70% of our meeting budget, but that's time and materials, so TerraHaven is required to pay us what it costs, right?"

"You're right that we charge for each hour we spend on those tasks, and that's all we charge. However, if you look back at the contract, you'll recall that TerraHaven placed a not-to-exceed condition on those budgets. This means we can't charge more than the $22,000 we budgeted." She paused before adding a qualifier, "Unless we submit a change order to increase the budget. We don't control much with meeting frequency and duration; we answer when the client calls. John understands this, but still appreciates seeing something in writing when we get close to the budget."

> This highlights the upper threshold for not-to-exceed conditions with time-and-materials tasks. Clients will expect advance notice and justification if the budget might be exceeded.

Veronica thought for a moment and then offered, "Well, if we're only halfway done with the work, but we've spent $15,400 of the $22,000, we'd probably want to ask John to revise the budget to near $30,800 to cover all future meetings, so an extra $8,800?"

Maya was impressed with how quickly Veronica was tracking the process. "That's a good and quick way to forecast it. One thing I'd suggest, though, is to think about this task differently from your site plan work. Your site plan is 50% complete, but I think we'll have many more meetings as we work through plan review and permits later in the project. This means we're maybe only through one-third of our meeting effort, based on how things are tracking."

Veronica nodded as she followed along.

Maya continued, "I'd like to just make one adjustment in the budget, and avoid hitting John with multiple change orders. So, I might recommend something closer to a new $45,000 budget for the meeting scope. As for time and materials, he knows he'll only end up paying for what we spend."

Satisfied with their progress, Maya made notes on the invoice proof to establish the recommended invoice for accounting. She then made a note to draft a change order to increase the meeting budget.

As they wrapped up, Maya mentioned a new request from John. "He wants a scope and fee for tenant fits. Our site plan uses some standard templates for each of the lots, but John says he expects us to provide some revisions to each of the seven lots, tailored specifically to the tenants he has lined up. We don't need final costs yet, and we don't have a well-defined scope, but John is looking for a rough idea of what it would cost to do this after the initial site plan is permitted. Do you think you could estimate the price, maybe based on your recent similar efforts for the project? Go ahead and take the cost report with you to consider how much your work has cost so far."

Veronica nodded as she grabbed the report to add to her own notes, eager to tackle this new challenge.

Maya glanced at her watch. "I've got another meeting I need to head to, but let's find time later to work on your budget estimate soon."

As Veronica left Maya's office, her mind still processing the mechanics of cost management, she shifted focus and started to think through her new task of establishing some new budget estimates.

* * *

A7 Review Questions

1. Evaluate the advantages and disadvantages of the fixed price structure for the site plan, environmental, and landscape tasks compared to the time and materials (T&M) approach for zoning coordination and meetings. How do these billing structures impact the management of the project and the relationship with the client?

2. Considering the scenario, how would you calculate the cost performance index (CPI) for the site plan task? Discuss the significance of this index in project management and how it reflects on ApexTech's performance in managing the Millbrook Logistics Park project.
3. Reflect on the potential risks associated with accelerating the project timeline (A6), as suggested by John. How might this affect the cost management of the project, particularly in relation to the fixed price and T&M tasks?
4. If you were Maya, how would you approach drafting a change order to increase the budget for meetings, considering the project's current status? Discuss the importance of communication and justification when proposing budget changes to a client.
5. Given the new request from John for tenant fit revisions, outline the steps Veronica should take to estimate the costs for this additional scope. Consider the information she has from the current project and how it can inform her budget estimation for this new task. What resources could Veronica reference to help build confidence in her budget estimate?

* * *

A8: Risks and Decisions

John Branson, the lead developer at TerraHaven responsible for the ambitious Millbrook Logistics Park project, had been working with his project team for the last several years. John constantly coordinated the project with his broker, financial analyst, market research professionals, equity investors, lenders, and a team of design professionals. On the site design side were engineers Maya and Veronica of ApexTech, and Amar at Avrum Planners.

> Clients are engaged with many other project stakeholders beyond the purview of the engineering team.

By this stage in the project, John had evaluated the market conditions and financial viability of the proposed logistics park and convinced his internal investment committee it was feasible. Initially, he endured a slow project and started negotiating with different landowners to assemble a large enough property to create the Millbrook Logistics Park. So far, he managed a few unexpected delays from land-owner negotiations and new TerraHaven policies regarding procurement of design services. Throughout the project, John obtained the necessary debt and equity commitments to purchase the properties and began paying the team's many invoices.

More recently, John worked closely with his broker to identify suitable tenants to pre-lease the space. Early in the process, John had secured locally based tenants for three of the seven new lots at Millbrook Logistics Park. The remaining four properties had received strong interest from creditworthy national tenants, but only on the condition that they could take earlier occupancy of the site. Unfortunately, speculations about the economy and local market conditions also introduced some risks.

Since working with his design points of contact, Maya and Amar, to accelerate the project timeline to capture tenant demand for the remaining four lots, he was also able to incorporate an early delivery rental premium into the leases with those tenants.

* * *

September 24, Year 3

John provided regular internal project updates to his regional manager at TerraHaven. Today was the last quarterly investment meeting before the end of TerraHaven's fiscal year, and John was on the agenda to provide a 'project spotlight' update for the team working in his region.

One of John's counterparts told him there had been rumors about some unexpected disruption on other projects across the company and that upper management seemed to be exercising an unusually high level of scrutiny. John recalled how his project was faced with implementing a new TerraHaven policy during the procurement stage, requiring formal RFPs to compare bids for design services. He wanted to give his regional manager confidence that he knew well how to manage his projects, including managing the risks.

This new requirement is what prompted the procurement process change identified in A2.

John walked into the TerraHaven conference room prepared to present the Millbrook Logistics Park to his team.

"John! Good to see you," boomed the Head of Regional Development, Bob Cohen, as he went to shake John's hands, "I'm glad you were able to make it up here to HQ for the meeting, it's been a while. You know everyone . . ." Bob motioned his hand across the room to John's counterparts, other development managers working on various projects in the region.

Most of John's counterparts had already arrived and were exuding the usual mix of seemingly contradictory emotions: irritated, enthusiastic, anxious, and confident, among others. It was clear that the growing market uncertainty was on everyone's mind today.

"Let's get right to it," continued Bob, hardly giving John time to sit down, "what's going on with your Millbrook Logistics Park? I saw budget and schedule changes flagged ages ago but haven't had time to follow up on it. What's the latest?" This was Bob's direct yet friendly way of putting John on notice that his recent project changes needed further explanation. Most notably, John had spent additional money on the project design schedule.

"Sure, happy to do a review and get everyone caught up," John began. "In fact, I think I can summarize my approach in very simple terms: work the numbers but don't get too greedy."

John paused for dramatic effect before continuing, "As per standard TerraHaven procedure, we had been tracking several risks: supply chain delays, community opposition, design

John is referencing the organizational assets that guide teams on risk identification and priorities based on institutional knowledge.

and construction issues – the usual. But one thing was always on our mind and became a top priority: the market.

"We've been incorporating input from the different team members to develop and continually update a projected timeline and budget while we anticipate potential disruptions from market changes. So far, the projected returns met – and continue to meet – our required threshold, with the usual assumptions about inflation, interest rates, market rents, etcetera.

"But as we began our lease-up efforts several months ago, the project reached an inflection point." John paused again to allow time for any questions on the standard launch of the project, but everyone just nodded, which he interpreted to mean it was safe to continue.

"We all work with specialized brokers to get the best insight into specific markets and tenant needs. In this case, I'm working with a fantastic broker who deals exclusively with warehouse and light industrial properties and tenants. I do my best to maintain a general understanding of the tenants' needs, but he's really been invaluable."

"Anyway, I don't mind saying that I didn't realize all the creditworthy national tenants suddenly gained interest in early occupancy. Three of our seven lots had tenants committed and then we completely seemed to lose all leasing momentum – until the broker brought us several national tenants who were interested, but only if we could commit to earlier delivery. Ten months before we originally planned! That would require pushing both our design team to deliver faster than we originally planned and asking the general contractor to compress the normal construction timeline. But you all know the value of a creditworthy national tenant.

> In A6, John requested that his design team complete the design work six months sooner. He also requested that his general contractor compress the construction timeline by four months, thus starting the project ten months sooner.

"So a decision had to be made to either advance the project timeline and absorb any associated costs or to stick with our original projections. It really came down to a risk evaluation."

John stood back up and was thankful to move around as he made his way to a whiteboard in the conference room. The meeting participants watched as he started to explain his strategy.

"We figured we had a decision: hold our original schedule or invest more money into the project to accelerate the schedule. In each case, we figured we could experience one of two results: the market would remain steady, or we'd be faced with some adverse market conditions."

> John has framed this out so that a decision tree could be used as a risk analysis tool, whether through formal methods or a generalized qualitative approach.

Someone in the meeting room spoke up, "Or a third possibility is that the market could improve, right?" Given how things were going, John didn't see that as a possibility but felt the need to acknowledge the comment, "Sure, that's a possibility too, but one we didn't invest too much time in." John continued, "Our goal, of course, was to complete our pre-leasing for the last four lots as quickly as possible.

"Mapping the relative risks and rewards of the two options was pretty straightforward. We started by evaluating the TerraHaven records to see how our prior projects have managed risks, especially years back during the last recession," he said as he grabbed a marker and started sketching out a few notes.

- *Hold Original Schedule*
  - *Pro*
    - *Certainty about the agreed timeline and TerraHaven budgets*
  - *Con*
    - *Weak tenant interest could jeopardize earning projections*
    - *Market softening could mean tenants push for lower rent*
    - *Competition from other development projects could pull tenants away*
- *Accelerate Schedule*
  - *Pro*
    - *Attract four new tenants with a commitment to earlier delivery*
    - *Potential to negotiate premiums with current tenants for early delivery*
    - *Earlier delivery means early rental income, reduced interest carry*
    - *Lock in higher rental rates before the market potentially drops*
  - *Con*
    - *Increased cost to expedite work, as indicated by the design team and our construction contractors*
    - *No guarantee that additional investments will yield early delivery*

After writing out his pros and cons list, John spoke up again, "Ultimately, we figured that the longer this project takes, the more susceptible we are to a potentially softer market. We didn't expect to avoid all risks with the acceleration, but I decided that early delivery would increase our likelihood of capturing the current market demand." John took another minute to draw a simple diagram (Figure A8.1) to explain the possible outcomes to his audience so he could point out what his team had considered:

> Not all risk analysis problems can be evaluated through a quantitative approach; instead, many rely on qualitative evaluation with the best judgment.

John turned away from the whiteboard and locked eyes with Bob. "It was clear that the *pros* outweighed the *cons*. What wasn't so obvious," he continued, "was how to determine what the likelihood of each outcome would be." John traced a few lines of his decision tree as he spoke, "You know how these things can't be quantified. It almost always comes down to making the best possible professional judgment call – with input from the team of course."

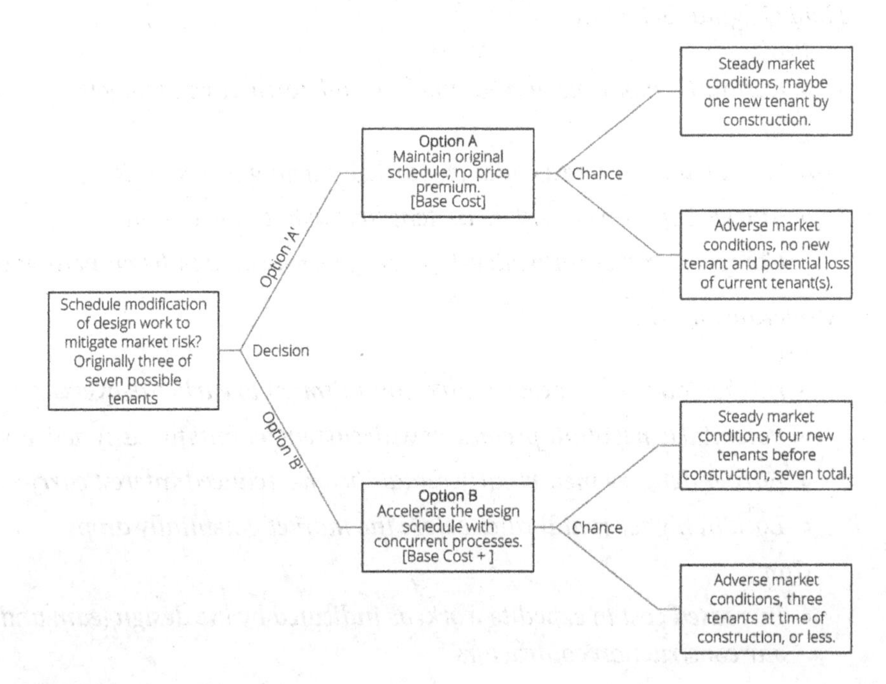

FIGURE A8.1

"Mmmm; Yea; Of course," came the general murmurs of agreement from the group.

One of John's fellow development colleagues raised a gentle voice of opposition. "You must have a different risk appetite down in your market. We would have stuck to the original schedule – like you said, we don't know what will happen in the future, and this seems to introduce a few *new* unknowns into the equation. But, we've got different equity partners and a different portfolio."

Bob acknowledged the perspective of one of the other managers but turned back to John, "So what finally swayed you to expedite? It sounds like you are paying a premium to move faster, both to your designers and your contract." Bob motioned to one of John's notes on the board. "I'm also guessing your design engineering team wasn't happy about that request," Bob pressed.

"It was the market. It seems like it always comes down to the market. In this case, there were just too many potential advantages from accelerating the timeline and too many compounding threats if we didn't.

All the market data had warned of a contraction in industrial and warehouse demand. Our financial analyst and I spent quite a bit of time running different scenarios through the pro forma and looking at some new sensitivity analyses, but the project returns almost always came out better *with* expediting.

Of course, at the time, we still had several unknown variables to tie up – what exactly was the increase in engineering costs? Would there be a subsequent cost increase from the contractor because of existing schedule conflicts that have to be resolved to move the timeline forward? Could we charge a premium to the tenants? Would the difference between those two numbers eat away at the project buffer we built into our assumptions? But standing here today, I'm pleased to say that all of that was resolved in our favor. The project is on schedule and may even exceed our original return requirements!"

John capped his marker, ready to sit back at the conference table and continue the conversation about his approach. Bob was fixed on John's last bullet though and interjected one last time, "But John, there's no guarantee that you'll actually deliver early right? I know we decided to invest some more money with the *expectation* that we'd finish early, but you're a few months from the site permit right? I like your strategy here, but we're not out of the woods yet are we?"

John wasn't surprised by this and remained confident, "You're right Bob, this hasn't had time to play out yet. But, I've worked with this team for years, especially the engineers – they always deliver, and they've been exceptional on this project. I'm also confident our general contractor will come through as well. I know this investment will pay off."

This action should consider how confidence can be portrayed and over-promising factors beyond the control of the project team.

A8 Review Questions

1. Identify at least three risks that John encountered during the Millbrook Logistics Park project. How would he assess these risks in terms of their potential impact on the project?
2. Discuss the decision John faced regarding accelerating the project timeline. What factors should be considered when making such decisions in project management, especially when dealing with uncertainties and external factors like market conditions?
3. How did John manage the expectations and concerns of different stakeholders, such as his regional manager, the design team, and the tenants? Provide examples from the scenario that illustrate effective stakeholder management.
4. Analyze the pros and cons list John created for holding the original schedule versus accelerating the schedule. How does this reflect a cost-benefit approach to risk management in project management? How could John's initial decision tree diagram be modified – how might he determine the expected monetary value (EMV) for his choices?
5. What project management strategies did John employ to mitigate risks and ensure the success of the Millbrook Logistics Park project? How did these strategies align with standard project management practices?

A9: Difficult Conversations

Several years into a new development project, ApexTech, the engineering firm responsible for the design, reached a significant milestone by submitting a site plan to Millbrook County for approval and land disturbance permits. This submission was the culmination of years of effort, largely due to an extended rezoning process required for the project, Millbrook Logistics Park.

Recently, TerraHaven, the project's developer, opted to expedite the design schedule to mitigate the risk of a softening market and the desire to secure additional tenants. This hastening of the schedule led to additional fees for ApexTech, which TerraHaven agreed to, valuing the advantage of commencing construction about six months earlier than initially planned. Since the site plan's submission in December of the previous year, ApexTech engineers Maya Mendez and Veronica Statton had shifted their focus to other projects, under the assumption that the site plan was progressing through the County's review process.

* * *

January 5, Year 4

Maya, the project manager at ApexTech, called Veronica into her office with a sense of urgency. She had submitted the site plan drawings to Millbrook County a month ago and had been expecting an email with confirmation of approval. Instead, she opened her inbox to find an unsettling message. As Veronica walked into the room, Maya motioned her over to her monitor to read it herself.

Veronica was initially quiet as she did a quick read before she spoke up. "We submitted this over a month ago. How is the County just now telling us about this?" She paused when she saw the CC line. "John is copied on this, I bet he's through the roof," Veronica noted as she scanned through the email once more.

Maya stared at the monitor, shaking her head at the ramifications. "Yeah, he's not going to be happy about a rejection and the delays it'll cause. Honestly, I'm not sure what to tell him about all this yet." She grew more frustrated at the thought of how much work was going to have to be redone. "We aren't responsible for the geotechnical engineering on this project. They aren't even our subconsultant. This is something TerraHaven should be managing. And maybe something Avrum Planners should have submitted already? I don't see how it's our fault."

TO: Mendez, Maya
CC: Branson, John

Letter of Noncompliance | 3365-SP-001 Millbrook Logistics Park

Maya Mendez

Your site plan application Millbrook Logistics Park 3365-SP-001 has been rejected for the following reasons:

- A geotechnical report has not been provided for this project. Therefore, the plans cannot be reviewed, as per the Millbrook County ordinance.

You may resubmit your site plan after the noted condition has been completed. Your plans are being returned to you.

Although it's critical to understand the cause of this disruption, Veronica is proactively focusing on progress here.

Veronica spoke up, "Well, that's good news, right? Or, maybe at least puts us in a better position? Regardless of blame, what does this mean for our timeline? Do we know if a geotechnical report has been completed? I always thought it was submitted separately, or maybe as part of the building plans, right?"

Maya started answering Veronica's questions, "We want to help John keep his schedule. After all the extra effort, overtime, and risk of accelerating things, it's frustrating that a month was wasted." Maya continued, "As for *when* a report is submitted and by *whom* – it can vary by jurisdiction. It seems Millbrook County wants it during the initial plan processing for some reason."

Maya's computer pinged with a notification of another email. It was from John this time, forwarding the County's email.

TO: Mendez, Maya

Fwd: Letter of Noncompliance | 3365-SP-001 Millbrook Logistics Park

What happened? We can't afford delays! Come to my office ASAP!!!

–J

"Well that's not surprising," she noted as she read John's message. Maya shut her eyes briefly and took a deep breath. "Veronica, I need you to figure out more of what's going on with the County. Hopefully, there's a way around this requirement for the geotechnical report. Let's see if we can encourage the County to start reviewing things while we figure out the issue here. Maybe see if they can help us rush it." Veronica nodded and started to head out before Maya made one last comment. "And Veronica, this is a phone call and probably a trip to the County, not an email. We need answers now." Veronica nodded with assurance, "Right. I'll see what I can find out." As she turned to leave, she began mentally preparing for her conversation with the County.

Maya recognizes the urgency and critical nature of this situation and is requesting that Veronica use direct methods of communication.

***

Later that day, at the TerraHaven office, Maya strode quickly into a conference room to see John pacing around the large table. As she entered and met eyes with John, she immediately spoke up, "We're addressing the issue, John. Veronica is on it as we speak," she assured him.

John was quiet, but it was abundantly clear he was worried and in a foul mood. He stared sternly at Maya for a moment, before walking over to a whiteboard in the room and writing out a number.

$32,258.07

He capped the marker and slammed it down on the table. He pointed forcefully to the number, "Do you know what this is Maya? Hm?" He said it with a tone of anger and desperation.

Maya was quick to register John's irritation, and while she didn't appreciate the unprofessional attitude, she didn't want to escalate the situation with a defensive response. She understood the situation was legitimately problematic, and responded as calmly as possible. "No, John. What is it?"

John snatched the marker back up from the table and vehemently circled the dollar amount several times. "This is my *daily* cost for this project. Taxes, loans, brokers, attorneys, designers. This is my *daily* cost. I pay the ApexTech invoices every month, and your team makes money." He paused for a moment, then continued again with the same desperation in his voice, "Me? I'm years away from collecting rent on this project." He turned back to the number and spoke again as he stared at it. "A month delay Maya? That costs me more than your entire engineering fee." He wasn't quite at the point of yelling, but Maya knew her tone had to remain level in order to keep from exacerbating things.

There was a long silence as they stood in the room collecting their thoughts. A moment passed before Maya steadily voiced her understanding of John's frustrations, and what actions she was taking to keep things moving forward. "John, I get it. This was totally unexpected and definitely put us in a bind. But I want you to know that we are well aware of the stakes. While we're *here*, I have Veronica working with the County to find a solution. Something, anything, we can do to get things back on track as quickly as possible." Maya paused to gauge John's reaction.

John seemed pacified knowing Maya was handling things to the best of her ability, but was still uneasy. "Maya, I need answers, fast. I have complete confidence in Veronica – but Maya, you're managing this project. You're accountable. You should be down there strong-arming the County to fix this." Maya wanted to interject that an aggressive approach with the County would likely only cause further issues and delays; she knew from experience that plan reviewers (and most people) did not respond well to being *strong-armed*. She was also considering how an aggressive approach would affect her personal reputation with the County – a relationship that was newly formed. Instead, she listened to John as he

> What seems like the right solution for an immediate problem could cause long-term challenges. In A3, it was noted that ApexTech also has a new contract with County transportation department, which means Maya should consider the ApexTech county-client relationship as well.

transitioned back to venting, "You need to talk to the people in charge there. Remind them of all the jobs we're trying to create for them – let alone the tax revenue they'll be getting. This should be their top priority!"

Maya didn't have any answers for John yet, but she trusted that Veronica was doing everything she could to find a creative solution at the County.

* * *

While Maya was meeting John, Veronica was trying to make progress with the County staff. After a call to the County, Veronica was invited to come to the County office for an impromptu meeting to talk through the site plan rejection. Veronica had managed to do a bit of quick preparatory research on the geotechnical report requirements and process before she made her way to the County office to find some answers and solutions.

Veronica arrived and was led into a County conference room, where she met and shook hands with Carl Winslow, the County planner who was in charge of the initial in-processing of the site plan. "Great to meet you Carl – I don't know if you remember, but you helped me with a few questions when we first got started on this project." Carl smiled as he greeted Veronica and they both started to sit before he said, "Oh that's right! It was right after the holidays, and you had some questions about our plan processing. I suppose this project has been focused on schedules right from the very beginning."

> Veronica exchanged messages with Carl in A3 with a professional and polite demeanor that could help facilitate this conversation.

Seated next to Carl, Veronica set some plans on the table along with a few notes she had made about the County ordinance and process. She spoke as she was getting settled, "Yeah, I'm guessing all projects have a certain sense of urgency, so we're no different there really. We've had quite the history already with this project, but the site plan and permit are some of the last pieces before we can get shovels in the ground."

Carl leaned back a bit in his chair and motioned with his arms as he spoke, "So, what can I do for you? Do you have the geotechnical report to submit?"

Veronica still wasn't sure what Maya had uncovered about the progress of the project's geotechnical report and could only imagine how the conversation with John was going. "I'm honestly not sure yet, Carl. The geotechnical work is being done by a different company, so we've had only limited interaction with them. We should hear from the client soon, but I wanted to make sure I understood the issue and some options as we dig into it."

Carl sat back up as he explained the permitting process to Veronica, "Sure. So, before we can review any site plan that includes buildings and building foundations, we need a geotechnical report. This may be unique to our County, but we have a lot of marine clays here, which cause stability and foundation support issues. We also need to keep an eye out for naturally occurring asbestos, which shows up in the Green Stone Rock formations of our bedrock. We like to see if a project site has any of these specific soil conditions before we start our review." Carl pulled one of Veronica's site plans closer and made a broad sweeping motion over an area with a proposed building, "See right here, hypothetically, if there's a swath of marine clays in this building footprint, then it will introduce a lot of potential issues."

Veronica nodded as she made a note, then pulled out some County soil maps she had printed right before the meeting. She placed them on the table and slid them over to Carl as she spoke, "Yes, I was reading about that with the ordinance. From the County soils map, it doesn't look like we have any of those issues near our site. The only exception would maybe be some marine clays near the western side by the natural pond, but we aren't building there. Would we still need the geotechnical report then?"

> This addresses Maya's question about why the other design team, Avrum Planners, may not have handled this during the rezoning work.

Carl studied the County soils map and compared it to the site plan. "You're right, it doesn't *look* like you'll have issues. This is probably why it wasn't flagged during your rezoning, but the County soils map is a starting point. Your client's geotech will need a field survey and a report to verify there's no issue. I'm sorry, but we always require a geotechnical report before we can start reviewing a site plan. My hands are pretty much tied on that point."

Veronica recognized that this was a hard rule and Carl wasn't going to budge on it. However, she'd come prepared with an alternative and broached the idea with Carl. "Thanks, Carl. That helps to clarify things, and I hope there's a report available soon that we just need to drop off with you today or tomorrow. Still, we're trying to catch up for the month we lost, and thinking about how we can get creative to help the client start work.

In your email a few years ago, about review timing, you mentioned that a rough grading plan, an RGP, can get reviewed faster. Is that a potential option here?"

Carl leaned back again in his chair as he answered, "Well, not really. The RGP is faster, yes, but you can't propose any buildings with an RGP, so you'd still need a site plan eventually. The RGP takes us about six weeks to review as compared to a site plan that takes more like three months. Still, all you get with an RGP is site clearing, utilities, and roads – nothing vertical."

Veronica had expected this possible response though, and was prepared to offer some ideas to Carl. "Well, let's say we submit an RGP," she suggested. "Could the developer still start work on the site, and we just submit a site plan with a geotechnical report later?" Veronica questioned before continuing, "I'm guessing it will be a year or more before they're ready for vertical construction anyway."

Carl seemed a bit surprised by Veronica's work-around. Not that he was old-fashioned, but he typically preferred that things were completed exactly in line with the published ordinance. Though he had to admit that he appreciated that she had done her research and was trying to solve the problem with some extra creativity. He responded after a moment's thought. "Veronica, I get it, I used to work in the private sector too and understand everything was due yesterday. You're right, that could all work. But, you'll need to revise your plans to meet the RGP requirements, and then you'll have a new and separate set of review fees to pay. I can't change any of that."

"I completely understand," Veronica smiled. At least it seemed like she was making some positive progress. "I'll just need to talk this over with our client and the project manager. But, I know they're laser-focused on schedule, so they might see value in this idea."

After Veronica and Carl finished talking through the process, Veronica thanked him for the meeting, packed up her plans, and went back to the office to share everything she had learned with Maya.

A9 Review Questions

1. Evaluate Maya and Veronica's initial response to the County's rejection email. How does it seem they handled communication between stakeholders, and what could have been done differently to manage the situation more effectively?

2. Discuss the implications of Maya's assumption that TerraHaven or Avrum Planners would manage the geotechnical report. How does this assumption reflect on the risk management practices within ApexTech?
3. Analyze Maya's interaction with John after receiving the County's rejection. How did she manage John's expectations and emotions, and what communication techniques did she employ to keep the situation under control?
4. Evaluate Veronica's approach to finding a solution with the County planner, Carl. How did her preparation and understanding of the County's process contribute to identifying a potential workaround with the rough grading plan (RGP)? How does it contrast with the discussion between Maya and John?
5. Considering Veronica's discussion with Carl about the RGP, what are the risks and benefits of pursuing this alternative path? How should ApexTech weigh these factors against the project's timeline and the client's urgency? What would the message to John look like to present these options?

A10: Retrospective

The Millbrook Logistics Park transitioned from design into construction several years ago. Originally a production engineer for ApexTech, Veronica had been promoted into a managerial role and was taking on more responsibility for the project during the construction administration phase. Still, she enjoyed updating her supervisor, Maya, about the status of the project and felt inclined to reflect on how the project ended up, and what was learned as they encountered several disruptions along the way.

* * *

April 23, Year 6

It was late in the evening as Veronica packed her bags and was heading out for the day when she noticed the light on in Maya's office. She walked over and stood in Maya's doorway to see Maya marking up some plans on her desk. Maya looked up and smiled, seeing that Veronica was headed out for the evening "Everything done? All the world's problems solved?" she joked with Veronica.

"Not quite everything," Veronica laughed. "But at least John seems happy with how things are going."

"Oh yes, our Millbrook Logistics Park. Projects tend to get quiet for us after permits. I know we did a bit of early construction administration for TerraHaven, but now that the early site work is complete it's quieted down a bit. How is everything going with that now? You've been doing well pretty much running that project for the last year or so."

Veronica took the compliment and generally felt good about the project. "It's going well. I've been supporting the construction administration here and there, though less now that the major work is done. We've made a few revisions based on new tenant requirements, but they're all pretty standard. I've enjoyed managing those new tasks."

"That's great to hear." Maya put down her pen and sat back from the plans, motioning for Veronica to take a seat. "So, let me ask you: What do you feel like you've learned about management with this project?"

Veronica took her bag off her shoulder and lowered it to the floor while she sat. She had been thinking about this for a while and appreciated the discussion with Maya, curious to hear her advice. "It's hard to explain. It's been very different from the technical work, which tends to have one correct solution

Project managers often spend most of their time on communication, which is influenced by each party's leadership styles and relationships.

we can solve. It's much more about leadership, power, and relationships than I realized. It's much heavier on communication too - I feel like that's all I'm doing some days now, is just transmitting information between stakeholders."

"Welcome to my world," Maya quipped, but she genuinely wanted to hear more about Veronica's perspective.

"The biggest revelation though, seems to be this endless requirement for adaptability." Veronica paused a moment to collect her thoughts. "The way I work with our other design engineers, other departments, the County staff, Avrum Planners, with John, each requires a different tactic." She continued, "It has a lot to do with authority, or power in each case. Even when I know what's best for the project, I often don't control what other people do, so it's a matter of convincing them what's right."

This underscores the importance of adaptability in project management and the nuances of exerting influence and authority in a multi-stakeholder environment.

Maya continued to follow along and was curious to hear more, "When you say 'convincing' what do you mean?"

"Well, I can *tell* our ApexTeam team what to do, so long as it's reasonable; and it helps them to understand why we're doing what we're doing," Veronica explained. "But, I can't tell Avrum what to do, nor the County, and certainly not John."

"Certainly not John," Maya chuckled back. "So, what's different there?"

"To some degree, over these last several years, I think Amar and John have learned to trust me, especially more recently," Veronica elaborated. "So, convincing them to see things the way I do seems more about their trust in my advice, and less about giving direct orders. John's also been more open about his upcoming prospective projects, seeking my engineering perspective on feasibility; and the Avrum Planners team is upfront about wanting to work together more in the future as well."

Maya smiled, impressed. "I agree with that. So, what does your observation about adaptability mean then?"

Veronica shifted a bit and crossed her arms. "To be frank? It means our original project plan was obsolete in the first month. It seems like even after every change we made to it, it was out of date by the following month." Maya nodded knowingly and let Veronica continue uninterrupted. "I suppose it's expected, but it's a lot of additional effort. More than I realized or anticipated."

> There's a dynamic nature to project management, where project plans must be continuously updated to reflect changing circumstances and requirements.

"Yes, it's surprising for sure; constantly refining the project plans is the nature of our work, I suppose. How do you feel about being adaptive in your communication and leadership?" Maya asked.

Veronica looked down, thinking momentarily as she had also been contemplating this for a bit, "It's hard to say. In these last few years, I've realized the *need* to adapt. I've learned to shift my communication style based on who I'm working with, and whether I have any formal authority over them – and even when I do, maybe because I 'outrank' them here at ApexTech, an authoritative approach never feels effective, and has never really felt right to me.

"I like when the team is committed to the project on their own. Which seems to be motivated by what's interesting and exciting about the project, what's new and complex about the work, or how a project benefits the community – depending on who I'm working with. So, I make a point to discuss this regularly with the team to help keep us focused on the core objectives.

"What I still find challenging though is maintaining my own individuality behind the adaptation. If I'm constantly changing my leadership style to best suit everyone else, it's not really *my* style anymore. I end up just operating in a way that pleases everyone, even when I feel like a more direct approach would be most efficient. Sometimes I just need people to trust me, especially when we don't really have time for *convincing* someone that I might be right; I know I am, and I know what needs to be done."

> This illustrates the challenge of balancing adaptability in leadership with maintaining one's personal style and authenticity.

Maya was interested to hear this perspective and took everything in. She had heard from Veronica's peers and from the client that they enjoyed working with her and appreciated her technical knowledge, as well as her demeanor. Still, it sounded like Veronica might be going too far to adapt for everyone else. "Veronica, it's great that you've learned to be adaptive in your leadership style, but it's meant to have a limit. The right style should be a blend of your personality and the needs of the team. You're also meant to be flexible to changing project conditions, so it's important not to lose your own voice.

"If you've established a good rapport with your team, as you have, then they'll trust you and understand when it's time to focus and get the work done when the project demands it." Veronica took her advice. "You're right though," Maya amended. "You still need to reflect and be attentive to the reaction of your leadership style, because it's unlikely your peers would feel comfortable speaking up and telling you something is wrong if they felt their voices weren't being heard."

Veronica wanted to heed Maya's advice, but was still uncertain. "It's hard to tell though. There's no textbook solution to the *right* way to work with your team and client. I know there's a lot of things I should avoid, but how do I know when I've found the correct approach, with our team or with a client?"

Maya kind of smirked at this, "Well, John would certainly let you know if there's something he doesn't like."

Veronica laughed, "Oh, I know he would. I can't really see him adapting his style for anyone at all. He's candid, but also sometimes lets his emotions, or his excitement, get the better of him. He also doesn't gain much by catering to our needs, does he? We're not his client."

Maya felt the need to interject, "That's right, but from the very beginning we're thinking about whether we want to work on a project and with a client. I know John can be a bit difficult, and we don't see much of his side of things when it comes to risks and stresses. But, he does understand the need for mutual respect in a professional relationship. If he didn't, we wouldn't work for him, and he does appreciate our work, your work, just as much as we like working on his project." She took a moment to think back to a more challenging conversation she'd had with John, "Remember a few years ago, when we had a tense conversation with him about the County review delays."

Veronica's face noticeably darkened as she remembered the dilemma, "Oh, I'm not going to forget about that anytime soon."

Maya agreed and continued, "Right, and while John was very 'excited' at the time, he did eventually call to apologize, in his own way. His approach is candid, as you said, and he was reasonably upset about the situation. But initially, his attitude was unprofessional and not really constructive. I gently said as much in a later meeting, after things had calmed down, and he acknowledged his actions. I like John, and I respect him. We want to work with him on his projects, and he also benefits from a good relationship with us."

Veronica still seemed sore about that series of events that led to John's frustration, "By the way, I'm still not sure what we could have done differently to avoid all that, and I'm still not convinced it was our fault that report was missed."

Maya shrugged, "It doesn't matter at this point. You and I can decide where we think the fault lies, and even note as much in our lessons learned register for ApexTech, but the immediate *right* action was about finding a solution, and John will remember that. For our next project, we hope that we maintain the integrity of that lesson and apply it to our contract language to clarify accountability, change our communication plans, or maybe change our efforts to better research local policies – whether explicitly included in an ordinance or just by understanding expectations."

Maya pressed on, "What's important is that we don't lose these lessons. It's too easy for information to disappear because we're unorganized and undisciplined. There's value in our information, our institutional knowledge. As you said, there's no perfect solution. We improve our leadership and management skills through our experiences and then by sharing those experiences. It's critical for us to have a reliable resource on these lessons or we end up being forced to rely on whatever information we can find. The more resources you have and the more diverse resources you can explore, the more you can maintain that individualism you mentioned. Too few resources or too much of the same content, and you'll miss the truly novel ideas. I think part of what has helped us make this project successful was how we responded to any disruptions, or honestly, how many times *your* creative solutions kept things on track."

Veronica was listening but shifted her gaze out the office window as she processed it all. Maya could see she was lost in her own thoughts, and felt like a bit of reassurance was warranted. "Veronica, you've done a great job on this project and your others. I doubt you'll see a perfect project where everything goes according to the plan you created on day one. The point is that you've learned to be resilient in your strategy, and you know the right level of adaptation for project conditions and for the stakeholders." Maya straightened up in her chair, prepared to get back to her work, but not before adding, "And besides all that, why are you here so late? Managing your time and well-being for life outside of work is even more important than your skill in managing projects. It's about long-term sustainability in our careers. Go home, relax, go have some fun!"

Veronica smiled, "Yes, I need to get out of here – and so do you!" Maya shrugged knowingly. "I appreciate the chat, though, Maya; it helps to put it all in perspective, including the reminders about what's really important."

A10 Review Questions

1. How does Veronica's experience with different stakeholders illustrate the need for adaptability in leadership styles? Discuss the pros and cons of maintaining one's individual leadership style versus adapting to the needs of different stakeholders.
2. Veronica mentions that much of her work involves transmitting information between stakeholders. How does effective communication play a crucial role in project management, especially in complex projects like the Millbrook Logistics Park?
3. In the scenario, Veronica navigates different levels of authority and influence with various stakeholders. How does understanding and managing power dynamics contribute to successful project management?
4. Maya emphasizes the importance of learning from experiences and sharing knowledge. Discuss the value of institutional knowledge in project management and how it contributes to resilience and innovation in future projects.
5. The scenario ends with a reminder about the importance of work-life balance. Consider the importance of addressing work-life balance for project managers and team members, and what impact it has on overall project success and team morale.

Bibliography

The following documents have been referenced during the production of this book and serve as a list of extended resources available. While we have covered many essential aspects of civil engineering project management, the scope of this field is vast. We encourage readers to continue exploring and learning from these valuable resources to deepen their understanding and stay updated with the latest practices and methodologies:

- **A Guide to the Project Management Body of Knowledge (PMBOK® Guide) – Sixth Edition**
  - Project Management Institute. (2017). *A Guide to the Project Management Body of Knowledge (PMBOK® Guide)* (6th ed.). Project Management Institute, Inc.
- **A Guide to the Project Management Body of Knowledge (PMBOK® Guide) – Seventh Edition**
  - Project Management Institute. (2021). *A Guide to the Project Management Body of Knowledge (PMBOK® Guide)* (7th ed.). Project Management Institute, Inc.
- **Agile Practice Guide**
  - Project Management Institute. (2017). *Agile Practice Guide*. Project Management Institute, Inc.
- **Black's Law Dictionary, 11th Edition**
  - Garner, B. A. (2019). Black's Law Dictionary (11th ed.). Thomson Reuters.
- **Complex System of Systems**
  - Haimes, Y. Y. (2012). *Systems-Based Guide to System of Systems Engineering (SoSE) Design and Management*. CRC Press.
- **Consumer Research Insights on Brands and Branding: A JCR Curation**
  - Keller, K. (2020). Consumer research insights on brands and branding: A JCR curation. *Journal of Consumer Research* 46(5): 995–1001.

Management Essentials for Civil Engineers: A Practical Guide to Business, Communication, Ethics, and Risk, First Edition. Cody A. Pennetti, C. Kat Grimsley, and Brian M. Grindall.

- **Corporate Brand Design: Developing and Managing Brand Identity (1st ed.)**
  - Foroudi, M.M., and Foroudi, P. (eds.). (2021). *Corporate Brand Design: Developing and Managing Brand Identity* (1st ed.). Routledge.
- **Developmental Sequence in Small Groups**
  - Tuckman, B. W. (1965). Developmental sequence in small groups. *Psychological Bulletin* 63(6), 384–399.
- **Four Failures in Project Management**
  - Cooper, K. G. (1998). Four failures in project management. In J. Pinto (ed.), *Project Management Handbook.* Newtown Square, PA: Project Management Institute.
- **Guide to the Systems Engineering Body of Knowledge (SEBoK), v. 1.9.1**
  - Pyster, A., Olwell, D., Hutchison, N., Enck, S., Anthony, J., & Henry, D. (eds.). (2017). *Guide to the Systems Engineering Body of Knowledge (SEBoK)*, v. 1.9.1. The Trustees of the Stevens Institute of Technology.
- **Guide to Writing Requirements**
  - INCOSE Requirements Working Group. (2015). *Guide to Writing Requirements*. INCOSE.
- **Hierarchical Leadership Versus Self-Management in Teams: Goal Orientation Diversity as a Moderator of Their Relative Effectiveness**
  - Nederveen Pieterse, A., Hollenbeck, J.R., van Knippenberg, D. et al. (2019). Hierarchical leadership versus self-management in teams: goal orientation diversity as a moderator of their relative effectiveness. *The Leadership Quarterly* 30(6): 101343.
- **How Toxic Workplace Environment Effects the Employee Engagement: The Mediating Role of Organizational Support and Employee Wellbeing**
  - Rasool, S. F., Wang, M., Tang, M., et al. (2021). How toxic workplace environment effects the employee engagement: the mediating role of organizational support and employee wellbeing. *International Journal of Environmental Research and Public Health* 18(5): 2294.
- **INCOSE Systems Engineering Handbook: A Guide for System Life Cycle Processes and Activities (Fifth Edition)**
  - INCOSE. (2023). *INCOSE Systems Engineering Handbook: A Guide for System Life Cycle Processes and Activities* (5th ed.). Wiley.

- **INCOSE Systems Engineering Vision 2025**
  - INCOSE. (2014). *INCOSE Systems Engineering Vision 2025*. INCOSE.
- **INVEST in Good Stories, SMART Tasks**
  - Wake, B. (2003). INVEST in Good Stories, and SMART Tasks. XP123. https://xp123.com/invest-in-good-stories-and-smart-tasks/
- **Leadership versus Management: How They Are Different, and Why**
  - Toor, S., and Ofori, G. (2008). Leadership versus management: how they are different, and why. *Leadership and Management in Engineering* 8(2): 61–71.
- **On the Quantitative Definition of Risk**
  - Kaplan, S., and Garrick, B. J. (1981). On the quantitative definition of risk. *Risk Analysis* 1(1): 11–27.
- **Organizational Project Management Maturity Model (OPM3) – Third Edition**
  - Project Management Institute. (2013). *Organizational Project Management Maturity Model (OPM3)* (3rd ed.). Project Management Institute, Inc.
- **Patterns of Team Adaptation: The Effects of Behavioural Interaction Patterns on Team Adaptation and the Antecedent Effect of Empowering Versus Directive Leadership**
  - Rico, R., Uitdewilligen, S., and Dorta, D. (2021). Patterns of team adaptation: the effects of behavioural interaction patterns on team adaptation and the antecedent effect of empowering versus directive leadership. *Journal of Contingencies and Crisis Management*: n.p. Web.
- **Practice Standard for Earned Value Management – Second Edition**
  - Project Management Institute. (2011). *Practice Standard for Earned Value Management* (2nd ed.). Project Management Institute, Inc.
- **Practice Standard for Project Configuration Management**
  - Project Management Institute. (2007). *Practice Standard for Project Configuration Management*. Project Management Institute, Inc.
- **Practice Standard for Project Estimating**
  - Project Management Institute. (2020). *Practice Standard for Project Estimating*. Project Management Institute, Inc.

- **Practice Standard for Project Risk Management**
  - Project Management Institute. (2009). *Practice Standard for Project Risk Management*. Project Management Institute, Inc.
- **Practice Standard for Scheduling – Second Edition**
  - Project Management Institute. (2011). *Practice Standard for Scheduling* (2nd ed). Project Management Institute, Inc.
- **Practice Standard for Work Breakdown Structures – Third Edition**
  - Project Management Institute. (2013). *Practice Standard for Work Breakdown Structures* (3rd ed.). Project Management Institute, Inc.
- **Solving for Project Risk Management: Understanding the Critical Role of Uncertainty in Project Management**
  - Smart, C. (2020). *Solving for Project Risk Management: Understanding the Critical Role of Uncertainty in Project Management*. Routledge.
- **The Art of System Architecting (3rd ed.)**
  - Rechtin, E., & Maier, M. W. (2009). *The Art of System Architecting* (3rd ed.). CRC Press.
- **The Mythical Man-Month: Essays on Software Engineering (Anniversary Edition)**
  - Brooks, F. P. (1995). *The Mythical Man-Month: Essays on Software Engineering* (anniversary edition). Addison-Wesley Professional.
- **Transparency and Control in Email Communication: The More the Supervisor Is Put in cc, the Less Trust is Felt**
  - Haesevoets, T., De Cremer, D., De Schutter, L. et al. (2021). Transparency and control in email communication: the more the supervisor is put in cc, the less trust is felt. *Journal of Business Ethics* 168: 733–753.
- **Transformational Leadership Versus Shared Leadership for Team Effectiveness**
  - Tran, T.B., and Vu., A.D. (2021). Transformational leadership versus shared leadership for team effectiveness. *Asian Academy of Management Journal* 26(2): 143–171.
- **Win More Work: How to Write Winning AEC Proposals**
  - Powell, A. (2017). *Win More Work: How to Write Winning AEC Proposals*. ArchMark.

Index

Management Essentials for Civil Engineers: A Practical Guide to Business, Communication, Ethics, and Risk, First Edition. Cody A. Pennetti, C. Kat Grimsley, and Brian M. Grindall.

D

E

P